PARKS AND PLATES

The Geology of
Our National Parks,
Monuments,
and Seashores

PARKS AND PLATES

The Geology of Our National Parks, Monuments, and Seashores

Robert J. Lillie

OREGON STATE UNIVERSITY

W. W. NORTON AND COMPANY

New York • London

W. W. Norton & Company has been independent since its founding in 1923, when William Warder Norton and Mary D. Herter Norton first published lectures delivered at the People's Institute, the adult education division of New York City's Cooper Union. The Nortons soon expanded their program beyond the Institute, publishing books by celebrated academics from America and abroad. By mid-century, the two major pillars of Norton's publishing program — trade books and college texts — were firmly established. In the 1950s, the Norton family transferred control of the company to its employees, and today — with a staff of four hundred and a comparable number of trade, college, and professional titles published each year — W. W. Norton & Company stands as the largest and oldest publishing house owned wholly by its employees.

Editor: Leo Wiegman
Project Editors: Thomas Foley and Mary Kelly
Director of Manufacturing, College: Roy Tedoff
Editorial Assistants: Robert Bellinger and Sarah Mann
Copy Editor: Andrew Saff
Managing Editor: Marian Johnson
Photo Researcher: Neil Ryder Hoos
Front Cover Photo: © David Muench/Corbis

Design and Layout by Anna Palchik
Illustrations by Robert J. Lillie
Composition by TSI Graphics
Manufacturing by R. R. Donnelley

ISBN 0-393-92407-6

W. W. Norton & Company, Inc., 500 Fifth Avenue, New York, N.Y. 10110
www.wwnorton.com

W. W. Norton & Company Ltd., Castle House, 75/76 Wells Street, London W1T 3QT

DEDICATION

To Carolyn,
*   You've been a special friend*
for all these years.

PARKS LIST

The following is a list of National Park Service and other sites covered in *Parks and Plates*. The numbers are used on maps and tables throughout the book. Table 1.1 lists the parks according to their tectonic settings.

Abbreviations for National Park System Areas: N & SP, National & State Park; NHP, National Historic Park; NHS National Historic Site; NL, National Lakeshore; NM, National Monument; NMem, National Memorial; NP, National Park; NPres, National Preserve; NRA, National Recreation Area; NRes, National Reserve; NRR, National Recreation River; NRRA, National River and Recreation Area; NS, National Seashore; NSR, National Scenic Riverway; NVM, National Volcanic Monument; WSR, Wild and Scenic River.

Acadia NP (59)
Agate Fossil Beds NM (93)
Alibates Flint Quarries NM (94)
NP of American Samoa (87)
Aniakchak NM (53)
Apostle Islands NL (22)
Arches NP (111)
Assateague Island NS (28)
Badlands NP (95)
Bandelier NM (16)
Big Bend NP (71)
Big South Fork NRRA (60)
Bighorn Canyon NRA (131)
Biscayne NP (29)
Black Canyon of the Gunnison NP (112)
Blue Ridge Parkway (61)
Bluestone NSR (62)
Bryce Canyon NP (113)
Buck Island Reef NM (83)
Cabrillo NM (77)
Canaveral NS (30)
Canyon de Chelly NM (114)
Canyonlands NP (40)
Cape Cod NS (31)
Cape Hatteras NS (32)
Cape Krusenstern NP (73)
Cape Lookout NS (33)
Capitol Reef NP (115)
Capulin Volcano NM (17)
Carlsbad Caverns NP (18)
Cedar Breaks NM (41)
Channel Islands NP (78)
Chattahoochee River NRA (63)
City of Rocks NRes (1)
Colorado NM (116)
Crater Lake NP (49)

Craters of the Moon NM (88)
Cumberland Island NS (34)
Curecanti NRA (117)
Cuyahoga Valley NP (96)
Death Valley NP (2)
Delaware Water Gap NRA (64)
Denali NP (141)
Devils Postpile NM (3)
Devils Tower NM (132)
Dinosaur NM (118)
Dry Tortugas NP (36)
Effigy Mounds NM (97)
El Malpais NM (119)
El Morro NM (120)
Everglades NP (37)
Fire Island NS (35)
Florissant Fossil Beds NM (133)
Gates of the Arctic NP (74)
Gauley River NRA (65)
Glacier NP (139)
Glacier Bay NP (142)
Glen Canyon NRA (121)
Golden Gate NRA (79)
Golden Spike NHS (4)
Grand Canyon NP (42)
Grand Portage NM (23)
Grand Staircase-Escalante NM (122)
Grand Teton NP (5)
Great Basin NP (6)
Great Sand Dunes NM (134)
Great Smoky Mountains NP (66)
Guadalupe Mountains NP (19)
Gulf Islands NS (38)
Hagerman Fossil Beds NM (89)
Haleakalā NP (85)
Hawai`i Volcanoes NP (86)

Hot Springs NP (70)
Indiana Dunes NL (98)
Isle Royale NP (24)
Jewel Cave NM (135)
John Day Fossil Beds NM (90)
Joshua Tree NP (7)
Katmai NP (54)
Kenai Fjords NP (48)
Keweenaw NHP (25)
Kings Canyon NP (56)
Kobuk Valley NP (75)
Lake Chelan NRA (144)
Lake Clark NP (55)
Lake Mead NRA (8)
Lake Meredith NRA (99)
Lassen Volcanic NP (50)
Lava Beds NM (9)
Mammoth Cave NP (100)
Mesa Verde NP (123)
Mississippi NRRA (101)
Missouri NRR (102)
Mojave NPres (10)
Mount Rushmore N Mem (136)
Mt Rainier NP (51)
Mt St Helens NVM (52)
Natches Trace Parkway (103)
Natural Bridges NM (124)
Navajo NM (125)
Newberry NVM (11)
Niobrara NSR (104)
Noatak NPres (76)
North Cascades NP (145)
Obed WSR (67)
Olympic NP (45)
Oregon Caves NM (46)
Organ Pipe Cactus NM (12)

CONTENTS

PREFACE

S and-covered beaches on the East Coast. Inspiring peaks in the Appalachians and Rockies. Active volcanoes in the Pacific Northwest and Hawai`i. Dramatic landscapes are often a visitor's first glimpse of a national park, an impression that can remain special for a lifetime. How do the landscapes of national parks form, and why are they similar in some parks, yet so different in others?

National parks are places where dramatic geologic events are taking place. People can go to parks in California to observe changes in the landscape associated with earthquakes. They can see volcanic eruptions and their products in the Pacific Northwest, Alaska, and Hawai`i. In national seashores they see effects of erosion and deposition. Through examples available for public viewing in parks, this book examines the scientific basis for understanding these and other natural processes, as well as society's need to preserve and protect outstanding geologic features.

Plate tectonics is a simple yet eloquent way to visualize geologic processes. Park visitors might use it to recognize, for example, why volcanoes in the Cascade Mountains look so much like those in national parks on the Alaskan Peninsula (same processes in a

◀ **Sunset Crater Volcano National Monument, Arizona.**
Gas-charged lava erupted high into the air, raining down as solid particles that piled up as a cinder cone.

similar tectonic setting). Or why mountains in national parks in the Cascades differ from those in the Appalachians (different processes in different tectonic settings). Or why there is so much earthquake and volcanic activity in parks in the West (at active plate boundaries), compared to few earthquakes and no volcanism in national seashores on the East Coast (far away from plate boundaries).

Parks and Plates illustrates that the landscapes of our national parks result from movements of large plates of Earth's outer shell. Mountains, volcanoes, shorelines, and various types of rocks develop through interactions along plate boundaries, or where a plate moves over a hotspot. The book is intended for visitors to national parks, monuments, and seashores; college and university courses on Geology of National Parks; and interpretation, natural resource, and research staffs in parks. It is written and illustrated at the level of an Introductory Geology textbook, so that no prior knowledge of geology is needed. Each part of the book

begins with basic questions about a particular type of plate boundary or hotspot (for example, a divergent plate boundary). Each chapter within that part illustrates and discusses national parks from a specific setting formed along that type of plate boundary (for example, a continental rift zone).

The "national park idea" has been described by some as our country's greatest contribution to society. Yellowstone was established as the world's first national park in 1872. The National Park Service now administers more than 380 sites, and the list of countries preserving areas as parks continues to grow. This book helps park visitors realize that, like Yellowstone, many of our national parks, monuments, and seashores were established because of their inspiring geologic features.

– Robert J. Lillie
Professor of Geology
Oregon State University

ACKNOWLEDGMENTS

I'm thankful to many people and organizations for their help and encouragement during my work in national parks and in the development of this book. The Geological Resources Division of the National Park Service made my visits to parks possible and provided much help and encouragement in the course of the project. I am especially grateful to Judy Geniac, Bob Higgins, and Jim Wood for financial, logistical, and editorial support, and for always being there to help—their commitment to promoting geology in parks is contagious!

Many individuals from parks have been especially generous with facilities and hospitality. I would like to express thanks to Marsha McCabe, Kent Taylor, Steve Mark, and Tom McDonnough (Crater Lake National Park), Marianne Mills (Badlands National Park), Allyson Mathis and Judy Hellmich (Grand Canyon National Park), Michael Smithson and Kathy Steichen (Olympic National Park), Ted Stout and Doug Owen (Craters of the Moon National Monument), Sherri Forbes (Mount Rainier National Park), Vicki Ozaki and Cathy Cook (Redwood National and State Parks), Patty Lockamy (Blue Ridge Parkway), Anita Davis (Sunset Crater Volcano National Monument), Jim Gale (Hawai`i Volcanoes National Park), Sharon Ringsven (Haleakalā National Park), and Mary Janes and H. D. Simpson (Gulf Islands National Seashore).

The enthusiasm and help on public outreach in parks from Carolyn Driedger, Melanie Moreno, and Tom Sisson of the U.S. Geological Survey is much appreciated. Grant Kaye of the Oregon State University Geosciences Department did a superb job preparing the digital relief maps for the book.

I have been fortunate to work with such a professional and good-natured editorial staff at W. W. Norton. Many thanks to Leo Wiegman, Roy Tedoff, Thom Foley, Andy Saff, Stephanie Hiebert, Marian Johnson, Rob Bellinger, Sarah Mann, and Mary Kelly for your hard work and for always keeping things on task! Thanks to Anna Palchik for a beautiful interior design and layout.

I am grateful to the many anonymous reviewers who provided insight that greatly improved the manuscript at various stages. A special thanks to Kathy Bricker for providing thorough, insightful, and helpful reviews of the entire book.

Many of the photos used in the book are public domain materials. I am grateful to the National Park Service and the U.S. Geological Survey for making those photos available. I extend special thanks to Bob Lawrence, Jo Ann Callahan, Randy Milstein, Emily Larkin, Ed Buchner, Cy Field, Mike Appel, and Hillary Senden for allowing me to use their outstanding photographs. Chris Warner of Earth Treks Climbing Center contributed a spectacular photo of K-2.

My own graduate students have been the source of a wealth of information, ideas, and general good humor. Thank you so much Robyn Green, Stacy Wagner, Becky Ashton, Jen Natoli, and Kim Truitt for your help and inspiration, and for contributing so much to the interpretation of geology in our national parks.

PART I

Earth Systems and Our National Parks

Our national parks reveal
processes that shape our planet.
Chapter 1 explains how plate
tectonics revolutionized thinking
about the Earth and shows how
plate-boundary processes result
in dramatic park landscapes. The
next chapter introduces basic
geologic concepts important to
understanding features in parks.

Why does Yellowstone National Park have geysers and hot springs? Why do the coastlines of the eastern and western United States look so different? How does the formation of the Sierra Nevada Mountains differ from that of the Appalachians? And why do volcanoes in Hawai`i look so different compared to those in the Pacific Northwest? Spectacular geologic features in national parks provide some of the answers to these and other questions about the Earth. The answers can be appreciated through an understanding of **plate tectonics**, an exciting way to view features on Earth's surface and the processes responsible for their formation. **Geology** is the study of the Earth. It incorporates more than just the identification of rocks. Geology involves the study of processes occurring within or on the Earth that make the planet come alive. Processes *within the Earth* include those responsible for **earthquakes**, **volcanoes**, and the formation of **mountain ranges**. The actions of wind, water, and ice occur *at Earth's surface,* resulting in **erosion**, **exposure** of older rocks, and the **deposition** of sediment. Park visitors can observe features at Earth's surface and understand them in terms of Earth's internal and external processes.

Plate tectonics has revolutionized the way we view large features on the surface of the Earth. Earth's internal processes were previously thought to operate in a vertical fashion, with continents, oceans, and mountain ranges bobbing up and down, without much sideways movement. But the acceptance of continental drift and other evidence for large lateral motions has changed all that. Now it's understood that Earth's internal processes can move large plates of Earth's outer shell great horizontal distances. Plate tectonics thus provides the "big picture" of geology; it explains how mountain ranges, earthquakes, volcanoes, shorelines, and other features tend to form where the moving plates interact along their boundaries.

National parks, monuments, seashores, and other areas administered by the National Park Service are established to preserve our cultural and natural heritage. Park landscapes include spectacular geologic

▲▲▲▲▲▲ THE EARTH IS A STAGE

The geological landscape is a grand stage. Episodes of cultural and natural history are performed on that stage. But the stage is not static. It changes in subtle ways during scenes, and more dramatically between acts. Geologists observe as the play unfolds, and try to imagine what happens behind the curtains.

◄ **[Overleaf]**
Crater Lake National Park, Oregon. U-shaped valley reveals that this was once a very high mountain, covered by glaciers.

features, such as the geysers of Yellowstone, the volcanoes of Hawai`i, and the granite peaks of Yosemite. National parks have two purposes: *1) to preserve features of aesthetic and historical importance; and 2) to make the features available for the education and enjoyment of the public.* Those two purposes are often at odds—the National Park Service is continually caught in the conflict between the need for preservation and the right of access. It is important to understand why certain geologic features are so unique, or so beautiful, or have such scientific significance, that society benefits from their preservation and access.

Alan Gussow writes "There is a great deal of talk these days about saving the environment. We must, for the environment sustains our bodies. But as humans we also require support for our spirits, and this is what certain kinds of places provide. The catalyst that converts any physical location—any environment if you will—into a place, is the process of experiencing deeply. A place is a piece of the whole environment that has been claimed by feelings. Viewed simply as a life-support system, the Earth is an environment. Viewed as a resource that sustains our humanity, the Earth is a collection of places. We never speak, for example, of an environment we have known; it is always places we have known—and recall. We are homesick for places, we are reminded of places, it is the sounds and smells and sights of places which haunt us and against which we often measure our present."

TEN QUESTIONS: Earth Systems and the Landscapes of Our National Parks

1. What role does geology play in national parks?

2. Why is tectonic activity concentrated over such narrow zones on Earth's surface?

3. Why are there plates?

4. Why do plates move relative to one another?

5. How fast do plates move?

6. What are hotspots?

7. What causes volcanic activity?

8. What causes earthquakes?

9. How do mountain ranges form?

10. Why is there so much more tectonic activity in the western United States than in the East?

Olympic National Park, Washington. The Needles are ocean crustal rocks that were lifted out of the sea as plates converge.

1

Plate Tectonics

"Plate tectonics helps us to understand the inspiring landscapes that attract us to national parks, and to compare geologic features in one park to features we see in others."

Areas are designated National Park Service lands because of their historical significance or natural beauty. The latter category includes areas of mountains, valleys, seashores, or rock formations, features commonly associated with large-scale, or tectonic, processes. The term **tectonics** originates from the Greek word *tektōn*, referring to a builder or architect. **Plate tectonics** suggests that large features on the Earth's surface, such as continents, ocean basins, and mountain ranges, result from interactions along the edges of large plates of the Earth's outer shell, or **lithosphere** (Greek *lithos*, "hard rock"). The plates, composed of the Earth's crust and uppermost mantle, ride on a warmer, softer layer of the mantle, the **asthenosphere** (Greek *asthenos*, "lacking strength"). The Earth's lithosphere is broken into a mosaic of seven major and several minor plates (Fig. 1.1). Relative motions between plates define three types of boundaries: **divergent**, where plates rip apart, creating new lithosphere; **convergent**, where one plate dives beneath the other and destroys lithosphere; **transform**, where plates slide past one another, neither creating nor destroying lithosphere (Fig. 1.2). Another large-scale feature is a **hotspot**, where a plate rides over a fixed "plume" of hot mantle, creating a line of volcanoes. Plate tectonics helps us to understand the inspiring landscapes that attract us to national parks, and to compare geologic features in one park to features we see in others. For example, volcanoes in parks in the Pacific Northwest and Alaska are similar because they formed at boundaries where plates converge, whereas a different volcanic type occurs in Hawaiian parks because those volcanoes formed over a hotspot.

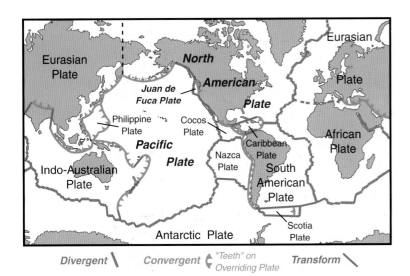

Divergent   Convergent   *"Teeth" on Overriding Plate*   Transform

FIGURE 1.1 Plate tectonic map of the world, showing the three types of plate boundaries. Most of the current tectonic activity in the United States involves interactions between the North American, Pacific, and Juan de Fuca plates. The western United States is near plate boundaries and thus has volcanoes, earthquakes, and developing mountain ranges, while the East Coast is far from plate boundaries and lacks active tectonic features.

In "The Earth Speaks," Steve Van Matre writes, "Have you listened to the Earth? Yes, the Earth speaks, but only to those who can hear with their hearts. It speaks in a thousand, thousand small ways, but like our lovers and families and friends, it often sends its message without words." Through the geologic features we see in national parks, the Earth speaks about processes that occurred long ago, as well as others that continue to shape the landscape.

THE WHOLE EARTH

The Earth was very hot when it formed 4.5 billion (4,500 million) years ago. In fact, it was so hot that it was entirely molten. The fluid nature of the early Earth helps explain some basic things about its form and gross layering. The Earth is a sphere, with a radius of about 4,000 miles (6,400 kilometers) that varies by only 12 miles (20 kilometers) from the equator to the poles. So why is the Earth such a nearly perfect sphere? The reason is gravity. Suspended in space with gravity working equally in all directions, the molten material organized itself into a sphere. The rotation about its polar axis makes the Earth bulge out slightly at the equator.

Gross Divisions of the Earth

Gravity also explains why the Earth is layered. In its early, molten state, different materials settled into three layers (left side, Fig. 1.3). Dense elements, mostly iron with some nickel, fell toward the center to form the **core**. Lighter compounds of silicon and oxygen (silicates) remained closer to the surface. Silicates rich in iron and magnesium formed the **mantle**, overlain by a thin **crust** of silicates containing light elements such as aluminum, calcium, potassium, and sodium. In the 1900s, seismologists discovered this classical division, based on *chemical composition*, by analyzing earthquake seismic waves that penetrated the Earth.

By the mid-twentieth century, improved seismic observations made it possible to study the Earth's interior in finer detail. Since forming, the Earth has had time to cool down substantially so that only a small portion of it is still liquid. The seismic data reveal that

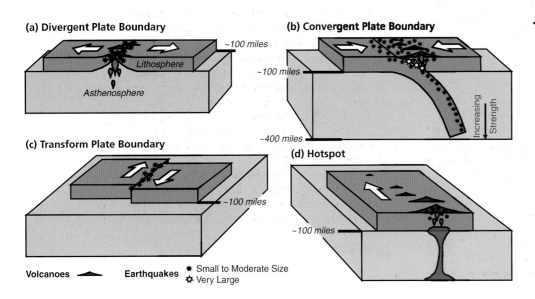

(a) Divergent Plate Boundary

(b) Convergent Plate Boundary

(c) Transform Plate Boundary

(d) Hotspot

Volcanoes ━━   Earthquakes   ✳ Small to Moderate Size   ✿ Very Large

◀ **FIGURE 1.2 Tectonic activity occurs at the three types of plate boundaries and at hotspots.** Volcanoes erupt in the zone where plates diverge, on the overriding plate where plates converge, and along a line where a plate rides over a hotspot. Only shallow earthquakes (less than 40 miles, 70 kilometers deep), of small to moderate size, occur at divergent and transform boundaries and at hotspots. The cold, brittle lithosphere may extend to great depths (up to 400 miles, 700 kilometers) at convergent boundaries, accompanied by a dipping zone of shallow to very deep earthquakes; the largest earthquakes occur at convergent boundaries where the two plates lock together for centuries, then suddenly let go (white stars).

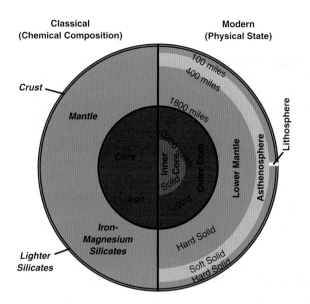

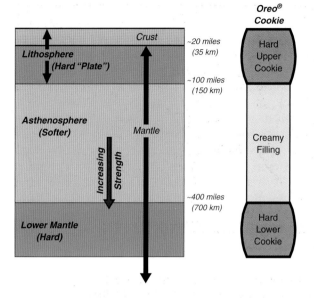

FIGURE 1.3 Gross layers of the Earth. Left: The classical division of the Earth is based on **chemical composition**, the heavier materials concentrated toward the center. **Right:** In modern times, the three chemical divisions are classified into five zones according to the physical state caused by temperature and pressure changes within the Earth.

FIGURE 1.4 Cross section of the upper 600 miles (1,000 kilometers) of the Earth, including the crust and part of the mantle. Increases in temperature and pressure with depth cause the mantle to exist in three different states. The uppermost mantle and crust comprise the cold, rigid plates of lithosphere. Hotter mantle below forms the somewhat softer asthenosphere. Pressure increase with depth causes the asthenosphere to increase in strength, to the more solid lower mantle. Lithospheric plates can be compared to a hard Oreo® cookie, riding on the soft, creamy filling (asthenosphere). The lower cookie (lower mantle) does not move.

the Earth's three chemical layers are differentiated into five zones based on physical state (right side, Fig. 1.3). Changes in physical state occur because both temperature and pressure increase downward in the Earth. The temperature is still hot enough at depths of 1,800 to 3,200 miles (2,900 to 5,100 kilometers) that the iron of the **outer core** is liquid. Below that, to a depth of 4,000 miles (6,400 kilometers) at the Earth's center, pressure is so great that the **inner core** is solid.

The iron/magnesium silicates of the mantle also show stratification due to the increasing temperature and pressure (Fig. 1.4). The outermost mantle and crust are cold and hard (like butter in a refrigerator), forming the rigid lithosphere. Below about 100 miles (150 kilometers), the mantle is warmer, so that the asthenosphere is a softer solid (like butter left on a dinner table). Still deeper, from 400 to 1,800 miles (700 to 2,900 kilometers), pressure is so great that the **lower mantle** is a hard solid. The changes in strength of the mantle create a unique situation where plates of lithosphere ride over the softer zone of asthenosphere, similar to sliding a hard Oreo® cookie over the creamy filling (Fig. 1.5).

Continental versus Oceanic Crust

Imagine you're a space traveler seeing Earth for the first time. You're immediately struck by a startling and beautiful aspect of the planet: about two-thirds of its

EARTH: THE ORIGINAL RECYCLER

Nowadays we're concerned with recycling. We save household materials for curbside pickup, donate newspapers to local organizations, and bring bottles and cans back for deposits at grocery stores. These habits may seem small, but they have significant impact when everyone does them. On a grand scale, the Earth recycles. New plate material is manufactured at divergent plate boundaries, particularly where crust forms at mid-ocean ridges. But the oldest oceanic crust that's still intact formed only about 200 million years ago. That's because plates are recycled back into the deep mantle, where they subduct at convergent plate boundaries. The Earth is thus teaching us an important lesson—that recycling is important on all scales!

surface is blue and most of the rest is brown. You realize that the blue area is water and the brown land. But
how is it that the surface of continents lies above sea
level, while the floor of the ocean is a couple of miles
below? The answer lies in the fact that crust of the
continents is so much *thicker* than crust of the oceans.
Although both crust and mantle are solid, crust can be
thought of as "floating" on the more dense mantle.
Because the crust of continents is thicker, it is *more
buoyant* than the crust of the ocean (Fig. 1.6). If one
envisions Earth's mantle as a swimming pool, continental crust might be thought of as a soccer ball, which
sticks up higher out of the water than a smaller tennis
ball (the oceanic crust).

The great difference in thickness of **continental
crust** compared to **oceanic crust** is an important factor
in how plate tectonics operates, and in how the landscapes of national parks developed. National seashores
on the Atlantic and Gulf coasts of the United States lie
in the zone of transition between thick continental and
thin oceanic crust, known as passive continental margins. In the Pacific Northwest, the Juan de Fuca Plate,
capped by thin oceanic crust, is descending (subducting) beneath thick, more buoyant continental crust of
the North American Plate. Parks along such an active

Sliding Plate over
Asthenosphere

Divergent
Plate Boundary

Convergent
Plate Boundary

Transform
Plate Boundary

FIGURE 1.5 Simulating plate boundaries with Oreo® cookies.

FUN WITH FOOD!

Sliding Plates and Oreo® Cookies

The three types of plate boundaries (Fig. 1.2) can be
demonstrated with Oreo® cookies (Fig. 1.5). The upper
cookie is the lithosphere, the creamy filling the asthenosphere, and the lower cookie the lower mantle
(Fig. 1.4). (Be sure to get the "Double Stuf"® variety,
which has adequate asthenosphere!). First, carefully re-
move the upper cookie (a "twisting" motion is required). Slide the upper cookie over the creamy filling
to simulate motion of a rigid lithospheric plate over the
softer asthenosphere. Next, break the upper cookie in
half. As you do so, listen to the sound it makes. What
does that sound represent? (An *earthquake*. Message: it
takes cold, brittle lithosphere to make earthquakes—
earthquakes do not occur in the soft, flowing asthenosphere.)

a. To simulate a *divergent plate boundary,* push down
   on the two broken cookie halves and slide them
   apart. Notice that the creamy filling between the

two broken "plates" may tend to flow upward, similar to the rising, decompression, and partial melting
of hot asthenosphere at mid-ocean ridges.

b. Push one cookie piece beneath the other to make a
   convergent plate boundary. Note that this is the only
   situation where the cold, brittle lithosphere extends
   to great depths, and hence the only place where
   deep earthquakes occur. The very largest earthquakes are at subduction zones where two plates
   get stuck together for centuries, then suddenly let
   go.

c. Simulate a *transform plate boundary* by sliding the
   two cookie pieces laterally past one another, over
   the creamy filling. You can feel and hear that the
   "plates" do not slide smoothly past one another, but
   rather stick and then let go, stick and then let go.
   The cracking sound you hear each time is like an
   earthquake occurring along the San Andreas Fault in
   California.

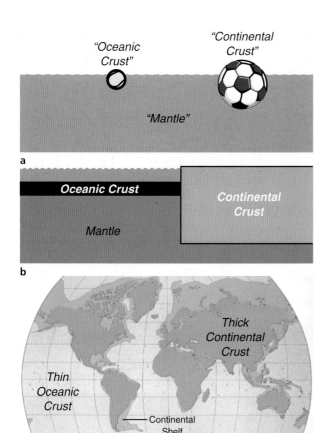

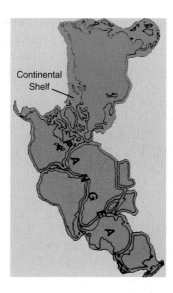

FIGURE 1.7 Continental jigsaw puzzle. When fit together along the edges of their shelves, the continents form the super-continent called Pangea.

FIGURE 1.6 Buoyancy of thick (continental) compared to thin (oceanic) crust. a) A bigger, more buoyant soccer ball sticks up higher, and rests lower, in a swimming pool than a smaller, less buoyant tennis ball. **b)** Thick (more buoyant) continental crust "floats" up higher on the mantle than the thin (less buoyant) oceanic crust. **c)** Relatively thick crust includes not only regions above sea level (orange), but also the continental shelves (dark blue) that lie beneath shallow water.

continental margin in northern California, Oregon, and Washington have a dramatically different, mountainous landscape.

CONTINENTAL DRIFT AND THE DEVELOPMENT OF PLATE TECTONIC THEORY

Ever since the first maps of the Atlantic Ocean were made in the sixteenth century, people noticed how Africa and South America fit together like pieces of a giant jigsaw puzzle. The fit is even more impressive if the continents are joined together along the edges of their continental shelves (Fig. 1.7). The resulting land

mass, called **Pangea**, represents a glimpse in time, about 250 million years ago, when most of the continental crust happened to be joined together. Prior to that time, the continents were apart; since then, they have drifted away from one another (imagine bumper cars being stuck together for a while, then flying apart!). This theory, called **continental drift**, was viewed with skepticism in the early twentieth century. It was thought impossible for blocks of crust to plow their way over Earth's mantle, which was known to be far more rigid and dense.

Two scientific advances in the mid-twentieth century resulted in information critical to acceptance of continental drift and plate-tectonic theory. First, the topography of the ocean floor was mapped in great detail during and after World War II (Fig. 1.8). It was discovered that the floors of the ocean basins are not flat. A continuous mountain chain circumscribes the globe near the centers of oceans, and in places, the ocean floor descends abruptly into deep-sea trenches. Second, a network of seismographs was installed around the world in the early 1960s, to detect nuclear tests during the Cold War. As a result, geologists could more precisely locate earthquakes and map the speed of seismic waves passing through various regions of the Earth.

The new seismic data revealed three startling observations. First, earthquakes are not scattered throughout the oceans, but instead are confined to narrow, rather continuous bands (Fig. 1.9). The narrow zones of earthquakes outline the boundaries of moving plates.

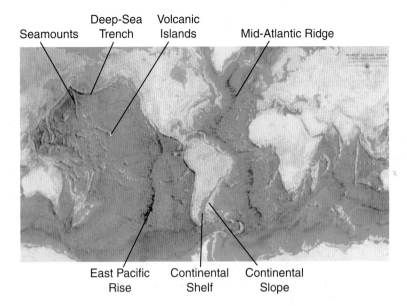

Seamounts   Deep-Sea   Volcanic
              Trench   Islands   Mid-Atlantic Ridge

East Pacific   Continental   Continental
    Rise   Shelf   Slope

FIGURE 1.8 World map showing features of the ocean floor. A system of mid-ocean ridges forms the broadest and longest mountain chain on Earth (the examples pinpointed are the Mid-Atlantic Ridge and the East Pacific Rise). Continental shelves extend outward to where the water deepens abruptly on continental slopes (the East Coast of South America). On the edges of some oceans, narrow deep-sea trenches occur (the Aleutian Trench). Lines of volcanic islands form long chains on the ocean floor, deepening progressively as underwater seamounts (the Hawai`i-Emperor Seamount Chain).

FIGURE 1.9 Locations of large earthquakes recorded over seven-year span, plotted without geographic reference. The earthquakes outline the plate boundaries shown in Fig. 1.1. Only shallow earthquakes occur at *divergent* boundaries (the Mid-Atlantic Ridge), *transform* boundaries (the San Andreas Fault), and *hotspots* (Hawai`i). Zones of earthquakes extending from shallow to intermediate to deep outline subducting plates along *convergent* boundaries (western South America).

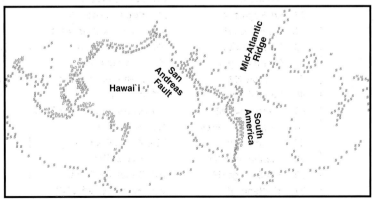

S = Shallow (< 40 miles)
M = Intermediate (40 - 200 miles)
D = Deep (> 200 miles)

Second, only shallow earthquakes (less than 40 miles, 70 kilometers deep) occur where plates diverge along mid-ocean ridges, while earthquakes at deep-sea trenches extend downward to as deep as 400 miles (700 kilometers) where plates converge (Fig. 1.2). Finally, seismic waves slow down as they travel through a zone about 100 to 400 miles (150 to 700 kilometers) deep, a sign that there is a relatively soft layer within the Earth's mantle. This last observation is the "Rosetta Stone" for plate-tectonic theory. It provides a means by which continents can drift apart. Instead of plowing directly over mantle, the continents are "passengers" on the tops of much thicker plates. The plates of crust and stiff mantle (lithosphere) move on the softer mantle layer beneath (asthenosphere).

TECTONIC SETTINGS OF OUR NATIONAL PARKS

The shaded-relief map of the United States highlights different tectonic settings (Fig. 1.10). Superimposed in red are the more than 380 National Park Service sites. Two patterns are apparent: 1) the western United States is much more rugged than the East; and 2) there is a lot more park land in the West than in the East. The dramatic topography in the West is due to its *youth*. Mountains and coastlines are continually being built and deformed because they are at or near active plate boundaries (Fig. 1.1). Features in the East, such as the Appalachian Mountains and Atlantic Coast, formed at plate boundaries, but that was a long time ago. Now the nearest plate boundaries are more than a thousand miles away, and the landscapes are wearing away. The reasons that more park land exists in the West are both physical and historical. Mountainous terrain is much less amenable to human settlement, agriculture, industry, and commerce. The first national park, Yellowstone, was considered unsuitable for human development. It was established in 1872, even though the National Park Service was not created until 1916. By the late 1800s, most of the land in the East was in private ownership, while much of the land in the West was (and still is) public. It was thus a much easier task to set aside vast regions of incredible scenery in the West.

Table 1.1 lists the lands of the National Park Service and a few other sites that lie within specific tectonic settings. The term **tectonic setting** relates to the processes within the Earth responsible for the topography and rocks of a region. The tectonic setting for a park depends on whether features in the park formed in association with a divergent, convergent, or

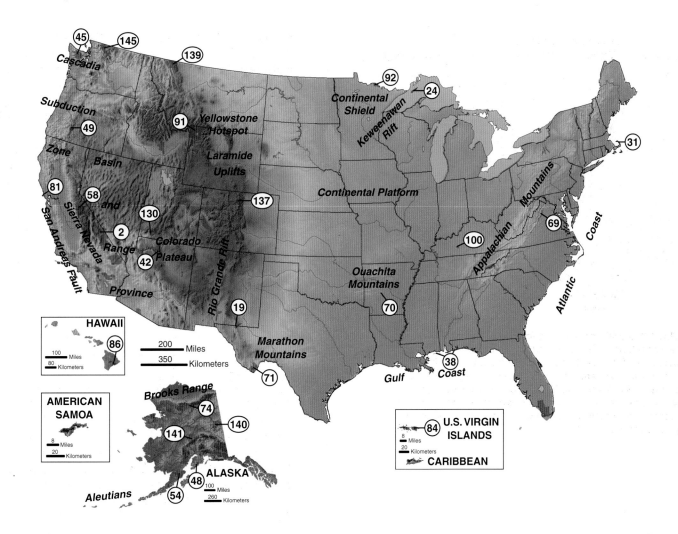

FIGURE 1.10 Shaded-relief map of United States, showing tectonic settings and National Park Service lands in red. See classification of parks by tectonic setting in Table 1.1. **DIVERGENT PLATE BOUNDARIES.** *Continental rifts:* Basin and Range Province; Rio Grande Rift; Keweenawan Rift. *Passive continental margins:* Atlantic Coast; Gulf Coast; western edge of Colorado Plateau. **CONVERGENT PLATE BOUNDARIES.** *Subduction zones:* Cascadia; Sierra Nevada; Aleutians–Southern Alaska. *Collisional mountain ranges:* Appalachian Mountains; Ouachita Mountains; Marathon Mountains; Brooks Range. **TRANSFORM PLATE BOUNDARIES.** San Andreas Fault; Caribbean. **HOTSPOTS.** *Oceanic:* Hawai`i, American Samoa. *Continental:* Yellowstone. **CONTINENTAL CRATON.** *Undeformed:* Continental Shield; Continental Platform. *Deformed:* Colorado Plateau; Laramide Uplifts. **ACCRETED TERRANES.** Southern/Southeast Alaska; Cordillera (Mountainous West).

Map Key

2.	Death Valley NP	74.	Gates of the Arctic NP & Pres.
19.	Guadalupe Mountains NP	81.	Point Reyes NS
24.	Isle Royale NP	84.	Virgin Islands NP
31.	Cape Cod NS	86.	Hawai`i Volcanoes NP
38.	Gulf Islands NS	91.	Yellowstone NP
42.	Grand Canyon NP	92.	Voyageurs NP
45.	Olympic NP	100.	Mammoth Cave NP
48.	Kenai Fjords NP	130.	Zion NP
49.	Crater Lake NP	137.	Rocky Mountain NP
54.	Katmai NP & Pres.	139.	Glacier NP
58.	Yosemite NP	140.	Yukon-Charley Rivers NPres.
69.	Shenandoah NP	141.	Denali NP & Pres.
70.	Hot Springs NP	145.	North Cascades NP
71.	Big Bend NP		

TABLE 1.1 Classification of National Park lands according to tectonic setting. Parks on the chart have landforms, rocks, or other features that tell a story about Earth's history and processes. Although each park appears only once on the chart, many individual park landscapes represent multiple tectonic episodes. Parks highlighted **with bold numbers** are located in Figure 1.10. NHP = National Historical Park; NL = National Lakeshore; NM = National Monument; NMen = National Memorial; NP = National Park; NPres = National Preserve; NRes = National Reserve; NRA = National Recreation Area; NRR = National Recreational River; NR= National River; NRRA = National River and Recreation Area; NS = National Seashore; NSR = National Scenic River; NST = National Scenic Trail; NVM = National Volcanic Monument; N & SP = National and State Parks; WSR = Wild and Scenic River. All of the sites are administered by the National Park Service, except Newberry and Mount St. Helens volcanic national monuments, which are under United States Forest Service jurisdiction, and Grand Staircase–Escalante National Monument, administered by the Bureau of Land Management.

DIVERGENT PLATE BOUNDARIES

CONTINENTAL RIFTS			PASSIVE CONTINENTAL MARGINS		
Active		**Ancient**	**Modern**		**Ancient**
Basin & Range Province	*Rio Grande Rift*	*Keweenawan Rift*	*Atlantic Coast*	*Gulf Coast*	*Colorado Plateau*
1. City of Rocks NRes, ID	16. Bandelier NM, NM	22. Apostle Islands NL, WI	28. Assateague Island NS, MD	36. Dry Tortugas NP, FL	40. Canyonlands NP, AZ
2. Death Valley NP, CA/NV	17. Capulin Volcano NM, NM	23. Grand Portage NM, MN	29. Biscayne NP, FL	37. Everglades NP, FL	41. Cedar Breaks NM, UT
3. Devils Postpile NM, CA	18. Carlsbad Caverns NP, NM	**24.** Isle Royale NP, MI	30. Canaveral NS, FL	**38.** Gulf Islands NS, FL/MS	**42.** Grand Canyon NP, AZ
4. Golden Spike NHS, UT	**19.** Guadalupe Mountains NP, TX	25. Keweenaw NHP, MI	**31.** Cape Cod NS, MA	39. Padre Island NS, TX	43. Petrified Forest NP, AZ
5. Grand Teton NP, WY	20. Petroglyph NM, NM	26. Pictured Rocks NL, MI	32. Cape Hatteras NS, NC		44. Walnut Canyon NM, AZ
6. Great Basin NP, NV	21. White Sands NM, NM	27. St. Croix NSR, MN/WI	33. Cape Lookout NS, NC		
7. Joshua Tree NP, CA			34. Cumberland Island NS, GA		
8. Lake Mead NRA, CA			35. Fire Island NS, NY		
9. Lava Beds NM, CA					
10. Mojave NPres, CA					
11. Newberry NVM, OR					
12. Organ Pipe Cactus NM, AZ					
13. Saguaro NP, AZ					
14. Sunset Crater Volcano NM, AZ					
15. Timpanogos Cave NM, UT					

CONVERGENT PLATE BOUNDARIES

SUBDUCTION ZONES			COLLISIONAL MOUNTAIN RANGES		
Active		**Ancient**	**Appalachian-Ouachita-Marathon**		**Northern Alaska**
Accretionary Wedge: Coast Ranges	*Volcanic Arc: Cascades*	*Volcanic Arc: Sierra Nevada*	*Appalachian Mountains*	*Ouachita Mountains*	*Brooks Range*
45. Olympic NP, WA	**49.** Crater Lake NP, OR	56. Kings Canyon NP, CA	59. Acadia NP, ME	**70.** Hot Springs NP, AR	73. Cape Krusenstern NP, AK
46. Oregon Caves NM, OR	50. Lassen Volcanic NP, CA	57. Sequoia NP, CA	60. Big South Fork NR&RA, KY/TN		**74.** Gates of the Arctic NP, AK
47. Redwood N&SP, CA	51. Mount Rainier NP, WA	**58.** Yosemite NP, CA	61. Blue Ridge Parkway, VA/NC	*Marathon Mountains*	75. Kobuk Valley NP, AK
	52. Mount St. Helens NVM, WA		62. Bluestone NSR & RA, WV	**71.** Big Bend NP, TX	76. Noatak NPres, AK
Southern Alaska			63. Chattahoochee River NRA, GA	72. Rio Grande WSR, TX	
48. Kenai Fjords NP, AK	*Southern Alaska*		64. Delaware Water Gap NRA, PA		
	53. Aniakchak NM, AK		65. Gauley River NRA, WV		
	54. Katmai NP, AK		66. Great Smoky Mountains NP, NC/TN		
	55. Lake Clark NP, AK		67. Obed WSR, TN		
			68. Russell Cave NM, AL		
			69. Shenandoah NP, VA		

TRANSFORM PLATE BOUNDARIES

CONTINENTAL	OCEANIC
San Andreas	**Caribbean**
77. Cabrillo NM, CA	83. Buck Island Reef NM, US Virgin Islands
78. Channel Islands NP, CA	**84.** Virgin Islands NP, US Virgin Islands
79. Golden Gate NRA, CA	
80. Pinnacles NM, CA	
81. Point Reyes NS, CA	
82. Santa Monica Mountains NRA, CA	

HOTSPOTS

OCEANIC	CONTINENTAL
85. Haleakalā NP, HI	88. Craters of the Moon NM, ID
86. Hawai`i Volcanoes NP, HI	89. Hagerman Fossil Beds NM, ID
87. NP of American Samoa	90. John Day Fossil Beds NM, OR
	91. Yellowstone NP, WY/ID/MT

GROWTH OF NORTH AMERICAN CONTINENT

CONTINENTAL CRATON		DEFORMED CONTINENTAL CRATON		ACCRETED TERRAINS	
Shield	Platform	Colorado Plateau	Foreland Structures	Alaska	Lower 48
92. Voyageurs NP, MN	93. Agate Fossil Beds NM, NB	111. Arches NP, UT	**Basement Uplifts**	**141.** Denali NP, AK	144. Lake Chelan NRA, WA
	94. Alibates Flint Quarries NM, TX	112. Black Canyon of the Gunnison NP, CO	131. Bighorn Canyon NRA, MT	142. Glacier Bay NP, AK	**145.** North Cascades NP, WA
	95. Badlands NP, SD	113. Bryce Canyon NP, UT	132. Devils Tower NM, WY	143. Wrangell-St. Elias NP, AK	146. Ross Lake NRA, WA
	96. Cuyahoga Valley NP, OH	114. Canyon de Chelly NM, AZ	133. Florissant Fossil Beds NM, CO		
	97. Effigy Mounds NM, IA	115. Capitol Reef NP, UT	134. Great Sand Dunes NM, CO		
	98. Indiana Dunes NL, IN	116. Colorado NM, CO	135. Jewel Cave NM, SD		
	99. Lake Meredith NRA, TX	117. Curecanti NRA, CO	136. Mount Rushmore NMem, SD		
	100. Mammoth Cave NP, KY	118. Dinosaur NM, UT/CO	**137.** Rocky Mountain NP, CO		
	101. Mississippi NRRA, MN	119. El Malpais NM, NM	138. Wind Cave NP, SD		
	102. Missouri NRR, SD/NB	120. El Morro NM, NM			
	103. Natchez-Trace Parkway MS/AL/TN	121. Glen Canyon NRA, UT	**Fold/Thrust Belts**		
	104. Niobrara NSR, NB	122. Grand Staircase-Escalante NM, UT	**139.** Glacier NP, MT		
	105. Ozark NSR, MO	123. Mesa Verde NP, CO			
	106. Pipestone NM, MN	124. Natural Bridges NM, UT	**Alaska**		
	107. Scotts Bluff NM, NB	125. Navajo NM, AZ	**140.** Yukon-Charley Rivers NPres AK		
	108. Sleeping Bear Dunes NL, MI	126. Pipe Spring NM, AZ			
	109. Tallgrass Prairie NPres, KS	127. Rainbow Bridge NM, UT			
	110. Theodore Roosevelt NP, ND	128. Tuzigoot NM, AZ			
		129. Wapatki NM, AZ			
		130. Zion NP, UT			

---

transform plate boundary or a hotspot, and whether the plates involved were capped by thick continental or thin oceanic crust.

Divergent Plate Boundaries

Where plates diverge, they can rip a continent apart and eventually open an entire ocean (Fig. 1.11). A **continental rift zone** is developing in the western United States, where blocks of crust drop down as **rift valleys** separating long mountain ranges. The resulting Basin and Range Province includes Death Valley, Great Basin, Grand Teton, and Saguaro national parks (Fig. 1.12a). The Rio Grande Rift is an arm of the Basin and Range Province extending across westernmost Texas and New Mexico into Colorado. Earthquakes, fault-block mountains, and volcanism at Guadalupe Mountains National Park and Bandelier and White Sands national monuments are consequences of the ongoing continental rifting.

About a billion years ago, the central part of North America nearly ripped apart along the Keweenawan Rift. The process stopped, but not before the area looked something like the present East African Rift Zone or the Red Sea. Volcanic rocks found in Isle Royale National Park and Keweenaw National Historical Park are products of the ancient continental rifting (Fig. 1.12b).

If plate divergence continues, a new ocean forms between the two continental fragments. The continents slowly drift away from one another as the ocean widens at the active plate boundary located at a mid-ocean ridge. The continental edges are called **passive continental margins** because they lack the high levels of earthquake, volcanic, and mountain-building activity characteristic of continental margins that are right at plate boundaries. The beautiful beaches of the 11 national seashores on the Atlantic and Gulf of Mexico coasts lie along passive continental margins that developed as Europe, Africa, and South America drifted away from North America beginning about 200 million years ago (Fig. 1.12c). Some parks in the western United States, including Grand Canyon, contain similar sedimentary layering that was deposited along an ancient passive continental margin (Fig. 1.12d).

Convergent Plate Boundaries

Where plates converge, the one with the thinner oceanic crust will descend (subduct) beneath the one with the thicker continental crust (Fig. 1.13a). Two parallel mountain ranges form along the edge of the continent at a **subduction zone**. Sedimentary layers and hard crust are scraped off the top of the oceanic plate, uplifted, and deformed into a coastal range. The rocks and structures of Olympic and Redwood national parks

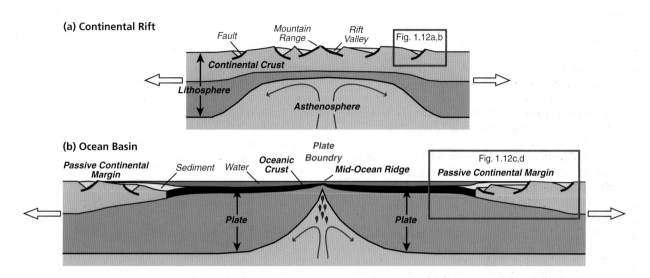

FIGURE 1.11 Plate divergence results in continental rifting and the development of a mid-ocean ridge between passive continental margins. a) Rift valleys form as the crust rips apart and thins. At Great Basin National Park in Nevada, the Snake Valley is a rift valley that dropped down along a fault at the front of the Snake Range (Fig. 1.12a). Isle Royale National Park in Michigan lies within the ancient Keweenawan Rift, where the North American continent nearly split apart 1.1 billion years ago (Fig. 1.12b). **b)** As the continental crust completely rips apart, the two fragments move away from one another as parts of different lithospheric plates. The transitions from thick continental to thin oceanic crust are passive continental margins because they are far away from the plate boundary at the mid-ocean ridge. Gulf Islands National Seashore in Florida and Mississippi lies above thick sedimentary layers deposited as the passive margin subsides (Fig. 1.12c). Similar strata, depostited on an ancient passive continental margin, are preserved as the spectacular layering at Grand Canyon National Park (Fig. 1.12d).

FIGURE 1.12 National Park Service lands ▶ highlight features created at divergent plate boundaries. a) Great Basin National Park in the Basin and Range Province of Nevada lies within a long mountain range bordering a continental rift valley (Fig. 1.11a). **b)** Isle Royale National Park in Lake Superior off the coast of the Upper Michigan Peninsula has dark lava flows that erupted as the ancient Keweenawan Rift developed over a billion years ago. **c)** Gulf Islands National Seashore in Florida and Mississippi lies along the northern Gulf of Mexico, a passive continental margin developed 150 millions years ago as South America drifted away from North America (Fig. 1.11b). **d)** The sedimentary layers in Grand Canyon National Park in Arizona were deposited along the edge of western North America as it subsided along a passive continental margin 250 to 525 million years ago.

(a) Great Basin National Park (Continental Rift Zone)

(b) Isle Royale National Park (Ancient Continental Rift)

(c) Gulf Islands National Seashore (Passive Continental Margin)

(d) Grand Canyon National Park (Ancient Passive Margin)

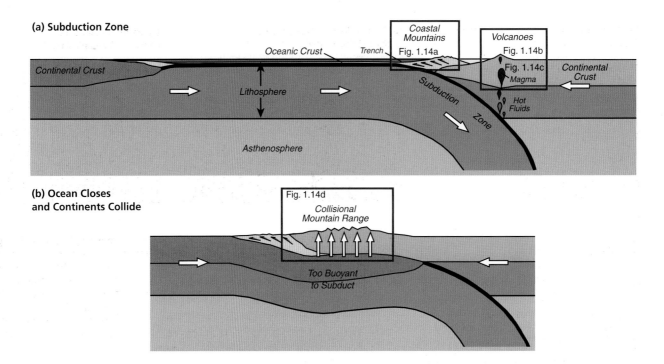

(a) Subduction Zone

(b) Ocean Closes and Continents Collide

FIGURE 1.13 Plate convergence results in subduction and the development of mountain ranges. a) A plate with thinner (less buoyant) oceanic crust descends (subducts) beneath a plate with thicker (more buoyant) continental crust. Two parallel mountain ranges form on the overriding plate: a coastal range (accretionary wedge) develops where oceanic material is scraped off the descending plate (Fig. 1.14a); a line of volcanoes (volcanic arc, Fig. 1.14b) forms above the region where the plate gets hot enough to dehydrate (that is, it loses water, or "sweats"). Magma-chamber rocks can be exposed if subduction stops and the volcanoes erode away (Fig. 1.14c). **b)** As plate convergence closes the ocean basin, the thin (less buoyant) oceanic crust can subduct, while thick (more buoyant) continental crust cannot. A collisional mountain range develops as blocks of thick continental crust crash together, deforming oceanic and continental rocks and uplifting a broad region to great elevation as the crust thickens (Fig. 1.14d).

were manufactured in the ocean and were later uplifted as the Juan de Fuca Plate subducted beneath the Pacific Northwest (Fig. 1.14a). Farther inland, the top of the plate extends to about 50 miles (80 kilometers) deep, where it is so hot that fluids (mostly water) are driven from its crust. The fluids rise, melting rock and generating magma that erupts at the surface, forming explosive volcanoes like those in Mount Rainier, Crater Lake, and Lassen Volcanic national parks (Fig. 1.14b). The spectacular granitic rocks of Kings Canyon, Sequoia, and Yosemite national parks were once parts of magma chambers formed at an ancient subduction zone; since the subduction has stopped, the volcanoes have eroded away, exposing the hardened magma-chamber rock (Fig. 1.14c). Another subduction zone extends from southern Alaska to the Alaska Peninsula and Aleutian Islands, where the Pacific Plate is going beneath the North American Plate. Kenai Fjords National Park lies within the coastal range part of the system, while active volca-

noes occur in Katmai, Lake Clark, and other parks farther inland.

Sometimes entire oceans close via subduction as plates converge (Fig. 1.13b). Blocks of thick continental crust collide, making high and spectacular mountains like the Himalayas in Asia and the Alps in Europe. Such a **collisional mountain range** formed about 300 million years ago when Europe, Africa, and South America crashed into North America. From their lofty heights, the mountains have since eroded to the more modest Appalachians, which include Shenandoah and Great Smoky Mountains national parks and the Blue Ridge Parkway. The Appalachian collision zone continues southwestward as the Ouachita and Marathon mountains in Arkansas and Texas, where Hot Springs and Big Bend national parks contain rocks of the closed ocean. Gates of the Arctic National Park and Preserve shows more youthful topography from continental collision that formed the Brooks Range in northern Alaska 100 million years ago (Fig. 1.14d).

**(a) Redwood National and State Parks
(Accretionary Wedge)**

Uplifted
Marine
Layers

**(b) Mount Rainier National Park
(Active Volcanic Arc)**

Mount
Rainier   Composite
Volcano

**(c) Yosemite National Park
(Ancient Volcanic Arc)**

Half Dome

Granitic
Magma
Chamber

**(d) Gates of the Arctic National Park
(Collisional Mountain Range)**

Brooks Range

**FIGURE 1.14 National Parks reveal features formed by convergent plate boundary processes.
a)** Redwood National and State Parks in the northern California Coast Range contain deformed layers of sandstone, shale, and basalt that were originally deposited in the ocean, then lifted out of the sea as part of the accretionary wedge. **b)** Mount Rainier National Park in the Cascade Mountains (volcanic arc) of Washington State is a pile of lava flows, ash, pumice, and mud flows that erupted as hot fluids rose through the overriding plate. **c)** The granitic rock found in Yosemite National Park in California cooled within magma chambers feeding ancient volcanoes that eroded away after subduction stopped. **d)** Gates of the Arctic National Park and Preserve in northern Alaska contains part of the Brooks Range, formed as continents collided about 100 million years ago.

Transform Plate Boundaries

A broad zone of deformation develops where plates slide horizontally past one another at a **transform plate boundary** (Fig. 1.15). Earthquakes along the San Andreas Fault in California result from transform plate motion as the Pacific Plate moves northward past North America. Point Reyes National Seashore, Pinnacles National Monument, and Channel Islands National Park contain long mountain ranges that were sheared up along the plate boundary (Fig. 1.16). In the Caribbean Sea, Virgin Islands National Park lies where oceanic crust is being deformed as the Caribbean and North American plates slide past one another.

Hotspots

Some chains of volcanoes lie within the interiors of plates, rather than along the edges (Fig. 1.17). The volcanoes are progressively older away from the largest

Transform Plate Boundary

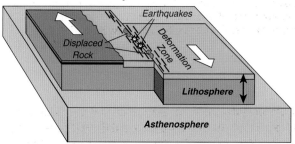

Earthquakes
Displaced
Rock
Deformation Zone
Lithosphere
Asthenosphere

FIGURE 1.15 Plates shearing past one another at a transform plate boundary result in crustal deformation, shallow earthquakes, and large lateral displacement of rocks.

**Point Reyes National Seashore
(Transform Plate Boundary)**

Sedimentary/Metamorphic Rocks
Tomales Bay
San Andreas Fault
Granite

FIGURE 1.16 Displaced rock in national park land along ▶ transform plate boundary. Granite in Point Reyes National Seashore has moved more than 250 miles (400 kilometers) northward along the San Andreas Fault, the main fault accommodating the transform motion between the Pacific and North American plates.

(a) Oceanic Hotspot

(b) Continental Hotspot

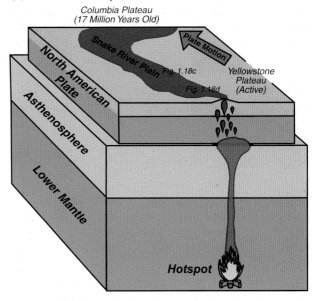

FIGURE 1.17 Chains of volcanoes form over hotspots. a) Hawai'i Hotspot Track. As the Pacific Plate moves over the hotspot, volcanic islands form; the islands are older in the direction of plate motion. Away from the hotspot, volcanic activity ceases and the islands stop growing. The islands subside as the plate cools and are further reduced in height by erosion. The Big Island of Hawai'i, including Hawai'i Volcanoes National Park, is directly above the hotspot, where it experiences extensive volcanic activity. Haleakalā National Park, on the island of Maui, has moved off the hotspot, but still has some eruptions. **b)** Yellowstone Hotspot Track. Volcanic rocks (red pattern) formed as the North American Plate moved southwestward over a hotspot. John Day Fossil Beds National Monument, Craters of the Moon National Monument, and Yellowstone National Park represent different stages of passage of the plate over the hotspot.

**(a) Haleakalā National Park
(Passed Over Oceanic Hotspot)**

**(b) Hawai'i Volcanoes National Park
(Oceanic Hotspot)**

**(c) Craters of the Moon National Monument
(Passed over Continental Hotspot)**

FIGURE 1.18 National park landscapes formed above hotspots. Hawai'i Hotspot Track:
a) Haleakalā National Park is on the island of Maui, which developed as that part of the Pacific Plate moved over the hotspot about a million years ago. **b)** Most of the current volcanic activity occurs in Hawai'i Volcanoes National Park, where the Big Island of Hawai'i is forming directly above the hotspot. Yellowstone Hotspot Track: **c)** Craters of the Moon National Monument in Idaho is now covered with basalt lava flows and cinder cones, formed after the area now occupied by the Snake River Plain moved over the hotspot. **d)** Hot springs, geysers, and rhyolite lava flows in Yellowstone National Park signify that the northwestern corner of Wyoming lies directly above the hotspot.

**(d) Yellowstone National Park
(Continental Hotspot)**

and most active volcano. Earthquake and volcanic activity in Haleakalā and Hawai`i Volcanoes national parks is the result of the Pacific Plate moving northwestward over a hotspot rising from the deep mantle (Fig. 1.18a,b). The spectacular geysers and hot springs in Yellowstone National Park and the volcanic rocks in Craters of the Moon National Monument result from southwestward movement of North America over another hotspot (Fig. 1.18c,d).

Building the North American Continent

The parks mentioned in the preceding sections fit nicely into the plate-tectonic scheme because their landscapes can be explained in terms of ongoing or past activity at or near a plate boundary or hotspot. Plate-tectonic interpretations of other parks are not as straightforward because they involve more complicated, and sometimes more controversial, concepts. The geologic features in those parks can be appreciated by examining the history of development of the North American continent (Fig. 1.19).

Continents tend to grow outward. The **craton** is the internal portion of the continent that has undergone very little deformation over the last several hundred million years. It consists of a region of very old rocks, known as the continental **shield**, and the adjacent continental **platform**, an area covered by a thin veneer of sedimentary layers (Fig. 1.20a,b). Those layers were deposited by shallow seas that periodically covered portions of the continent.

In places, the edge of the North American craton has been subjected to deformation and uplift. Some of our best-known national parks, such as Grand Canyon, Zion, and Mesa Verde, lie on the **Colorado Plateau**, a region that has been gradually uplifting during the past few million years (Fig. 1.20c). Other parts of the craton have been broken into smaller blocks and thrust upward. The **Laramide Uplifts** form some of the spectacular scenery in Rocky Mountain National Park, Florissant Fossil Beds National Monument, and other parks in the frontal ranges of the Rocky Mountains. Forces causing the uplifts may relate to times when the oceanic plate was subducting at a low angle, so that the

(a) Ancient North America

(b) Low-angle Subduction

(c) Terrane Accretion

◄ **FIGURE 1.19 Building the North American continent.** The craton is the ancient nucleus of the continent. It consists of a continental shield of very old rocks, covered on its edges by thin sedimentary layers on the continental platform. Deformation along the edge of the craton has uplifted strata on the Colorado Plateau, and created **foreland structures**. The continent has been growing outward as accreted terranes slam into the edge the craton. **a)** About 500 million years ago, the craton was bounded by passive continental margins, with seas periodically rising and depositing sedimentary layers over the platform region. The eastern margin was later deformed during continental collisions as the Appalachian Mountains; part of the western passive margin and platform were uplifted as the Colorado Plateau. **b)** Part of the craton was compressed and squeezed upward as foreland structures known as Laramide Uplifts, perhaps when the oceanic plate was subducting at a low angle. **c)** Plate convergence has brought in blocks of crust that accreted to the edge of the craton, a process that continues to build the continent westward.

FIGURE 1.20 National park lands record the history of development of the North American continent. a) Very old igneous and metamorphic rocks of Voyageurs National Park in northern Minnesota hint at ancient plate-tectonic processes that formed the continental shield. **b)** Flat-lying sedimentary layers at Badlands National Park in South Dakota record the advance and retreat of shallow seas across the continental platform, as well as deposits due to uplift and erosion of the nearby Black Hills. **c)** Layers in Canyonlands National Park in Utah represent deposition of marine and continental strata that were exposed when a large block on the edge of the craton, the Colorado Plateau, recently uplifted. **d)** Rocky Mountain National Park is in the Front Range of Colorado, where compression due to low-angle subduction caused igneous and metamorphic ("crystalline basement") rocks of the craton to be thrust upward. **e)** Rocks of North Cascades National Park in Washington State formed elsewhere, were transported via plate movement, and were accreted to the edge of the continent.

(a) Voyageurs National Park (Continental Shield)

(b) Badlands National Park (Continental Platform)

(c) Canyonlands National Park (Uplifted Platform)

(d) Rocky Mountain National Park (Laramide Uplift)

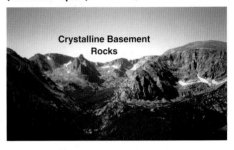

(e) North Cascades National Park (Accreted Terranes)

top of the descending plate was not deep enough to dehydrate and produce volcanoes. Instead, convergence of the two plates caused compression beneath the interior of the continent, uplifting dramatic mountain ranges along faults (Fig. 1.20d).

Much of western North America consists of pieces of crust that formed elsewhere and then slammed into the edge of the continent. Such **accreted terranes** form the landscapes of North Cascades National Park, as well as some of the spectacular Alaskan parks, including Denali and Wrangell–St. Elias (Fig. 1.20e).

Effects of more than one type of tectonic event are apparent in many parks. For example, the Grand Canyon owes its great depth to erosion accompanying uplift of the Colorado Plateau, but the western portion of the park has faulting and volcanic rocks characteristic of continental rifting in the adjacent Basin and Range Province. The spectacular layering so characteristic of the canyon is a product of an earlier time, when western North America was part of a passive continental margin. Big Bend National Park lies at the juncture of three tectonic provinces: it is at the end of the Appalachian/Ouachita/Marathon continental collisional zone, but was later subjected to compression forming Laramide Uplifts, and then to Basin and Range continental rifting.

FURTHER READING

GENERAL

Chronic, H. 1986. *Pages of Stone: Geology of Western National Parks and Monuments*. Seattle: The Mountaineers. 4-vol. series.

Elwood, B. B. 1996. *Geology of America's National Park Areas*. Upper Saddle River, NJ: Prentice Hall. 372 pp.

Harris, A. G., E. Tuttle, and S. P. Tuttle. 2004. *Geology of National Parks*, 6th Ed. Dubuque: Kendall/Hunt. 882 pp.

Kiver, E. P., and D. V. Harris. 1999. *Geology of U.S. Parklands*, 5th Ed. New York: John Wiley. 902 pp.

McPhee, J. 1998. *Annals of the Former World*. New York: Farrar, Straus and Giroux. 696 pp.

Moores, E. M. (editor). 1990. *Shaping the Earth: Tectonics of Continents and Oceans, Readings from Scientific American*. New York: Freeman. 206 pp.

Van Matre, S., and B. Weiler (editors). 1983. *The Earth Speaks*. Greenville, WV: The Institute for Earth Education. 187 pp.

Wilson, J. T. (editor). 1976. *Continents Adrift and Continents Aground: Readings from Scientific American*. New York: Freeman. 230 pp.

TECHNICAL

Bally, A. W., and A. R. Palmer (editors). 1989. *The Geology of North America: An Overview*. Boulder, CO: Geological Society of America, Decade of North American Geology, v. A. 619 pp.

Cox, A. (editor). 1973. *Plate Tectonics and Geomagnetic Reversals: Readings with Introductions by Allan Cox*. San Francisco: Freeman. 702 pp.

Dewey, J. F., and J. M. Bird. 1970. Mountain belts and the new global tectonics. *Journal of Geophysical Research*, v. 75, p. 2625–2647.

Kay, M. 1951. *North American Geosynclines*. Boulder, CO: Geological Society of America, Memoir 48. 143 pp.

Kearey, P., and F. J. Vine. 1996. *Global Tectonics*, 2nd Ed. Oxford: Blackwell Science. 333 pp.

Le Pichon, X. 1968. Sea-floor spreading and continental drift. *Journal of Geophysical Research*, v. 73, p. 3661–3697.

Moores, E. M., and R. J. Twiss. 1995. *Tectonics*. New York: Freeman. 415 pp.

Strahler, A. N. 1998. *Plate Tectonics*. Cambridge, MA: Geo-Books. 554 pp.

Sullivan, W. 1992. *Continents in Motion: The New Earth Debate*, 2nd Ed. New York: American Institute of Physics. 430 pp.

Vine, F. J. 1966. Spreading of the ocean floor: New evidence. *Science*, v. 154, p. 1405–1415.

Wegener, A. L. 1924. *The Origin of Continents and Oceans* (trans. by J. G. A. Skerl). London: Methuen. 245 pp.

WEBSITES

National Park Service: www.nps.gov

Park Geology Tour:
www2.nature.nps.gov/grd/tour/index.htm

Park Geology Photos:
www2.nature.nps.gov/grd/edu/images.htm

Individual Park Photos:
www.nps.gov/pub_aff/imagebase.html

U.S. Geological Survey: www.usgs.gov

Geology in the Parks:
www2.nature.nps.gov/grd/usgsnps/project/home.html

National Park Photos:
libraryphoto.er.usgs.gov/parks1.htm

U.S. Forest Service: www.fs.fed.us

Geology:
www.fs.fed.us/geology/mgm_geology.html

**Bureau of Land Management:
www.blm.gov/nhp/index.htm**

Environmental Education:
www.blm.gov/education

**National Association for Interpretation:
www.interpnet.com**

National Science Teachers Association: www.nsta.org

Plate Tectonics:
www.scilinks.org/retrieve_outside.asp?sl=
92635699108810331055

University of California–Santa Barbara

Plate Tectonic Animations:
transfer.lsit.ucsb.edu/geol/projects/emvc/
cgi-bin/dc/list.cgi?lis

**National Center for Science Education:
www.ncseweb.org**

2

Geologic Features and Processes

▲▲▲▲▲▲ *"Understanding the processes responsible for the spectacular mountains, valleys, and coastlines of national parks helps us to appreciate the importance of geologic features, and to devise ways to make them accessible."*

What do geologists notice, measure, and think about? Thomas Nolan remarks that, "Geologists habitually think in three dimensions, instinctively use the fourth dimension of time, and appreciate the inevitability of change."

The preceding chapter described how the landscapes in national parks, monuments, seashores, and other areas form at plate boundaries and hotspots. Various types of rocks, geologic structures, volcanoes, mountain ranges, and earthquakes in national parks result from geologic processes accompanying the plate movements (Fig. 2.1). This chapter presents an in-depth look at those processes.

FIGURE 2.1 **Geologic features and processes are evident in Cabrillo National Monument, California.** The sedimentary layers exposed on the cliffs tell of past episodes of erosion, transportation, and deposition of rock particles, then compaction and hardening into rock. Tilting of the lower layers indicates deformation associated with uplift of the strata out of the sea. The waves pounding on the cliffs cause renewed erosion, and beach sand represents ongoing deposition. The dynamic coastline is one indication of shearing of the continental edge associated with the San Andreas transform plate boundary.

National parks are often about superlatives: Yellowstone has the largest concentration of geysers and hot springs in the world; Crater Lake the deepest lake in the United States; Hawai`i Volcanoes one of the most volcanically active places on Earth. The National Park Service has two important, yet often conflicting, missions: 1) to preserve areas of outstanding physical, historical, or cultural significance; and 2) to make those areas accessible to the public. The conflict between preservation and access of geologic features raises important environmental, natural resource, and political questions. Are some geologic features so special that they demand protection, with little or no access by the public? But if so little access is allowed, then what's the point of preservation? Understanding the processes responsible for the spectacular mountains, valleys, and coastlines of national parks helps us to appreciate the importance of geologic features, and to devise ways to make them accessible.

GEOLOGIC TIME

Some geologic processes in parks, such as earthquakes, volcanic eruptions, or the erosion of coastlines, are evident because they involve change we can observe during our lifetimes. But many geologic features form through very slow processes occurring over incomprehensibly long periods of time.

Geology's most important contribution to human thought may be that, according to scientific observation, the Earth is about 4.5 billion years old. The great length of geologic time is difficult to appreciate, given that the span of a human life is only a few decades. For perspective, consider that by some accounts, based on cultural and religious beliefs, the Earth is about 6,000 years old. Consider that 4.5 billion years is about *one million* times as long as 6,000 years. With only 6,000 years available, it is necessary for processes to occur very quickly to form continents, oceans, mountains, valleys, and other features. With a million times as much time to work with, however, those features could develop through slow processes acting over long periods of time. Whether one accepts the Earth as very old (4.5 billion years) or relatively young (6,000 years), the controversy affects thinking far beyond geology and other sciences. The debate encompasses history, philosophy, and religion, involving ideas at the very heart of the nature of people and the universe. Features in many national parks, such as the layering of Grand Canyon in Arizona or the fossils at John Day in Oregon, provide important observations that help us understand the nature of geologic time.

Geologic Time Scale

Table 2.1 is a simplified version of the geologic time scale. The divisions relate to forms of life that occurred at various times in the geologic past. The 4.5 billion

---

DEEP TIME

Large amounts of both distance and time are hard to imagine. Our bodies occupy about one square foot of space, and five or six feet of height. We can relate that to the dimensions of our house and the town we live in. But distance becomes harder to grasp when we consider a trip across the country or to Europe—our bodies and homes seem so small. We've traveled to the moon and may think it's far away, yet a lunar round-trip is only 500,000 miles, equivalent to going around the Earth 20 times (some of those '57 Chevys and Volkswagen buses have done that!). But think about Mars. A round-trip journey there would be about 1,000 times as far as a moon excursion. And a trip to the nearest star (Alpha Centauri) is about 100,000 times the Mars journey, or 100 million moon trips. Imagine going to the nearest galaxy (Andromeda), equivalent to 10 trillion moon trips! It is clear that, based on the size of our bodies and our living space, we cannot begin to envision such distance. The concept of **deep space** sums up our inability to comprehend distances so large.

Likewise it is difficult to imagine geologic time. We can understand a day of our lives, and maybe even the idea that our lifetime covers about 30,000 days. We might be surprised to find that the 6,000 years of recorded history is equivalent to only about 75 of our lifetimes. We can imagine that. But what about 5 million years ago, when our human-like ancestors first came on the scene? That would be equivalent to 60,000 lifetimes. And what about the heyday of the dinosaurs 200 million years ago, roughly equivalent to 3,000,000 of our lifetimes? Now consider that the 4.5 billion years since the Earth formed spans about 60,000,000 human lifetimes. So, analogous to our inability to imagine large distances, we can think of **deep time** as symbolizing the futility of trying to comprehend how long the Earth has been around.

(4,500 million) years of Earth history are first broken into four **eons**. In the *Hadean* and *Archean* eons, spanning nearly half of geologic time (from 4.5 to 2.5 billion years ago), there were only very primitive bacteria and algae. The *Proterozoic* Eon, from 2.5 billion to 543 million years ago, saw the development of primitive aquatic plants. In the *Phanerozoic* Eon, beginning 543 million years ago, life abruptly flourished and evolved to the complex plants and animals we know today.

The Phanerozoic Eon is broken into finer divisions called **eras**, which in turn are divided into **periods** and **epochs**. The first period of the Phanerozoic Eon is the Cambrian. Rocks older than the Phanerozoic Eon (that is, older than the Cambrian Period) are sometimes collectively termed **Precambrian**. At the beginning of the *Paleozoic* (meaning "time of early life") Era, marine invertebrates suddenly flourished, followed progressively by the development of fish and early land plants. In the *Mesozoic* ("middle life") Era, trees ap-

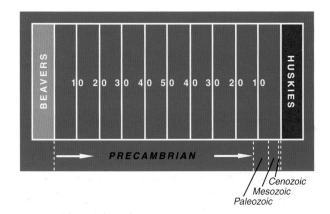

FIGURE 2.2 The geologic time scale (Table 2.1) compared to the length of a football field.

peared, along with dinosaurs and the first birds and mammals. The *Cenozoic* ("modern life") Era saw the rise of mammals and, very recently, humans.

We can visualize the enormous span of geologic time, and where our own lives fit in, by putting the 4.5 billion years between the goal lines of a 100-yard football field (Fig. 2.2). The span of very primitive life (Precambrian) would stretch all the way across midfield to the opponent's *12 yard line*! Extinction of the dinosaurs (end of the Mesozoic Era, 65 million years ago) would be between the 1 and 2 yard lines. The appearance of the earliest human ancestors (5 million years ago) would be *4 inches* from the goal line. The beginning of recorded history (about 6,000 years ago) would be *5 one-thousandths of an inch* (.005 inch) from the goal line, less than the thickness of the "dimples" on the leather ball!

How Old Is This Rock?

The rocks preserved in national parks provide a remarkable account of geologic history. They record events such as seas coming in and out of a region, or the presence of nearby volcanoes. Age dating techniques help us determine the sequence of events and when they occurred.

ABSOLUTE AGE DATING. To appreciate how old rocks are and the rates at which geologic processes occur, we must know the actual age of Earth materials. **Absolute age dating** techniques rely on the fact that certain things happen, at known rates, after a rock or sedimentary layer forms (Table 2.2). Isotopic methods are based on knowing the time it takes for half of a radioactive element to decay to another form. The length of time is known as the **half-life**. For example, the isotope uranium-235 decays to lead-207, with a half-life of

TABLE 2.1 Geologic time scale. Note that, in this portrayal, the Phanerozoic Eon is broken into much finer divisions than the three earlier eons, known collectively as the Precambrian.

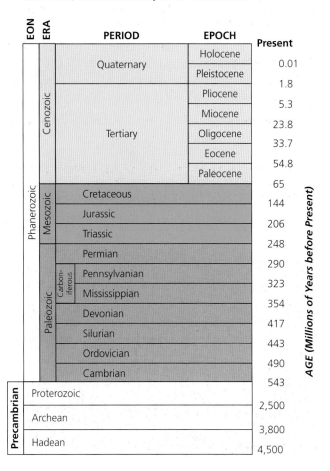

EON	ERA	PERIOD	EPOCH	AGE (Millions of Years before Present)
				Present
Phanerozoic	Cenozoic	Quaternary	Holocene	0.01
			Pleistocene	1.8
		Tertiary	Pliocene	5.3
			Miocene	23.8
			Oligocene	33.7
			Eocene	54.8
			Paleocene	65
	Mesozoic	Cretaceous		144
		Jurassic		206
		Triassic		248
	Paleozoic	Permian		290
		Carboniferous — Pennsylvanian		323
		Carboniferous — Mississippian		354
		Devonian		417
		Silurian		443
		Ordovician		490
		Cambrian		543
Precambrian		Proterozoic		2,500
		Archean		3,800
		Hadean		4,500

about 700 million years. If a mineral within a rock contains half uranium-235 and half lead-207, then the rock formed about 700 million years ago; if the mineral has only one-fourth uranium-235 and three-fourths lead-207, then the rock formed 1,400 million (1.4 billion) years ago.

RELATIVE AGE DATING. The sequence of geologic events can aid our understanding of the geologic history of a region, even if we don't know exactly when each event occurred. Such **relative age dating** is analogous to observing the members of an extended family. Even if we don't know the exact ages of each family member, it's clear that the grandmother is older than the mother, and the mother is older than the daughter. A basic principle of geology is known as **superposition**. As we cook pancakes and place one on top of another, we know that the one cooked first is at the bottom of the stack, and that the pancakes get younger upward. Superposition assumes that, *in a sequence of sedimentary layers, the oldest layers are at the bottom and the youngest at the top*. (Exceptions occur where deformation overturns layers.) At Grand Canyon National Park, we know that the layers at the bottom of the canyon were formed before those higher up on the canyon walls (Fig. 2.3).

Fossils, the preserved remains or other evidence of plants and animals, are used extensively to determine the relative ages of rock layers. Species generally exist for only a few million years before becoming extinct. If we know that a certain species lived before or after other species, we can thus determine that sedimentary layers are older or younger than other layers. We can then establish the sequence of geologic events, such as structural deformation or volcanic activity, that affected the layers.

Reading Geologic History

Sedimentary layers can be viewed as pages of a book, telling a sequential geologic story. But there are places where the geologic story is incomplete because of nondeposition or erosion, as if some of the pages were

	TECHNIQUE	MATERIAL DATED	DATES MATERIAL FORMED THIS MANY YEARS AGO
Isotropic	**Uranium to Lead**	Minerals	1 Million to 4.5 Billion
	Rubidium to Strontium	Minerals	60 Million to 4.5 Billion
	Potassium to Argon	Minerals	10,000 to 3 Billion
	Uranium Series Disequilibrium	Minerals, Shell, Bone, Teeth, Coral	0 to 400,000
	Carbon-14	Minerals, Shell, Wood, Bone, Teeth, Water	0 to 40,000
Radiation Exposure	**Fission Track**	Minerals, Natural Glass	500,000 to 1 Billion
	Thermo-luminescence	Minerals, Natural Glass	0 to 500,000
	Electron Spin Resonance	Minerals, Tooth Enamel, Shell, Coral	1,000 to 1 Million
Other	**Geomagnetic Polarity**	Minerals	780,000 to 200 Million
	Amino Acid Racemization	Shells, Other Biocarbonates	500 to 300,000
	Obsidian Hydration	Natural Glass	500 to 200,000
	Dendro-chronology	Tree Rings	0 to 12,000
	Lichenometry	Lichens	100 to 9,000

TABLE 2.2 Age dating methods. Each technique relies on the fact that certain things happen, at known rates, after a specific type of Earth material forms. There is a limited range of dates for which each technique is reliable.

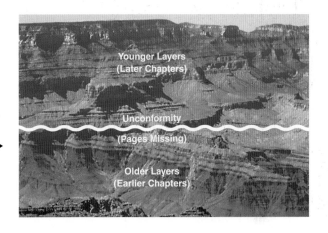

FIGURE 2.3. The rock layers of Grand Canyon National Park are like the pages of a book, recording parts of the past 2 billion years of Earth history. A hike down into the canyon is through flat-lying, younger layers, then older strata that were tilted and their tips eroded away. The unconformity between the two sequences is like pages missing from the book.

(a) Angular Unconformity

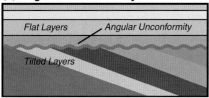

Flat Layers   Angular Unconformity

Tilted Layers

(b) Disconformity

Younger Layer   Disconformity

Older Layer

(c) Nonconformity

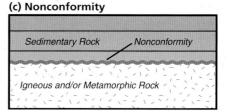

Sedimentary Rock   Nonconformity

Igneous and/or Metamorphic Rock

FIGURE 2.4 Types of unconformities. a) At an angular unconformity the lower rock layers are tilted at a different angle than the overlying layers. **b)** A disconformity separates rock layers that are parallel. **c)** A nonconformity is a surface where sedimentary layers are deposited on top of igneous or metamorphic rocks.

never written, or were written but later ripped out of the book. An **unconformity** is the surface that represents the missing pages of the geologic history book.

Unconformities preserved in national parks tell us important things about the natural history of a region (Fig. 2.4). An **angular unconformity** separates a tilted sequence of layers from horizontal layers above. It indicates a sequence of four geologic events: 1) sedimentary layers were deposited; 2) deposition stopped and the region was exposed to erosion; 3) structural deformation (folding or faulting) tilted the layers and their ends were truncated by erosion; 4) another sequence of sedimentary layers was deposited on top of the tilted layers. Sedimentary layers commonly, but not always, are deposited in oceans or shallow seas. Periods of deposition (events 1 and 4) commonly coincide with advance of the sea as the region subsided or the sea level rose. A period of non-deposition (event 2) relates to uplift of the region above the sea, or a drop in sea level. A **disconformity** separates sequences of parallel layers and implies a similar history, but without the episode of deformation and tilting. Disconformities can tell us about seas periodically advancing over continental shelf and platform regions, then retreating and leaving those regions subject to erosion. At a **nonconformity**, sedimentary strata are deposited on top of igneous or metamorphic rocks. A nonconformity often represents a profound amount of uplift and erosion, because igneous and metamorphic rocks can form many miles below the surface.

The coastal areas of Olympic National Park contain some spectacular angular unconformities. That is, the strata below the unconformity surface are tilted at an angle, whereas those above are nearly horizontal (Fig. 2.5a). The angular unconformities in Olympic illustrate that older sandstone and shale layers were tilted as they were lifted out of the sea, then partially eroded. Flat-lying glacial deposits later covered some of those truncated layers. The angular unconformities thus represent the missing time interval between deposition of the sandstone and shale layers (about 20 million years ago) and the glacial material (less than a million years ago). Disconformities at Grand Canyon National Park illustrate that the region at times underwent episodes of invasion by the sea accompanied by deposition of sedimentary layers, then a period of erosion as the sea moved out, followed by deposition of more sedimentary layers as the sea once again invaded (Fig. 2.5b). At the bottom of Grand Canyon, the Colorado River has carved down to igneous and metamorphic rocks that formed during a period of ancient mountain building 1.7 billion (1,700 million) years ago (Fig. 2.5c). In places, sedimentary layers deposited only 540 million years ago rest directly on top of those rocks. The nonconformity separating the two sequences is thus a surface representing more than a billion years of missing geologic history. During that time, the region must have experienced an enormous amount of uplift, as erosion removed about 10 miles (16 kilometers) of rock.

CRUSTAL DEFORMATION AND MOUNTAIN BUILDING

Huge forces occur where plates interact along their boundaries. Those forces result in earthquakes, the deformation of rock into geologic structures, and the uplift of mountain ranges.

Forces that deform the crust into mountains depend on the type of plate boundary present, and whether oceanic or continental crust caps each plate. The crust responds to the forces by faulting and folding, and in places rises to great heights to maintain a state known as **isostatic equilibrium**.

(a) Angular Unconformity

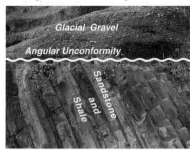

(b) Disconformity

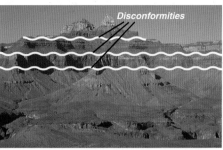

(c) Nonconformity

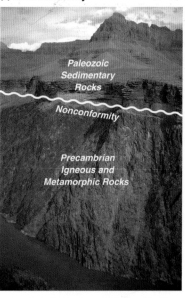

FIGURE 2.5 Unconformities in national parks. a) Beach 3 north of Kalaloch in Olympic National Park. The tilted sandstone and shale layers below the angular unconformity are about 20 million years old, whereas the glacial gravel layers above are less than 1 million years old.
b) A disconformity is harder to identify because it occurs parallel to layering. The missing time it represents could be very small or substantial. Sedimentary strata at Grand Canyon National Park contain many significant disconformities, recording the advance and retreat of shallow seas.
c) A prominent nonconformity in the lower part of Grand Canyon is an erosional surface between old igneous and metamorphic rocks of the Inner Canyon, and younger sedimentary layers above.

Isostatic Equilibrium

The term **isostasy** is from the Greek *isos* ("equal") and *stasis* ("standing"). A broad region of high topography, such as a mountain range or a plateau, represents an enormous weight of rock. That downward force must be balanced by an equal upward force somewhere beneath the mountains or plateau. **Isostatic uplift** results when the upward force from below exceeds the weight of the mountains or plateau, and the region rises just enough so that the additional topographic weight balances the upward force. Buoyancy of low-density material within the Earth commonly provides the upward force. The situation is similar to taking a beach ball into a swimming pool and placing it under your belly. The buoyancy of the ball lifts just enough of your body out of the water to weight down the ball. You and the beach ball are then in a state of isostatic equilibrium.

There are two situations where broad mountain ranges form to compensate low-density material at depth. At divergent plate boundaries, where the lithosphere is thin, the underlying asthenosphere is shallow (Fig. 2.6a). Asthenosphere has relatively low density because it is hot, expanded mantle. It provides the buoyancy that elevates the topography at mid-ocean ridges and continental rift zones. National parks in the Basin and Range Province, a continental rift zone in the western United States, are at high elevation because of hot mantle beneath the region. Similarly, some of the high elevation of Yellowstone and Hawai`i

Volcanoes national parks can be attributed to the fact that these parks lie above hotspots.

Where the crust is thick, a low-density "root" sticks downward into the denser mantle (Fig. 2.6b). The resulting buoyancy causes uplift of the crust until just enough topography is created to weight down the root. The average elevation of 2½ miles (4 kilometers) in the Himalayan Mountains and Tibetan Plateau results from the crust doubling its thickness as India collides with Asia. Park lands in ancient continental collision zones, like the Appalachian Mountains in the eastern United States and Brooks Range in northern Alaska, have significant topography because the crust is still thicker than crust of the surrounding regions.

Geologic Structures

Geologic structures, such as the breaking of rocks along faults or buckling into folds, commonly form where rocks are stressed along plate boundaries. The landscapes of many national parks are the topographic expressions of these structures. In some parks, smaller examples of faults and folds are exposed on rock faces.

FAULTS. A **fault** is a break between two blocks of rock, along which the two blocks slide relative to one another. Faulting is **brittle** deformation of rock, occurring primarily in the cold, upper portion of the crust.

(a) Elevated Asthenosphere

Continent before Rifting

Continental Crust

Lithosphere

Mantle

Asthenosphere

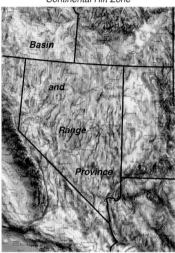

Basin and Range
Continental Rift Zone

Basin

and

Range

Province

Continental Rift Zone

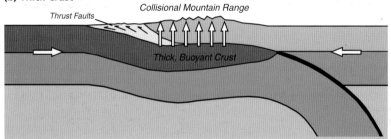

Normal
Fault

Fault-Block
Mountains

Rift
Valley

Hot, Buoyant Mantle

(b) Thick Crust

Thrust Faults

Collisional Mountain Range

Thick, Buoyant Crust

Himalayan Continental Collision Zone

Himalayas

Tibetan
Plateau

FIGURE 2.6 Regions of high topography formed by isostatic uplift. a) Hot mantle (shallow asthenosphere). Left: As a plate rips apart, the hot asthenosphere rises and expands like a hot-air balloon, elevating topography at a continental rift zone. Right: The Basin and Range Province reveals high elevations due to the presence of hot, buoyant mantle beneath. **b)** Thick crust. Left: As the crust thickens during continental collision, it rebounds upward, building just enough topography to compensate its buoyancy. Left: The Himalayan Mountains and Tibetan Plateau result from the collision of the Indian subcontinent with Asia.

The warmer, lower crust tends to fail in a **ductile** fashion, flowing slowly rather than breaking.

There are three main types of faults, described by how one fault block moves relative to the other (Fig. 2.7). The types of faults can be described by envisioning the layers of rock before and after faulting. In the nineteenth century, coal miners in Pennsylvania recognized faults offsetting the strata along the walls of mine shafts. They described the type of fault motion by standing across the sloping fault surface. The block where they hung their helmets they called the **hanging wall**, the one at their boots the **foot wall**. A fault where the hanging wall block moved downward, relative to the foot wall, was termed a **normal fault**. A **reverse fault** occurred where the hanging wall moved upward, relative to the foot wall. A special kind of reverse fault, where the fault surface slopes at a low angle, is called a **thrust fault**. The term **strike-slip fault** was later applied to the situation where one block slid laterally past the other.

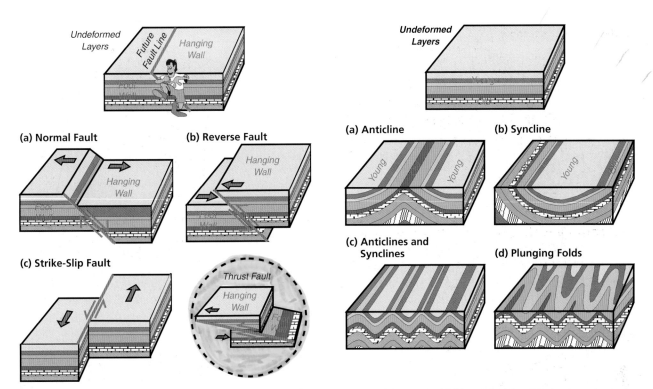

FIGURE 2.7 Types of faults. When people stand across the fault, their heads are next to the hanging wall, their feet next to the foot wall. **a)** Normal fault. The hanging wall moves down, relative to the foot wall. **b)** Reverse fault. The hanging wall moves up, relative to the foot wall. (A thrust fault is a low-angle reverse fault.) **c)** Strike-slip fault. One block slides laterally past the other.

FIGURE 2.8 Types of folds. a) Anticline. The layers are bent upward so that, on the eroded surface, the older rocks are at the center. **b)** Syncline. Layers bend downward, exposing younger rocks at the center. **c)** Series of anticlines and synclines. The surface often develops parallel ridges and valleys. **d)** Series of plunging anticlines and synclines. Characteristic "V-shaped" patterns develop on the eroded surface.

The type of faulting often relates to the type of stress that occurs at a plate boundary (Fig. 1.2). Plates tend to pull apart at divergent plate boundaries, forming normal faults. They compress at convergent plate boundaries, so that blocks are thrust over one another along reverse faults. Where plates shear past one another along transform plate boundaries, faults are predominately strike-slip. Although this simple pattern generally holds true, stress patterns are often complex, so that all three types of faults might be found along a given plate boundary.

FOLDS. A **fold** is the bending of rock layers. Folding can occur where rocks are cold and brittle, accompanied by cracks and small faults that allow the rocks to bend. Folds also form in hot, ductile rock, as the rock flows and bends instead of cracking.

As with faults, it helps to envision rock layers before and after folding (Fig. 2.8). In a normal sequence of rocks, the older layers are at the bottom, the youngest on top. An **anticline** is where the layers are folded upward; after some surface erosion, the older layers are at the center of the anticline, the younger ones toward the flanks. Where rock is bent downward, at a **syncline**, the younger layers appear at the center. Rock layers are often buckled into a series of anticlines and synclines, the same pattern you get when you push on a rug. In some places, such as the Valley and Ridge Province of the Appalachian Mountains, rock layers are folded in a more complex fashion, forming **plunging folds**. After erosion, distinct "V-shaped" patterns develop on the ground surface.

Mountain Ranges Formed by Crustal Deformation

Mountain ranges form in several ways due to deformation of Earth's crust: through pulling apart (**extension**) where plates diverge, crunching together (**compres-**

FIGURE 2.9 Mountain ranges formed at a subduction ▶ zone. The accretionary wedge (commonly a coastal mountain range) forms where sedimentary layers and hard crust of the ocean are scraped off the top of the subducting plate, compressed, and lifted out of the ocean along thrust faults. The volcanic arc forms above the position where the plate becomes so hot that water rises from its surface and melts rock in its path.

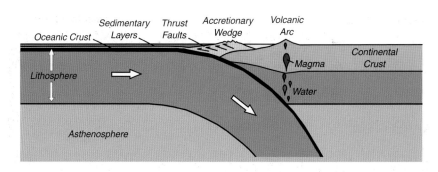

sion) where plates converge, and sliding by one another (**shearing**) at transform plate boundaries.

EXTENSIONAL MOUNTAIN RANGES. Where plates diverge, the cold, brittle part of the crust can fracture because of tensional stresses, forming a series of normal faults. **Fault-block mountains** thus form, each of the blocks separated by normal faults. The Basin and Range Province in the United States and Mexico is an example of block faulting in a continental rift zone. The rift valleys (or "basins") are down-dropped, hanging-wall blocks, while the mountain ranges are foot-wall blocks (Fig. 2.7a). Such topography is apparent in Great Basin, Grand Teton, Saguaro, and Death Valley national parks.

COMPRESSIONAL MOUNTAIN RANGES. Sedimentary layers and hard crust can be scraped off the cold portion of the down-going oceanic plate at a convergent boundary. The material is squeezed upward by compression between the two plates, resulting in a structural mountain range parallel to the volcanic chain (Fig. 2.9). The overall shape is that of a wedge that is attached, or "accreted," to the crust of the overriding plate. Such a structural mountain range, consisting of numerous folds and reverse (thrust) faults, is called an **accretionary wedge.** The Coast Range in Washington, Oregon, and northern California collectively forms an accretionary wedge. Olympic National Park, Oregon Caves National Monument, and Redwood National and State Parks showcase sedimentary strata that were deposited in the sea, then were scraped off and uplifted where the Juan de Fuca Plate plunges eastward beneath North America.

The highest elevations on Earth occur in **collisional mountain ranges** that form when an ocean basin closes because of plate convergence (see the previous discussion on isostasy). Rocks are compressed in the vise between the colliding continents, folding the sedimentary layers and thrusting them upward along reverse faults. Flying over portions of the Appalachian Mountains, an ancient continental collision zone that has been heavily eroded in the last 300 million years, one sees spectacular examples of plunging anticlines and synclines. The ancient continental collision zone in the eastern United States includes not only Acadia, Shenandoah, and Great Smoky Mountains national parks and the Blue Ridge Parkway, but also Hot Springs National Park in the Ouachita Mountains of Arkansas, and a portion of Big Bend National Park that extends into the Marathon Mountains in Texas. In northern Alaska, Gates of the Arctic National Park and Preserve is in the Brooks Range, a continental collision zone formed about 100 million years ago.

SHEARING MOUNTAIN RANGES. At transform plate boundaries, mountains develop as the plates slide by one another. The San Andreas Fault in California reveals long, sheared-up mountain ranges and valleys formed as the Pacific Plate moves northward relative to the North American Plate (Fig. 1.15). The shearing action is accompanied by numerous strike-slip faults, many that form the long mountain ranges apparent in Point Reyes National Seashore, Pinnacles National Monument, and Channel Islands National Park.

EARTH MATERIALS

The scenery of national parks includes forests, grasslands, rivers, lakes, and beaches. Those features develop on rocks that in many places are exposed and allow our inspection (Fig. 2.10). Rocks exposed in national parks tell us important things about processes that occur at plate boundaries and hotspots: hot, molten liquid cools and solidifies to form **igneous rock**; ice, wind, and water erode and transport parti-

(a) Igneous Rock

(b) Sedimentary Rock

1 Foot

(c) Metamorphic Rock

FIGURE 2.10 The three basic types of rocks are represented by the four most common rocks in Olympic National Park. a) Igneous. Basalt just beyond the tunnels on the Hurricane Ridge Road. **b)** Sedimentary. Sandstone interbedded with thin layers of shale on Beach 3 north of Kalaloch. **c)** Metamorphic. Slate two miles up the Obstruction Point Road.

cles of rock that settle and harden into layers of **sedimentary rock**; preexisting rocks recrystallize in a solid state to form **metamorphic rock**.

Three Basic Types of Rocks

Sometimes the terms *rock* and *mineral* are confused. A **mineral** is a substance that has the following properties: 1) Naturally occurring, inorganic solid. 2) Definite chemical composition. 3) Specific crystalline structure. 4) Definite physical properties that result from the

TABLE 2.3 Classification of igneous rocks.

		CHEMICAL COMPOSITION			
		% SILICA (SiO$_2$)			
		70%	60%	50%	40%
TEXTURE	**Fine-Grained Extrusive ("Volcanic")**	Rhyolite	Andesite	Basalt	
	Coarse-Grained Intrusive ("Plutonic")	Granite	Diorite	Gabbro	Peridotite

chemical composition and crystalline structure. Take, for example, the mineral quartz. Its chemical composition is SiO$_2$, meaning the crystal molecules have two atoms of oxygen (O) for every one atom of silicon (Si). Quartz forms long, six-sided crystals that come to a point. One of its physical properties is that it has hardness of 7, meaning it can scratch the mineral orthoclase (hardness 6), but not topaz (hardness 8).

A **rock** is a consolidated mixture of minerals. Generally (but not always), a rock is made of more than one kind of mineral. The mineral grains are held together as a rigid solid; either they become interlocked as they form, or the particles are cemented together in some natural way. For example, after sediment is deposited and buried, silicon and oxygen atoms can precipitate from groundwater, gluing grains of quartz and other minerals together as sandstone.

IGNEOUS ROCKS. Igneous rocks solidify from **magma**, which is molten rock that may contain gases and suspended solid material. The terms *magma* and *lava* are often used interchangeably, but differ in important ways. All melted Earth material is magma, while **lava** is magma that poured out on Earth's surface, forming **extrusive (volcanic)** rocks. When magma solidifies below Earth's surface, **intrusive (plutonic)** rocks result.

Igneous rocks are classified according to two parameters: texture and chemistry (Table 2.3). *Texture* refers to the *size of the mineral grains*, which is a func-

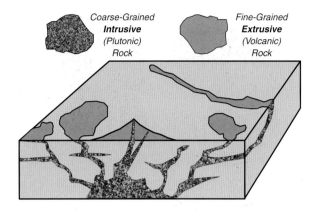

Coarse-Grained
Intrusive
(Plutonic)
Rock

Fine-Grained
Extrusive
(Volcanic)
Rock

FIGURE 2.11 Texture of igneous rocks. Magma may cool below Earth's surface, forming coarse-grained, intrusive rocks. Fine-grained, extrusive rocks form where magma pours out on Earth's surface as lava flows or other eruptive material. Volcanism refers to the fiery action at Earth's surface that Roman mythology attributed to the god Volcanus, while plutonism pertains to Pluto, Roman god of the underworld.

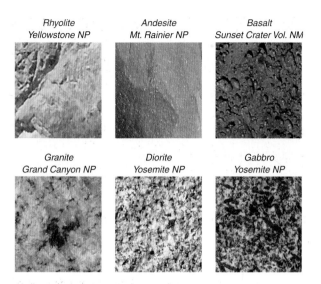

Rhyolite
Yellowstone NP

Andesite
Mt. Rainier NP

Basalt
Sunset Crater Vol. NM

Granite
Grand Canyon NP

Diorite
Yosemite NP

Gabbro
Yosemite NP

FIGURE 2.12 Examples of igneous rocks on national park lands. As in Table 2.3, fine-grained (extrusive) rocks are on top, coarse-grained (intrusive) rocks below. Silica content decreases from left to right.

tion of how quickly the magma cooled (Fig. 2.11). Intrusive rocks are *coarse-grained* because they cooled slowly within the Earth, where mineral crystals had time to grow large. Magma that encounters air or water at Earth's surface cools quickly, so that extrusive rocks have *fine-grained* minerals that you generally cannot see with your naked eye.

The chemistry of igneous rocks is commonly related to the amount of silica (silicon and oxygen) contained in the rock minerals. **Silica** is the same material that makes up window glass and, in its pure form, is the mineral quartz (SiO_2). Rocks with high silica content tend to be lightweight and light in color (Fig. 2.12). Coarse-grained (intrusive) **granite** and fine-grained (extrusive) **rhyolite** are commonly pink-to-white-colored igneous rocks that are high in silica ($\approx 70\%$ of the total rock mass is SiO_2). As the silica content decreases, minerals generally have higher percentages of heavier elements like iron, forming rocks that are darker and more dense. Intermediate silica ($\approx 60\%$ SiO_2) rocks are coarse-grained **diorite** and fine-grained **andesite** that are commonly gray. Low-silica ($\approx 50\%$ SiO_2) igneous rocks are black, including intrusive **gabbro** and extrusive **basalt**. When rock is very low in silica ($< 40\%$ SiO_2), it has a lot of iron and magnesium, so that it forms olive-green-colored **peridotite**, the dense,

coarse-grained rock comprising Earth's mantle. Crust has more silica and less iron, and is thus less dense, than mantle. Oceanic crust, made of gabbro and basalt, is somewhat denser than continental crust, which has rocks closer to the composition of granite.

SEDIMENTARY ROCKS. Most sedimentary rocks form through the cycle of: 1) erosion, chemical, or biological activity that forms particles (sediment); 2) transportation of the sediment by water, wind, or ice; 3) deposition of the sediment in oceans, lakes, or streams; 4) compaction and cementation of the sedimentary particles as they are buried beneath more sediment. Eroded particles of rocks are known as **clasts. Clastic sedimentary rocks** (for example, sandstone and shale) are composed of such particles, while **nonclastic sedimentary rocks** (such as limestone and rock salt) result from particles that precipitate out of solutions during biological or chemical activity.

Types of sedimentary rocks are best understood by imagining their environments of deposition (Table 2.4; Fig. 2.13). Gravel is found on the beds of fast-moving streams. When buried and cemented, it forms the sedimentary rock **conglomerate. Sandstone** is made of sand-sized grains, mostly quartz, that were deposited in streambeds or on beaches. In lakes or deeper parts

HOW FORMED				
			ERODED ROCK PARTICLES ("CLASTIC")	BIOLOGICAL OR CHEMICAL ACTIVITY ("NON-CLASTIC")
GRAIN SIZE	Fine	*Mud*	Shale	
		Silt	Siltstone	Lime-stone
		Sand	Sandstone	Rock salt
	Coarse	*Gravel*	Conglomerate	

TABLE 2.4 Classification of sedimentary rocks.

Shale Glacier NP

Limestone Grand Canyon NP

Sandstone Blue Ridge Pkwy

Conglomerate Blue Ridge Pkwy

of the ocean, where water is quieter, finer particles of silt and mud accumulate. Burial, compaction, and cementation turn them into **siltstone** and **shale**. In warm climates, shell fragments of marine organisms dissolve in seawater, precipitating out as calcium carbonate. This fine lime mud eventually turns into the sedimentary rock **limestone**. In dry climates where water circulation in an embayment or lake is restricted, high evaporation can leave deposits of **rock salt**.

METAMORPHIC ROCKS. A metamorphic rock forms by recrystallization of some or all of the minerals in an existing rock; the recrystallization occurs when the rock is subjected to increased temperature or pressure, but while the rock is still solid. Three general factors determine what type of metamorphic rock forms: 1) the original ("parent") rock; 2) the heat and pressure the rock endured; and 3) the presence (or absence) of fluids within the rock. The classification of metamorphic rocks is extensive and technical. Metamorphism, nonetheless, can be understood on a simple level if one imagines what happens to a specific parent rock as it is buried deeper and deeper within the Earth (that is, as it encounters increasing temperature and pressure). For example, as the sedimentary rock shale is initially buried, it begins to compress and heat up, forming the low-grade metamorphic rock **slate** (Table 2.5; Fig. 2.14). With deeper burial, larger crystals develop (especially the flaky mineral mica), forming intermediate-grade **phyllite** and then **schist**. Finally, at great depth where the temperature and pressure are high, large crystals of blocky minerals such as quartz and feldspar form, resulting in a highly-metamorphosed rock called **gneiss** (pronounced "nice"). If the rock is buried deeper, it becomes so hot that it melts, forming magma.

FIGURE 2.13 Examples of sedimentary rocks on national park lands. As in Table 2.4, fine-grained rocks are on top, with grain size increasing downward. Clastic rocks are on the left, non-clastic to the right.

LEAVERITE

Wonderful examples of rocks, minerals, and fossils occur on national park lands, some with names like rhyolite, andesite, quartzite, muscovite, and coprolite. But national parks, monuments, seashores, and other areas are set aside to protect such treasures. By law, no material, including plants, animals, rocks, minerals, or fossils, may be removed from park lands without special permission. So please admire these precious materials, but "leav-'er-ite where you found 'er!"

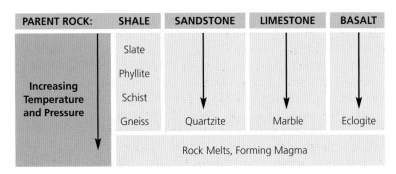

TABLE 2.5 Classification of metamorphic rocks.

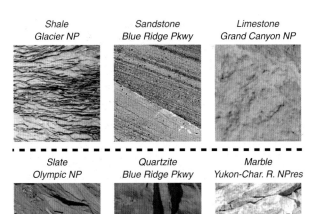

FIGURE 2.14 Metamorphic rocks in national parks. Rocks are arranged as in Table 2.5, with parent rocks on the top of each column, and metamorphic grade increasing downward.

VOLCANISM

Many of our national parks and monuments were established because of their spectacular volcanic features. The parks are natural classrooms where you can learn how magma erupts, the types of volcanoes and volcanic rocks that form, and how volcanoes affect people.

Why Rocks Melt

Contrary to popular conceptions, magma intruding through Earth's crust does *not* originate from the molten core. The iron/nickel liquid of Earth's outer core is far too dense to rise to the surface. The source of magma is much shallower, at depths of the lower crust and upper mantle.

Melting of rock generally occurs under two circumstances. 1) Hot material that was solid because of enormous pressure rises to where the pressure is less (**decompression melting**). 2) Material that was cold near Earth's surface is pushed to depth, where it heats up and releases water; the water rises and melts rock in its path (**hydration melting**).

DECOMPRESSION MELTING. Decompression melting can be understood by considering a pressure cooker. Under high pressure, water in the cooker remains liquid at temperatures considerably above its normal boiling point (212°F; 100°C). Removing the lid from the cooker suddenly releases the pressure, causing the hot water to flash to steam. Similarly, hot mantle that is solid under high pressure will begin to melt ("flash to liquid") when it rises and decompresses at divergent plate boundaries and hotspots (Fig. 2.15a,c).

The mantle is mostly iron and magnesium, with about 40% silica, forming the rock peridotite (Table 2.3). Under pressure, the peridotite of the hot mantle is solid. But as the asthenosphere part of the mantle rises and the pressure drops, it begins to melt. The magma that "bleeds off" is a bit richer in silica than the mantle, so it forms gabbro and basalt (50% silica). Where thin (and relatively low-silica) oceanic crust caps tectonic plates in divergent or hotspot settings, the magma does not have to melt much crustal material; the basalt/gabbro composition remains intact. Vast amounts of basaltic magma pour out along mid-ocean ridges, creating the longest mountain range on Earth (≈35,000 miles, or 60,000 kilometers, long; Figs. 2.16a and 1.8). The

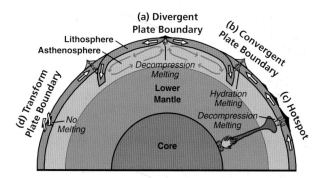

FIGURE 2.15 Reasons for volcanism. a) Drop in pressure on the asthenosphere as it rises at a divergent plate boundary causes some of the already hot material to melt. b) Heating of the crust of the descending plate at a convergent plate boundary releases hot fluids. Those fluids, mostly water, rise and cause rock to melt. c) At a hotspot, the deep mantle is heated, perhaps because there is a "bump" on the outer core. The hot mantle expands and rises in a solid state, until it is so shallow that the drop in pressure causes melting. d) At a transform plate boundary, material stays at about the same depth. There is no appreciable rise in temperature or drop in pressure; hence there is little or no volcanism.

Hawaiian Islands and their subsea extension, the Hawaiian Ridge and Emperor Seamount Chain, demonstrate how volcanic mountains developed over a hotspot (Fig. 2.16c). Hawai`i Volcanoes and Haleakalā national parks show different stages of volcanic activity related to the northwestward movement of the Pacific Plate over the hotspot. In continental rift or continental hotspot settings, rising magma must melt its way through thicker continental crust, enriching the silica content. Yellowstone National Park has rhyolite lava flows and very explosive eruptions because it lies directly above a continental hotspot.

FIGURE 2.16 Types of volcanic mountain chains. (Compare with tectonic settings in Fig. 2.15.) a) A mid-ocean ridge is a mountain range of overlapping shield volcanoes, formed at a divergent plate boundary with plates capped by oceanic crust (Mid-Atlantic Ridge). b) A volcanic arc is a chain of steep, composite volcanoes that forms above a subduction zone (Cascade Mountains in central Oregon). c) Plate motion over a hotspot leaves a chain of volcanoes that are youngest directly above the hotspot, and become older away from it (Hawai`i-Emperor Seamount Chain).

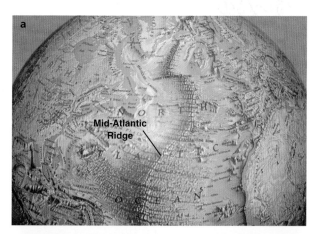

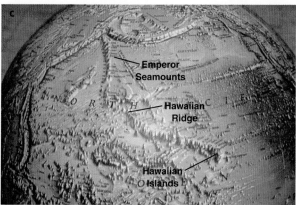

MELTING BY WETTING

It's easy to visualize hot water rising from a subducting plate, as the sedimentary layers and crust on top of the plate become so hot that water is driven off. In other words, the plate dehydrates. You might also imagine the water rising and melting rock in its path as it heats up the rock. But actually, that last part isn't quite right. The rising water is really not any hotter than the rock it rises through. But something else very important happens. Have you ever noticed that it can be a lot easier to cook food if you add some water? Rocks behave in much the same way. The presence of water can lower a rock's melting temperature. So it is really the rising water *wetting* the surrounding, already hot rock that is ultimately responsible for the generation of magma at subduction zones.

HYDRATION MELTING. We all know what happens when we get hot—we sweat! Subduction zone volcanism occurs for much the same reason (Fig. 2.15b). As you go deeper into the Earth, it gets hotter and hotter. At convergent plate boundaries, the sediment and crust on top of the subducting plate become so hot that they "sweat" fluids, primarily water. As the water rises, it melts some of the mantle and crust in its path. High-silica minerals are incorporated into the magma first, because they have low melting temperatures (Table 2.6). Magma at a convergent plate boundary may be enriched in silica, forming intrusive rocks with compositions from diorite (≈60% silica) to granite (≈70% silica), and extrusive (volcanic) rocks ranging from andesite (≈60% silica) to rhyolite (≈70% silica). In some situations, rocks with lower (≈50%) silica can also form (intrusive gabbro; extrusive basalt).

TABLE 2.6 Physical properties of igneous rocks and magmas. High-silica minerals tend to have lower melting temperatures than those with low silica. As magma cools to rock, the amount of silica influences the minerals formed and affects the rock's physical properties. High-silica minerals tend to be light in color and weight, forming rocks that have pink-to-white color and low density (granite, rhyolite). When rocks are low in silica, they have a larger proportion of heavy, dark minerals that are high in iron and magnesium; those rocks are therefore high density and are generally dark brown, black, or dark green (basalt, gabbro, peridotite). (Obsidian is an exception to the color scheme. Even though it has very high silica content, the refractive properties of the glass make it black.) Silica content tends to increase viscosity, which in turn affects the extent of lava flows, amount of trapped volatiles, and the nature of eruptions and volcanoes formed.

TEXTURE	CHEMICAL COMPOSITION (% SILICA)					
	70%	65%	60%	55%	50%	40%
Glass	Obsidian					
Fine-Grained	Rhyolite	Dacite	Andesite	Basaltic Andesite	Basalt	Komatiite
Coarse-Grained	Granite	Granodiorite	Diorite		Gabbro	Peridotite
Very Large Crystals	Pegmatite					

Common Minerals	Quartz Orthoclase-Feldspar		Amphibole Plagioclase-Feldspar		Pyroxene	Olivine
Color	Light	←		→	Black	Dark Green
Density (g/cm³)	2.7	←		→	3.0	3.3
Melting Temperature (°C)	800°	←		→	1200°	1400°
Viscosity of Magma	High	←		→	Low	
Extent of Lava Flows	Small Area	←		→	Large Area	
% Volatile Fluids	10%	←		→	1%	
Style of Eruption	Explosive	←		→	Quiet	
Types of Volcanoes	Composite Lava Dome	←		→	Shield Cinder Cone	

a) Low-Silica Lava

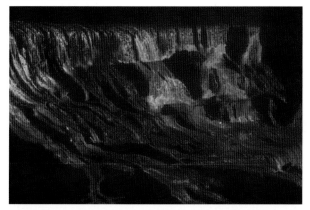

b) High-Silica Lava

FIGURE 2.17 The viscosity of lava measures its resistance to flowing. Silica is the "thickening agent" that increases the viscosity of lava. **a)** Low-silica lava in Hawaiʻi has low viscosity, capable of flowing over a long distance like a river of water. **b)** Lava that is rich in silica has high viscosity, behaving like thick paste. The sticky magma beneath Mount St. Helens would not allow gas to escape; built-up pressure was released through the violent explosion in 1980.

A chain of volcanoes forms on the overriding plate at a subduction zone (Fig. 2.9). Such chains of volcanoes are often curved, so that they are called volcanic arcs (for example, notice the Andes Mountains and Aleutian chain in Figs. 1.1 and 1.8). National parks on the Alaska Peninsula and in southern Alaska, as well as those in the Cascade Mountains in the Pacific Northwest, showcase composite volcanoes that are parts of volcanic arcs formed by subduction (Fig. 2.16b).

NO MELTING. Where plates slide past one another, materials do not deepen or shallow appreciably (Fig. 2.15d). There is no significant heating of cold crust, or decompression of hot mantle, that would induce melting. Volcanism is therefore not common at transform plate boundaries—young volcanic rocks are rare along the San Andreas Fault in California. (The movie *Volcano*, showing Los Angeles overrun by lava flows, is nonsense!)

Volcanoes and Their Eruptive Products

The style of volcanic eruption and the type of volcano formed depend on the temperature and amount of silica present in the erupting lava (Fig. 2.17; Table 2.6). Silica tends to be a thickening agent, much like flour added to pancake batter. The term **viscosity** refers to how a material *resists* flowing. Lava that is low in silica is thin and runny *(low viscosity)*—it flows freely like a fountain or river of water. High-silica lava is thick and pasty *(high viscosity)*—it flows sluggishly and sometimes explodes. **Volatile fluids** (water vapor and carbon dioxide) escape easily from fluid magma, but can be trapped under high pressure in high-silica magma, causing explosive eruptions.

LOW-SILICA LAVAS. Low-silica lava can be so fluid that it travels 10 to 30 miles per hour (15 to 50 kilome-

---

CAJUN COOKING

Growing up in the Cajun Country of South Louisiana, the author enjoyed some of the most delicious food in the world. His mother and grandmother made the best gumbos and étouffées you can imagine—muuuuuuh, ummmmmm! They always started by making a *roux*, a mixture of hot oil and flour. The more flour added, the thicker and more pasty the roux. Magma is similar, only the thickening agent is silica.

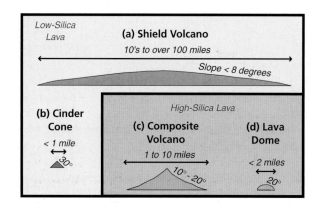

FIGURE 2.18 The size and shape of a volcano are influenced by the amounts of silica and gas in the erupting lava.
a) Low-silica (basalt) lavas produce a broad **shield volcano. b)** If basalt lava has a lot of gas, a central vent may erupt like a fire hose, forming a **cinder cone. c)** High-silica (andesite-to-rhyolite) lavas form a steeper **composite volcano. d)** A **lava dome** forms around a local eruption of high-silica lava.

ters/hour). Individual **lava flows** are commonly thin (< 20 feet or 6 meters), erupt from long fissures, and travel tens of miles. The resulting volcanoes are broad, with gentle slopes (Fig. 2.18a). From the air they look like giant warrior's shields, so they are called **shield volcanoes**. Shield volcanoes are predominately lava flows of basalt; the high amounts of iron and magnesium result in a dark, black landscape. The island of Hawai`i can be considered the tallest mountain on Earth. Starting from a depth of 18,000 feet (5,500 meters) below sea level, it rises to elevations over 13,000 feet (4,000 meters), as the giant shield volcanoes Mauna Loa ("Long Mountain") and Mauna Kea ("White Mountain") (Fig. 2.19a).

A **cinder cone** is a small volcano that forms from basaltic lava with high gas content (Fig. 2.18b). The lava is so fluid that the escaping gases result in fountains of hot liquid erupted through long fissures or central vents. As the material begins to cool and solidify, it falls to earth as particles, commonly sand-to-gravel-sized cinders. The pile of cinders steepens until it reaches a certain angle (the angle of repose), much like sand falling in an hourglass. The cinder cone can continue to grow, but the angle of repose

FIGURE 2.19 Examples of four kinds of volcanoes. a) Shield volcano, Mauna Kea, Hawai`i. **b)** Cinder cone, Wizard Island, inside Crater Lake, Oregon. **c)** Composite volcano, Mount Shasta, California. **d)** Lava dome, inside Mount St. Helens, Washington.

a

b

c

d

MORE FUN WITH FOOD!

Volcanic Sweets

The four types of volcanoes can be demonstrated with common goodies (Fig. 2.20.).

- The broad, flat appearance of a shield volcano is like a Vanilla Wafer.® Only get the chocolate variety, because basalt lava is dark!

- A steep, pointed composite volcano is like a Hershey's Kiss®. Again, be careful in your selection, as you don't want the chocolate in this case, but rather the Almond Vanilla to demonstrate light-colored andresite and rhyolite.

- A chocolate chip is like a cinder cone—dark, steep-sided, and much smaller than either a shield or composite volcano.

- A lava dome is like an M&M®—rather small and rounded on the top. The pink or red colors would best represent the high-silica, dacite-to-rhyolite lavas.

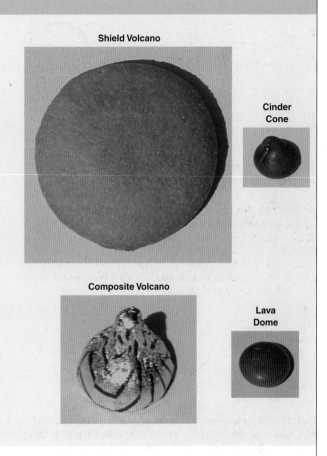

FIGURE 2.20 Common goodies represent the relative size, shape, and steepness of the four main types of volcanoes.

(commonly about 30º) remains the same. Sunset Crater Volcano National Monument, Capulin Volcano National Monument, and Wizard Island in Crater Lake National Park are all cinder cones that formed in the past 6,000 years (Fig. 2.19b). There are numerous young cinder cones in Haleakalā, Hawai`i Volcanoes, and other parks.

HIGH-SILICA LAVAS. Silica-rich lava generally does not flow more than a mile or so from its source. Individual andesite-to-rhyolite flows can be on the order of 100 feet (30 meters) thick. Because the lava is thick and pasty, gas may be trapped under high pressure, sometimes resulting in explosive eruptions such as that which occurred at Mount St. Helens in 1980. Such high-silica volcanoes are built from not only lava flows, but also stratified layers of ash, cinders, pumice, mud, and other materials; they are thus called **composite volcanoes** (or **strato-volcanoes**). The sticky lavas do not flow easily, so that steeper volcanoes result (Fig.

2.18c). Mount Shasta in the Cascade Mountains of northern California is a classic example of a steep-sided composite volcano (Fig. 2.19c). Composite volcanoes in national parks include Aniakchak in Alaska, Mount Rainier in Washington, Mount Mazama (which collapsed and is filled with Crater Lake) in Oregon, and Lassen Peak in California.

Where sticky lava concentrates in a small area, it tends to make small volcanoes with steep, rounded shapes (Fig. 2.18d). **Lava domes** are distinct by their light color, due to volcanic rocks with very high silica content (**dacite** to rhyolite). After the explosive eruption of Mount St. Helens, a lava dome formed in the resulting crater (Fig. 2.19d). The dome grew to about 1,000 feet (300 meters) high and a ⅔-mile (1-kilometer) diameter from 1980 to 1986. Other lava domes include Novarupta in Katmai National Park in Alaska, Llao Rock in Crater Lake National Park in Oregon, and O'Leary Peak near Sunset Crater Volcano National Monument in Arizona.

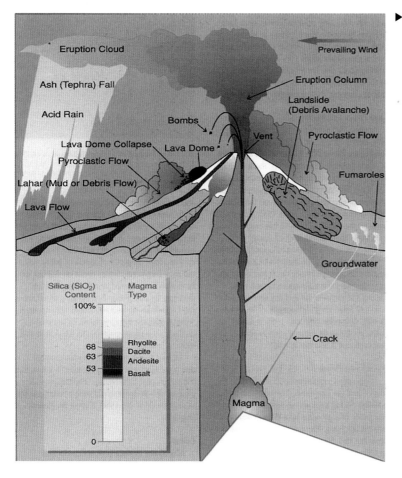

Silica (SiO₂) Content / Magma Type chart labels:
100%
68 — Rhyolite
63 — Dacite
    Andesite
53 — Basalt
0

▶ **FIGURE 2.21 Volcanic materials and hazards.** As the magma ascends, the pressure decreases and gases come out of solution. Gas bubbles expand rapidly to form a froth of molten rock, similar to the foam rising from a newly opened bottle of champagne. The bubble walls cool rapidly and harden. A mixture of gas and rock fragments, known as tephra, erupts into the atmosphere. The larger fragments, in which masses of bubbles are preserved, form pumice. The smaller fragments, mostly pieces of broken bubble walls and crystals, form volcanic ash. The mixture of hot volcanic gases, heated air, and rock fragments can be less dense than the surrounding atmosphere and rise tens of thousands of feet, forming an eruption column that may be carried by the wind as an eruption cloud. After the release of steam, magma pours out as lava flows. On steep volcanic slopes, the sharp-edged flows disintegrate and sweep across the surface as pyroclastic flows, often incorporating and melting glacial ice, producing volcanic mudflows known as lahars.

Volcanic Hazards

National park lands preserve not only a wide variety of materials erupted from volcanoes, but also the impressions of a dynamic Earth that volcanic events leave on the minds of human observers. Northwest native people, for example, witnessed eruptions of steam, ash, mudflows, and possibly lava flows from Mount Rainier during the past 5,600 years. Their stories of geologic events, while woven with mythic lore, are vivid in description. Their oral traditions tell of "blood running down the slopes" (lava flows); of the mountain losing its top and the ensuing landslide and mudslide wiping trees off valley slopes and filling the Puyallup Valley with bubble-filled stones; and of the Kent-Auburn Valley as an estuary that has since been filled in with sediment that originated from Mount Rainier mudflows.

Rapidly growing populations are infringing on volcanic areas, increasing the potential for disaster. More than 1,500 volcanoes have erupted in the world during the past 10,000 years. About 50 different volcanoes erupt every year. From 1980 to 1995, volcanic activity killed more than 29,000 people, caused about a million people to evacuate their homes, and resulted in economic losses of more than 3 billion dollars. One of the most volcanically active regions in the world is Hawai`i, evidenced by the fact that Kīlauea Volcano has been erupting continuously for the past 20 years! Volcanoes in the Cascade Mountain Range threaten communities in the Pacific Northwest, but most active composite volcanoes in the United States are where the Pacific Plate subducts beneath the Alaska Peninsula and Aleutian Islands.

Many national parks contain evidence of past volcanic events. Several have volcanoes that have erupted recently, or show signs that they will erupt in the not-too-distant future. Park areas help us appreciate volcanic hazards and how they affect our lives (Fig. 2.21).

PYROCLASTIC MATERIAL. The term **pyroclastic** (from the Greek *pur*, for "fire," and *klastos*, for "broken") refers to fragmented material that erupts from a volcano. Gas-rich magma can spew liquid and solid material high into the sky. As the liquid portion cools and solidifies, it rains down and blankets the landscape. Bombs and large blocks of pumice land within a mile or so of the vent; smaller pumice and scoria cover up to several miles, while wind carries fine ash far from the volcano, sometimes encircling the globe.

Unconsolidated accumulations of pyroclastic materials are known as **tephra** (Table 2.7). The nature of tephra depends on chemical composition as well as the size of particles. Fine-grained **volcanic ash** can originate from both high- and low-silica magmas. **Pumice** comes from high-silica lava, which has relatively low density. We all know that blocks of pumice float on water if the gas content is so great that numerous air pockets form in the rock. But if the magma is low in silica, then the rock mass is too dense for that to happen. Such material is called **scoria**, which can be large **volcanic bombs** or sand-to-pebble-sized **cinders**. Scoria is heavy and black because of its high iron content; red-colored iron oxide (rust) develops on blocks of scoria if hot water or gasses are present during the eruption.

Volcanic ash is not a by-product of combustion, like the ash from the burning of wood, leaves, or paper, but rather small jagged fragments of rock formed by the expansion of steam within rising magma. During an eruption, the force of escaping gas shatters solid rock. The ash soars into the atmosphere by force of the eruption because it is warmer and lighter than surrounding air. Once in the atmosphere, ash is blown by the winds, the heavier pieces falling out first and lighter pieces carried farther. Ash erupted from western volcanoes is prominent as far away as Badlands and Theodore Roosevelt national parks in the Dakotas and Agate Fossil Beds National Monument in Nebraska.

Volcanic ash sometimes surrounds organic material which can be radiocarbon dated. Ash layers are used as marker beds that "bracket" the ages of intervening layers, such as lahars, lava flows, and other volcanic ash where no organic material can be found. The Mazama Ash, which spewed forth from the volcano that collapsed 7,700 years ago and formed Crater Lake, blankets a large portion of the United States and Canada. It is an important marker bed for archaeologists, because layers below the Mazama Ash are older than 7,700 years, while those above are younger.

TABLE 2.7 Eruptive products of volcanoes.

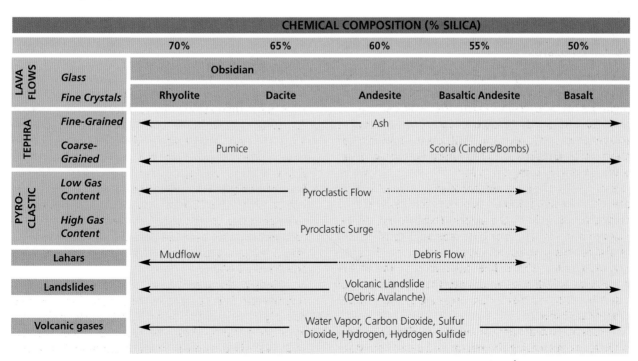

Sometimes a deadly mixture of hot ash, rock fragments, and gas moves down the slopes of a volcano during an explosive eruption or when the steep side of a lava dome collapses. Such **pyroclastic flows** can travel at speeds of more than 100 miles per hour (160 km/hr), toppling and burning everything in their path. The density of pyroclastic flows confines them to valleys. But relatively light, gas-rich versions that sweep up and over ridges are known as **pyroclastic surges**. The May 18, 1980, eruption of Mount St. Helens generated both varieties, destroying an area of over 200 square miles (500 km²), toppling trees more than 6 feet (2 meters) in diameter. The eruption of Italy's Mount Vesuvius in 79 A.D. is dramatic evidence that pyroclastic surges can choke and incinerate all people and animals in their paths.

VOLCANIC LANDSLIDES AND MUDFLOWS. Volcanic slopes can be precariously steep because of accumulation of eruptive material or injection of magma from below. If the rock has also been weakened by chemical alteration by hot water or gasses, then the potential is great for **volcanic landslides** (also known as **debris avalanches**). Once the conditions are ripe, shaking from an earthquake or saturation by heavy rainfall may trigger the landslide. In 1980, the north side of Mount St. Helens bulged out as magma moved upward; an earthquake on May 18 triggered the gigantic landslide that released pressure on trapped gasses and resulted in the huge lateral blast and mudflows.

Volcanoes can be the source of deadly mixtures of mud, sand, rock, and water. The Indonesian word **lahar** collectively describes such features, which are also called **mudflows** or **debris flows**. Lahars roar down a volcano's flanks at 20 to 40 miles per hour (30 to 60 kilometers/hour), and can continue for tens of miles farther down stream valleys. When more than half of their mass is rock debris, lahars can behave like wet concrete, carrying trees, boulders, houses, cars, and bridges. As they slow down, they become rock solid in a matter of minutes.

Lahars are associated with volcanic eruptions as well as other events. Their causes include snow and ice melted by lava or pyroclastic flows, flooding from heavy rain, sudden drainage of a lake near the volcano's summit, or landslides from oversteepened or chemically weakened slopes. The main volcanic hazard at Mount Rainier National Park is not lava flows, but rather lahars. The Osceola Mudflow 5,600 years ago and Electron Mudflow just 600 years ago were both large enough to flow down river valleys all the way to the Puget Sound region, which includes the cities of Seattle and Tacoma.

VOLCANIC GAS. About 90% of the gas emitted from volcanoes is water vapor. Significant amounts of carbon dioxide, sulfur dioxide, hydrogen, and hydrogen sulfide are also released. Gas escapes not only during eruptions, but also at quiet times through cracks and vents known as **fumaroles**. The gas can weaken rocks to a point where steep slopes fail, causing landslides that can feed devastating lahars. Hot gas is also a necessary ingredient for pyroclastic flows and surges.

Volcanic gasses can have other profound effects on areas near and away from their sources. Carbon dioxide is heavier than air, so it collects in low areas and suffocates people and animals. Sulfur dioxide gas reacts with water droplets to form acid rain, which corrodes metal and harms vegetation. If the sulfur dioxide reaches the stratosphere, it can react with water vapor to produce an aerosol of sulfuric acid. Such aerosols reflect sunlight and can lower Earth's average temperature a few degrees—snow fell in Boston the summer following the 1883 eruption of the Indonesian volcano Krakatau!

EARTHQUAKES

The magnificent landscapes characteristic of many national parks result, in part, from hundreds to thousands of movements along fault zones, accompanied by earthquakes. The Basin and Range Province, a rift in the continent that is developing into a divergent plate boundary, has shallow earthquakes that shake Great Basin and Death Valley national parks. Along the San Andreas Fault in California, a transform plate boundary, shallow earthquakes rattle Channel Islands National Park, Pinnacles National Monument, Point Reyes National Seashore, and Golden Gate National Recreation Area. In southern Alaska, the Pacific and North American plates converge, forming a subduction zone under Alaska. The 1964 Great Alaska Earthquake shook Katmai, Lake Clark, Denali, Kenai Fjords, and Wrangell-St. Elias national parks. In the Pacific Northwest, the Juan de Fuca Plate is subducting beneath the North American Plate. Although small-to-moderate-sized earthquakes have occurred there in historic time, it is thought that the plates have been locked together since the last large earthquake, 300 years ago. When the plates can no longer stand the accumulated strain, they will suddenly snap along their boundary, causing a large earthquake that may devastate the region, including Olympic, Mount Rainier, Crater Lake, Redwood, and Lassen Volcanic national parks.

What Causes Earthquakes?

Believe it or not, rocks are *elastic*—they behave like a rubber band or rubber ball. Under a small amount of stress (compression, pulling, or shearing), they gradually deform (compress, stretch, or bend). When the stress is removed, the rocks return to their original size and shape. But so much stress might be applied to a rock that it deforms permanently, either by ductile flow (like Silly Putty) or brittle cracking (like breaking a pencil). The level of stress where the material begins to flow or crack is called the **elastic limit**.

Rocks that are cold and brittle undergo **elastic rebound** when they are stressed beyond their elastic limit (Fig. 2.22). That is, the layers of rock initially bend in an elastic fashion, like stretching a rubber band or winding a clock spring. But when the elastic limit is reached, they suddenly break along a fault line, snapping to new positions that no longer line up together. The sudden snapping is an **earthquake** that releases energy in the form of seismic waves.

Two conditions are necessary for an earthquake to occur: 1) motion within the Earth so that material is stressed beyond its breaking point (elastic limit); 2) material that deforms in a snapping (brittle), rather

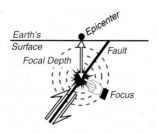

FIGURE 2.23 Cross-sectional view of a rupturing fault, illustrating terms used to describe the location of an earthquake.

than flowing (ductile), manner. Practically the only region of the Earth that can produce earthquakes is the lithosphere. In fact, earthquakes are generally confined to the cold, upper portions of lithospheric plates, which break like peanut brittle (Fig. 1.2). The lower, warmer regions of plates tend to bend easily without breaking.

Location of Earthquakes

The location of an earthquake can be described by its latitude, longitude, and depth. The **focus** is the actual "point" (relatively small volume) within the Earth where the earthquake energy was released (Fig. 2.23). The **epicenter** is the point on Earth's surface directly above the focus. The **focal depth** is the distance from the epicenter to the focus.

Earthquakes not only outline the boundaries of plates, but their depths reveal the distribution of cold, brittle material (Fig. 1.9). Earthquakes at divergent, transform, and hotspot settings are only shallow, within the upper 20 or so miles (30 kilometers) of Earth's surface (Fig. 1.2). Where plates converge fast enough, the subducting plate remains cold and brittle to considerable depth, much like a cube of ice remains cold and brittle for some time when it is pushed to the bottom of a cup of hot coffee. Earthquakes can occur to considerable depth (down to 400 miles; 700 kilometers) as the rigid top portion of the plate is stressed and contorted. The biggest earthquakes ever recorded, beneath Chile in 1960 and Alaska in 1964, occurred in regions where one converging plate had been locked against another for centuries, building enormous stress (white stars, Fig. 1.2b).

Strength of an Earthquake

Geologists use two terms to describe the strength of an earthquake: **magnitude** is quantitative, related to the amount of energy released by the earthquake; **inten-**

FIGURE 2.22 An earthquake occurs because of elastic rebound. a) Cross-sectional view of undeformed rock layers. **b)** Stress causes layers to bend elastically. **c)** When the elastic limit is reached, the rocks break along a fault. **d)** Earthquake seismic waves radiate from the broken rock as it suddenly snaps to new positions across the fault.

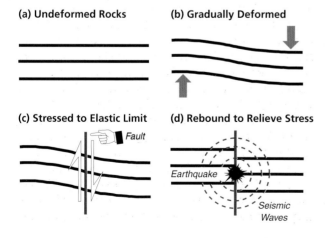

(a) Undeformed Rocks

(b) Gradually Deformed

(c) Stressed to Elastic Limit

Fault

(d) Rebound to Relieve Stress

Earthquake

Seismic Waves

(a) Amplitude of Seismic Waves   (b) Energy Released

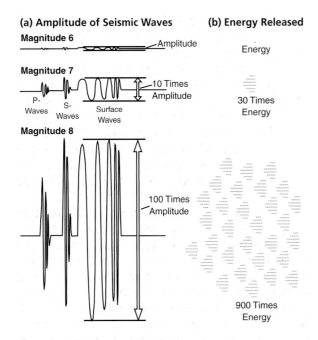

FIGURE 2.24 Earthquake magnitude is measured according to the size (amplitude) of seismic waves, which relates to how much energy was released by the earthquake.
a) Consider three earthquakes occurring on the San Andreas Fault in Los Angeles and recorded on the same seismograph in Seattle. The waves from a magnitude 6 earthquake would be recorded with a certain amplitude. Waves would appear 10 times that amplitude if the second earthquake were magnitude 7. Waves from a magnitude 8 earthquake would be 10 times as big as the magnitude 7, meaning 10 × 10, or 100 times as big as the waves from the magnitude 6 earthquake. **b)** It would take about 30 magnitude 6 earthquakes, all going off at the same time, to make the amplitude of the seismic waves grow by a factor of 10. Thus 30 times as much energy is released by a *magnitude 7* earthquake compared to a *magnitude 6*. And it would take about 900 (30 × 30) magnitude 6 earthquakes to equal the energy released by a magnitude 8!

sity is qualitative, describing the severity of shaking at a given location.

MAGNITUDE. The strength of an earthquake is commonly reported to the public in the familiar **Richter Scale**, which measures earthquake magnitude. Magnitude is based on precise measurements of the amplitude of seismic waves, as recorded by seismographs, and then corrected for the distances from each seismograph to the epicenter. Amplitude refers to how much the ground surface moves up and down as the seismic waves pass through a region. Magnitude is expressed on a logarithmic scale, whereby an increase in magni-

tude by one unit corresponds to a tenfold increase in amplitude of the seismic waves. For example, consider a magnitude 6 earthquake, compared to a magnitude 7 earthquake, occurring in exactly the same place (Fig. 2.24a). At a seismic station, the ground moves up and down 10 times as high for the magnitude 7 event as it did during the magnitude 6 ($10^7/10^6 = 10^1 = 10$). Now consider a magnitude 8 earthquake, which produces waves that are 10 times as large as the magnitude 7 earthquake. Compared to the magnitude 6 earthquake, the magnitude 8 would thus produce waves 100 times as large ($10^8/10^6 = 10^2 = 10 \times 10 = 100$).

Now let's consider how magnitude relates to the *amount of energy* released by an earthquake. It takes a thirtyfold increase in the amount of seismic energy to make the amplitude of seismic waves grow by a factor of 10. It would therefore take about 30 magnitude 6 earthquakes to equal the energy released by a single magnitude 7 earthquake (Fig. 2.24b). And about 900 (30 × 30) magnitude 6 earthquakes would have to go off all at the same time to equal a magnitude 8 earthquake!

The energy released by an earthquake relates to three parameters: 1) the surface area of the ruptured fault; 2) the amount of movement along the fault during the earthquake; and 3) the amount of friction along the fault. So the larger the area of rupture and fault movement, the larger the earthquake. Where hot asthenosphere shallows, the overlying plate is hot and only brittle enough to break along small faults. Earthquakes at divergent plate boundaries and hotspots are therefore only up to about magnitude 7.5 (Fig. 1.2a,d). The region is generally colder where plates slide past one another at transform plate boundaries, so that rocks break along bigger faults (Fig. 1.2c). Earthquakes as large as magnitude 8.5 can therefore occur in western California, when hundred-mile-long segments of the San Andreas Fault suddenly snap. A cold, brittle plate can be forced against another where plates converge (Fig. 1.2b). The plates may lock together and build stress for centuries, only to break along a fault zone up to 1,000 miles (1,500 kilometers) long. The resulting earthquakes can reach magnitude 9.5, similar to that of the largest ever recorded in South America in 1960 and Alaska in 1964. The subduction zone in the Pacific Northwest may be capable of producing earthquakes nearly as large.

INTENSITY. The intensity of an earthquake is based on effects at the surface, as witnessed by people. For a given location, intensity is reported as Roman numerals according to the **Mercalli Scale** (Table 2.8). Three

INTENSITY	OBSERVED EFFECTS
I	Not felt except by very few people, under special conditions. Detected mostly by instruments.
II	Felt by a few people, especially those on upper floors of buildings. Suspended objects may swing.
III	Felt noticeably indoors. Standing automobiles may rock slightly.
IV	Felt by many people indoors, by a few outdoors. At night, some are awakened. Dishes, windows, and doors rattle.
V	Felt by nearly everyone. Many are awakened. Some dishes and windows are broken. Unstable objects are overturned.
VI	Felt by everyone. Many people become frightened and run outdoors. Some heavy furniture is moved. Some plaster falls.
VII	Most people are in alarm and run outside. Damage is negligible in buildings of good construction.
VIII	Damage is slight in specially designed structures, considerable in ordinary buildings, great in poorly built structures. Heavy furniture is overturned.
IX	Damage is considerable in specially designed structures. Buildings shift from their foundations and partly collapse. Underground pipes are broken.
X	Some well-built wooden structures are destroyed. Most masonry structures are destroyed. The ground is badly cracked. Considerable landslides occur on steep slopes.
XI	Few, if any, masonry structures remain standing. Rails are bent. Broad fissures appear in the ground.
XII	Virtually total destruction. Waves are seen on the ground surface. Objects are thrown in the air.

TABLE 2.8 Mercalli Scale of earthquake intensity

factors tend to increase intensity at a given location: 1) magnitude of the earthquake; 2) proximity to the earthquake focus; and 3) loose soil as opposed to firm bedrock.

During an earthquake, people often report "hearing" the ground rumble a bit, like an approaching train, then "feeling" the ground shake violently. The reason is that an earthquake generates three types of seismic waves, and the waves travel at different speeds (Fig. 2.24). The fastest are **compressional waves** (P-waves), which are really sound waves traveling in the rock of the Earth. They move as the material compresses and expands. Moving at about half the speed are **shear waves** (S-waves), which contort the Earth in sideways motions, and then **surface waves**, which roll the land surface up and down and back and forth. People initially sense vibrations of the compressional waves, which commonly have smaller amplitudes and less severe ground motion than the later shear and surface waves. The more intense shear and surface waves cause the bulk of the damage.

Immediately after an earthquake, surveys are conducted of residents in the affected region to determine what people actually observed during and after the quake. Based on those surveys and other observations, intensities are determined for specific areas. Contour maps are drawn showing the observed intensity relative to the location of the earthquake epicenter (Fig. 2.25). Because seismic wave amplitudes get smaller with increasing distance from the earthquake source, intensities generally decrease with distance from the epicenter. Exceptions occur because of local ground conditions. Seismic waves are amplified in areas where there is unconsolidated sediment or landfill, locally increasing the intensity.

FIGURE 2.25 Earthquake intensity shows how severe the ground shaking was at a given location. It depends on the magnitude of the earthquake, the distance from the focus, and the local soil conditions. The contour map shows that intensity is generally highest near the epicenter, but can be enhanced in areas built on loose material, as in a sediment-filled valley.

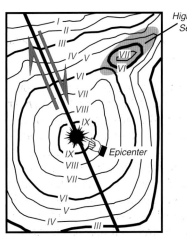

FURTHER READING

GENERAL

Bolt, B. A. 1988. *Earthquakes*. New York: Freeman. 282 pp.

Burchfield, J. D. 1990. *Lord Kelvin and the Age of the Earth*. Chicago: University of Chicago Press. 267 pp.

Chester, D. 1993. *Volcanoes and Society*. London: Edward Arnold. 288 pp.

Decker, R., and B. Decker. *Volcanoes in America's National Parks*. New York: Norton. 256 pp.

Grove, N. 1992. Crucibles of creation. *National Geographic*, v. 182, p. 5–41.

McPhee, J. 1998. *Annals of the Former World*. New York: Farrar, Straus and Giroux. 696 pp.

Oliver, J. 1996. *Shocks and Rocks: Seismology in the Plate Tectonics Revolution: The Story of Earthquakes and the Great Earth Science Revolution of the 1960's*. Washington: American Geophysical Union. 189 pp.

Zimmer, C. 2001. How old is it? *National Geographic*, v. 200, no. 3, p. 78–101.

TECHNICAL

AGI (American Geological Institute) and NAGT (National Association of Geoscience Teachers). 2003. *Laboratory Manual in Physical Geology*, 6th Ed. Upper Saddle River, NJ: Prentice Hall. 276 pp.

Boggs, S. 2003. *Petrology of Sedimentary Rocks*. Blackburn, NJ: Blackburn Press. 707 pp.

Cepeda, J. C. 1994. *Introduction to Minerals and Rocks*. New York: Macmillan. 217 pp.

Cohee, G. V., M. F. Glaessner, and H. D. Hedberg (editors). *Contributions to the Geologic Time Scale*. Tulsa: American Association of Petroleum Geologists. 388 pp.

Davis, G. H., and S. J. Reynolds. 1996. *Structural Geology of Rocks and Regions*. New York: Wiley. 800 pp.

Dewey, J. F., and J. M. Bird. 1970. Mountain belts and the new global tectonics. *Journal of Geophysical Research*, v. 75, p. 2625–2647.

Evernden, J. F., D. E. Savage, G. H. Curtis, and G. T. James. 1964. Potassium-argon dates and the Cenozoic mammalian chronology of North America. *American Journal of Science*, v. 262, p. 145–198.

Faure, G. 1986. *Principles of Isotope Geology*, 2nd Ed. New York: Wiley. 608 pp.

Francis, P. 1993. *Volcanoes: A Planetary Perspective*. Oxford, UK: Clarendon Press. 443 pp.

Gilluly, J. 1971. Plate tectonics and magmatic evolution. *Geological Society of America Bulletin*, v. 82, p. 2382–2396.

Hamblin, W. K., and E. H. Christiansen. 2001. *Earth's Dynamic Systems*, 9th Ed. Upper Saddle River, NJ: Prentice Hall. 735 pp.

Harland, W. B., R. L. Armstrong, A. V. Cox, L. E. Craig, A. G. Smith, and D. G. Smith. 1989. *A Geologic Time Scale*. Cambridge, UK: Cambridge University Press. 263 pp.

Hsu, K. J. 1982. *Mountain Building Processes*. New York: Academic Press. 263 pp.

Isaacs, B., J. Oliver, and L. R. Sykes. 1968. Seismology and the new global tectonics. *Journal of Geophysical Research*, v. 73, p. 5855–5899.

Jeffreys, H. 1976. *The Earth,* 6th Ed. Cambridge, UK: Cambridge University Press. 525 pp.

Lillie, R. J. 1999. *Whole Earth Geophysics: An Introductory Textbook for Geologists and Geophysicists*. Upper Saddle River, NJ: Prentice Hall. 361 pp.

Marshak, S. 2005. *Earth: Portrait of a Planet*, 2nd Ed. New York: Norton. 735 pp.

McDougall, I., and F. H. Chamalaun. 1966. Geomagnetic polarity scale of time. *Nature*, v. 212, p. 1415–1418.

Philpotts, A. R. 1990. *Igneous and Metamorphic Petrology*. Upper Saddle River, NJ: Prentice Hall. 494 pp.

Prothero, D. R., and F. Schwab. 1996. *Sedimentary Geology*. New York: Freeman. 575 pp.

Skinner, B. J., and S. C. Porter. 1987. *Physical Geology*. New York: Wiley. 750 pp.

Van der Pluijm, B. A., and S. Marshak. 2003. *Earth Structure: An Introduction to Structural Geology and Tectonics*, 2nd Ed. New York: Norton. 656 pp.

Watkins, N. D. 1972. Review of the development of the geomagnetic polarity time scale and discussion of prospects for its finer definition. *Geological Society of America Bulletin*, v. 83, p. 551–574.

WEBSITES

National Park Service: www.nps.gov

Geologic Resources Division:
www2.nature.nps.gov/grd/grdbroc.htm

Geoscientist-in-the-Parks Program:
www2nature.nps.gov/grd/geojob/index.htm

Teacher Materials:
www2nature.nps.gov/grd/edu/index.htm

U.S. Geological Survey: www.nps.gov

Geology:
geology.usgs.gov/index.shtml

Volcanoes:
volcanoes.usgs.gov
Cascade Volcanic Observatory: vulcan.wr.usgs.gov
Hawaiian Volcanic Observatory: hvo.wr.usgs.gov
Alaska Volcano Observatory:
www.avo.alaska.edu/avo4/index.htm
Historical Eruptions:
volcanoes.usgs.gov/Volcanoes/Historical.html
Volcano Monitoring:
volcanoes.usgs.gov/About/What/Monitor/
monitor.html
Monitoring Active Volcanoes:
pubs.usgs.gov/gip/monitor

Earthquakes:
earthquakes.usgs.gov
National Earthquake Information Center:
neic.usgs.gov/neis/states/states.html
Largest U.S. Earthquakes:
wwwneic.cr.usgs.gov/neis/eqlists/10maps_usa.html

Geologic Time:
pubs.usgs.gov/gip/geotime

Union of Two Maps: Geology and Topography:
tapestry.usgs.gov/ages/ages.html

U.S. Forest Service: www.fs.fed.us

Geology:
www.fs.fed.us/geology/mgm_geology.html

**Bureau of Land Management:
www.blm.gov/nhp/index.htm**

Environmental Education:
www.blm.gov/education

**National Association for Interpretation:
www.interpnet.com**

**National Center for Science Education:
www.ncseweb.org**

University of California, Berkeley

WEB Geological Time Machine:
www.ucmp.berkeley.edu/help/timeform.html

PART II
Divergent Plate Boundaries

Park features formed by divergent plate boundary processes reveal the progression of a continent ripping apart and forming an ocean basin. Chapter 3 presents structural and volcanic features found in parks in continental rift zones, both active and ancient. Chapter 4 then shows the beautiful shorelines that develop once a continent completely rips apart, and examines a continental margin formed in the distant past.

Where plates move away from one another, the lithosphere thins, so that the underlying, buoyant asthenosphere elevates a broad region. The elevated landscapes are continental rift zones or mid-ocean ridges, depending on whether the lithosphere is capped by thick continental or thin oceanic crust. Such regions are parts of a continuous process. Prolonged rifting can completely rip a continent apart, creating new ocean floor between the fragments. The edges of the continents that drift farther and farther apart are passive continental margins.

The region of northeastern Africa and Saudi Arabia illustrates how plate divergence can rip a continent apart and eventually open an entire ocean. Continental rifting is occurring in East Africa, producing long mountain ranges, valleys, and volcanoes such as Mt. Kilamanjaro. Many of the rift valleys have lakes because the valley floors are dropping faster than sediment can fill them. In fact, most of the world's

National park landscapes show stages of divergent plate boundary development similar to the East African region. **a)** Map of East Africa and the Arabian Peninsula. **b)** National park lands in the *Basin and Range Province* and *Rio Grande Rift* have rift valleys and volcanism very much like those found in *East Africa*. **c)** Parks in the *Keweenawan Rift* reveal a landscape that evolved to the stage between that of *East Africa* and the Red Sea. **d)** National seashores on the *Atlantic and Gulf coasts* lie along passive continental margins like those of the *Indian Ocean*.

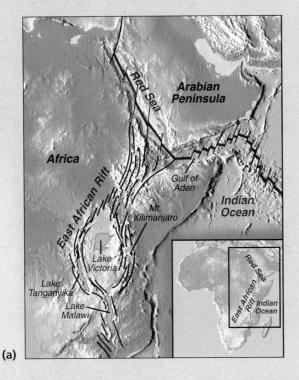

(a)

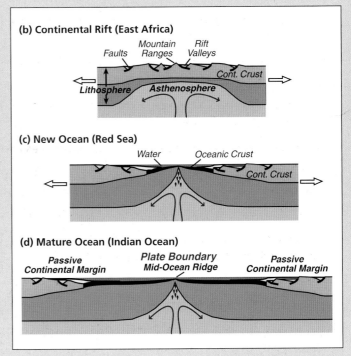

(b) Continental Rift (East Africa)

(c) New Ocean (Red Sea)

(d) Mature Ocean (Indian Ocean)

◀ [Overleaf]
Death Valley National Park, California. Long mountain ranges and deep valleys form as the North American continent rips apart.

deepest lakes are in continental rift valleys, including Lake Tanganyika and Lake Malawi in East Africa. The Red Sea is an early-stage ocean basin, with passive continental margins bordering Africa and Saudi Arabia, and a mid-ocean ridge down the center. The Indian Ocean is a more advanced ocean, with a well-developed mid-ocean ridge.

Plate divergence in North America has also created block-fault mountains and volcanic features where the continent rifts, and picturesque shorelines and layering where oceans are opening as continents drift apart. National park lands are located in the rugged landscape where active rifting is occurring in the Basin and Range Province and Rio Grande Rift, and where lava flows are preserved in the ancient Keweenawan Rift. The coastlines and barrier islands in the national seashores along the Atlantic Ocean and Gulf of Mexico are just the top portion of passive continental margin deposits—layering similar to that revealed in some western parks where canyons cut deeply into the Earth.

TEN QUESTIONS: Divergent Plate Boundaries

1. Why is North America above sea level, and the surrounding ocean floors below?

2. If the crust in the Basin and Range Province is so thin, then why is the topography so high?

3. Why are there so many types of volcanoes and volcanic products in continental rift zones?

4. Why are there long, north-south mountain ranges and valleys in the Basin and Range Province?

5. Why is Death Valley the lowest point in North America?

6. What types of earthquakes occur in the Basin and Range Province?

7. Why don't large earthquakes or other tectonic events occur along margins of the Atlantic Ocean or Gulf of Mexico?

8. How did the Atlantic Ocean and Gulf of Mexico form?

9. What factors influence the position of a coastline?

10. Why is it thought that the Colorado Plateau region was once part of a passive continental margin?

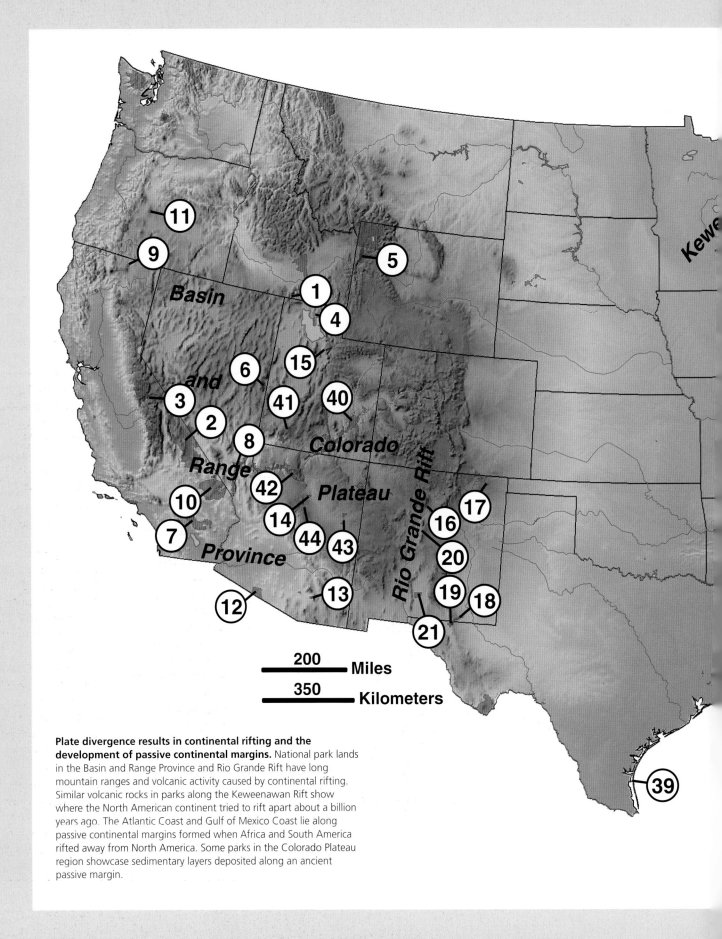

Plate divergence results in continental rifting and the development of passive continental margins. National park lands in the Basin and Range Province and Rio Grande Rift have long mountain ranges and volcanic activity caused by continental rifting. Similar volcanic rocks in parks along the Keweenawan Rift show where the North American continent tried to rift apart about a billion years ago. The Atlantic Coast and Gulf of Mexico Coast lie along passive continental margins formed when Africa and South America rifted away from North America. Some parks in the Colorado Plateau region showcase sedimentary layers deposited along an ancient passive margin.

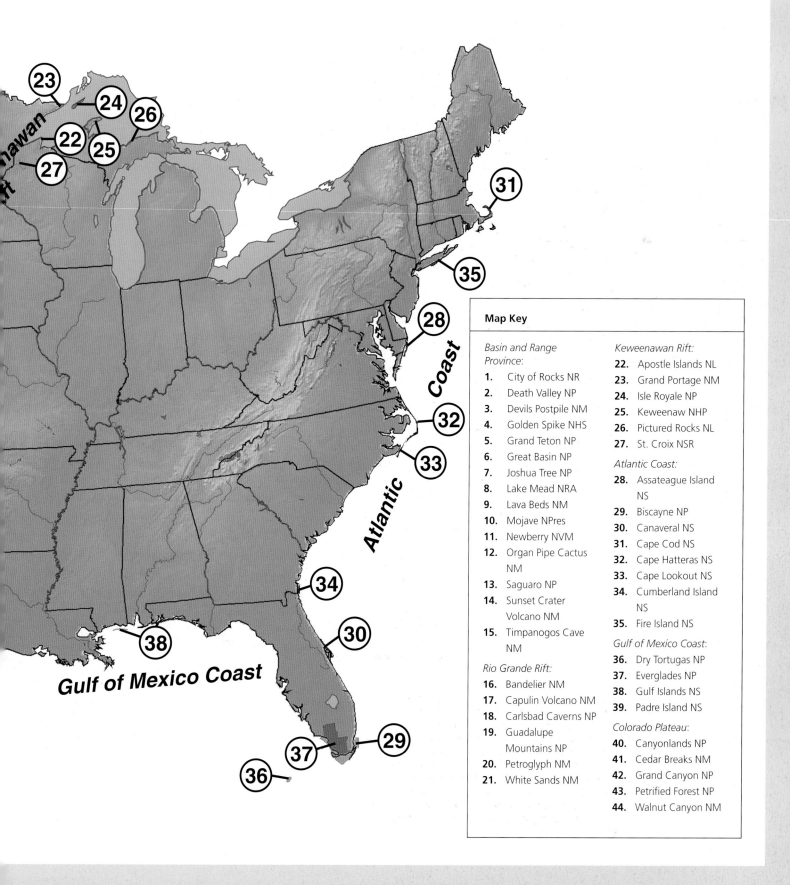

Map Key

Basin and Range Province:
1. City of Rocks NR
2. Death Valley NP
3. Devils Postpile NM
4. Golden Spike NHS
5. Grand Teton NP
6. Great Basin NP
7. Joshua Tree NP
8. Lake Mead NRA
9. Lava Beds NM
10. Mojave NPres
11. Newberry NVM
12. Organ Pipe Cactus NM
13. Saguaro NP
14. Sunset Crater Volcano NM
15. Timpanogos Cave NM

Rio Grande Rift:
16. Bandelier NM
17. Capulin Volcano NM
18. Carlsbad Caverns NP
19. Guadalupe Mountains NP
20. Petroglyph NM
21. White Sands NM

Keweenawan Rift:
22. Apostle Islands NL
23. Grand Portage NM
24. Isle Royale NP
25. Keweenaw NHP
26. Pictured Rocks NL
27. St. Croix NSR

Atlantic Coast:
28. Assateague Island NS
29. Biscayne NP
30. Canaveral NS
31. Cape Cod NS
32. Cape Hatteras NS
33. Cape Lookout NS
34. Cumberland Island NS
35. Fire Island NS

Gulf of Mexico Coast:
36. Dry Tortugas NP
37. Everglades NP
38. Gulf Islands NS
39. Padre Island NS

Colorado Plateau:
40. Canyonlands NP
41. Cedar Breaks NM
42. Grand Canyon NP
43. Petrified Forest NP
44. Walnut Canyon NM

3

Continental Rifts

▲▲▲▲▲▲ *"Much of the Western United States lies in a zone where the North American continent is ripping apart. The zone includes the Basin and Range Province . . . as well as the narrow Rio Grande Rift . . ."*

Spectacular mountain ranges, fault escarpments, and volcanic materials preserved on national park lands help us visualize continental rifting processes (Fig. 3.1). As a continent rips apart, it stretches, thinning the crust and entire lithosphere (Fig. 3.2). The region is raised to high elevation because the underlying asthenosphere is hot and buoyant. The decrease in pressure on the rising, hot asthenosphere generates magma that rises and pours out as extensive volcanic eruptions. The upper part of the crust deforms in a cold, brittle fashion, causing earthquakes and long mountain

FIGURE 3.1 Lava Beds National Monument in California illustrates the structure, volcanism, and stratigraphy associated with continental rifting in the Basin and Range Province. The steep escarpment on the left is part of the normal fault that bounds the west side of the Tule Lake rift valley. Young lava flows pond against the escarpment because the valley floor tilts westward as it moves downward. Tule Lake in the upper right illustrates that lake and river sediments are deposited on the subsiding valley floor.

(a) Continent begins to Rift Apart

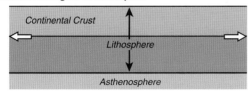

(b) Plate Stretches and Thins

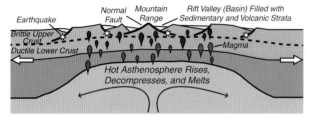

FIGURE 3.2 Continental rift zone. a) Plate divergence can pull a continent apart. **b)** The upper part of the crust is cold and brittle. It breaks along normal faults, causing earthquakes and blocks of mountain ranges separated by rift valleys. The hot, ductile asthenosphere rises like a hot-air balloon, elevating the topography and generating magma that leads to volcanic activity.

ACTIVE RIFTING IN THE WESTERN UNITED STATES

John McPhee writes in his book *Basin and Range*, "Basin. Fault. Range. Basin. Fault. Range. A mile of relief between basin and range. Stillwater Range. Pleasant Valley. Tobin Range. Jersey Valley. Sonoma Range. Pumpernickle Valley. Shoshone Range. Reese River Valley. Pequop Mountains. Steptoe Valley. Ondographic rhythms of the Basin and Range." Much of the western United States lies in a zone where the North American continent is ripping apart. The zone includes the **Basin and Range Province** in Nevada and portions of Oregon, Idaho, Wyoming, California, Utah, Arizona, and Mexico, as well as the narrow **Rio Grande Rift** extending from westernmost Texas through New Mexico and into Colorado (Fig. 3.3).

Structure

The topography of the Basin and Range Province and Rio Grande Rift is quite spectacular on a shaded-relief map (Fig. 3.4). Much of the region lies more than a mile

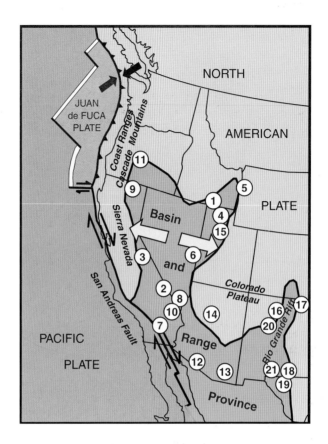

FIGURE 3.3 Plate tectonic map of the western United States. Active continental rifting is occurring in the *Basin and Range Province* and *Rio Grande Rift*. The Juan de Fuca Plate is subducting northeastward beneath the Pacific Northwest, forming the Coast Ranges and Cascade Mountains; the Sierra Nevada are the remnants of earlier subduction. The San Andreas Fault is a transform plate boundary where the Pacific Plate slides north-northwestward past the North American Plate. The Colorado Plateau is a relatively intact block of crust moving gradually upward. Continental rift faulting and volcanism have been progressing outward, eating away at the Cascade and Sierra Nevada mountains and the Colorado Plateau. Numbers refer to active continental rift parks indicated in Table 3.1.

(1½ kilometers) above sea level. The landscape reveals classic fault-block mountains, with "basins" (rift valleys) separated from the "ranges" (mountains) by normal faults (Fig. 2.7a). The ranges are made of older igneous, sedimentary, and metamorphic rocks, while the valleys contain young sedimentary deposits from rivers and lakes, with interlayered lava flows.

OVERALL ELEVATION. The hot, shallow asthenosphere associated with plate divergence contributes to the high elevation of the western United States. Overall el-

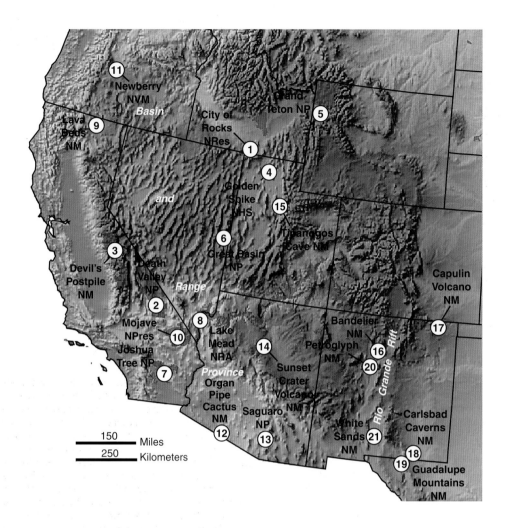

FIGURE 3.4 Shaded-relief map of the western United States.
National Park Service sites are shown in red, along with Newberry
National Volcanic Monument, a U.S. Forest Service site in Oregon.
Active continental rifting is revealed by north-south trending
mountain ranges in the Basin and Range Province, and by a promi-
nent depression along the Rio Grande Rift.

Map Key

1. City of Rocks NRes, ID	**13.** Saguaro NP, AZ
2. Death Valley NP, CA/NV	**14.** Sunset Crater Volcano NM, AZ
3. Devils Postpile NM, CA	**15.** Timpanogos Cave NM, UT
4. Golden Spike NHS, UT	**16.** Bandelier NM, NM
5. Grand Teton NP, WY	**17.** Capulin Volcano NM, NM
6. Great Basin NP, NV	
7. Joshua Tree NP, CA	**18.** Carlsbad Caverns NP, NM
8. Lake Mead NRA, CA	
9. Lava Beds NM, CA	**19.** Guadalupe Mountains NP, TX
10. Mojave NPres, CA	
11. Newberry NVM, OR	**20.** Petroglyph NM, NM
12. Organ Pipe Cactus NM, AZ	**21.** White Sands NM, NM

ACTIVE		ANCIENT
BASIN & RANGE PROVINCE	**RIO GRANDE RIFT**	**KEWEENAWAN RIFT**
1. City of Rocks NRes, ID	16. Bandelier NM, NM	22. Apostle Islands NL, WI
2. Death Valley NP, CA/NV	17. Capulin Volcano NM, NM	23. Grand Portage NM, MN
3. Devils Postpile NM, CA	18. Carlsbad Caverns NP, NM	24. Isle Royale NP, MI
4. Golden Spike NHS, UT	19. Guadalupe Mountains NP, TX	25. Keweenaw NHP, MI
5. Grand Teton NP, WY	20. Petroglyph NM, NM	26. Pictured Rocks NL, MI
6. Great Basin NP, NV	21. White Sands NM, NM	27. St. Croix NSR, MN/WI
7. Joshua Tree NP, CA		
8. Lake Mead NRA, CA		
9. Lava Beds NM, CA		
10. Mojave NPres, CA		
11. Newberry NVM, OR		
12. Organ Pipe Cactus NM, AZ		
13. Saguaro NP, AZ		
14. Sunset Crater Volcano NM, AZ		
15. Timpanogos Cave NM, UT		

TABLE 3.1 National parks in continental rift zones

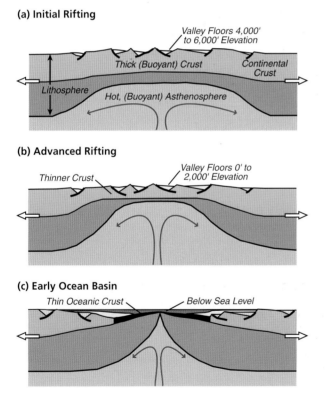

(a) Initial Rifting

Valley Floors 4,000'
to 6,000' Elevation

Thick (Buoyant) Crust

Continental Crust

Lithosphere

Hot, (Buoyant) Asthenosphere

(b) Advanced Rifting

Thinner Crust

Valley Floors 0' to
2,000' Elevation

(c) Early Ocean Basin

Thin Oceanic Crust

Below Sea Level

◀ **FIGURE 3.5 Stages of continental rifting affect overall elevation. a)** Initial rifting (high elevation). As the lithosphere rips apart, the underlying asthenosphere becomes shallow while the crust is still fairly thick. The buoyancy of the shallow asthenosphere and thick crust results in high overall elevation of the Basin and Range Province of southeastern Oregon, southern Idaho, northeastern California, northern Nevada, and Utah. **b)** Advanced rifting (lower elevation). As the continental crust thins, it is less buoyant. Elevations are lower in the southern part of the province in southern California, southern Nevada, Arizona, and Mexico. **c)** Early ocean basin (below sea level). When the continental crust completely rips apart, thin oceanic crust develops. The region drops below sea level, as seen in the Gulf of California.

evations in the northern part of the Basin and Range Province are higher than in the southern part, reflecting different stages of continental rifting (Fig. 3.5). In the northern part, valley floors are typically at least 4,000 feet (1,200 meters) above sea level, while in the south they are 2,000 feet (600 meters) or less, with Death Valley extending below sea level (Fig. 3.6).

The East Pacific Rise, a mid-ocean ridge system, extends right up the Gulf of California, toward the Salton Sea and Death Valley region of southern California (Fig. 1.8). Continental rifting in the Basin and Range Province can be considered an extension of the East Pacific Rise into the North American continent. The northern part of the province is at an early stage of continental rifting, when the crust is still fairly thick. But in the southern part, the crust is thinner, so that the region is at lower elevation. Farther south,

a   b   c

FIGURE 3.6 The elevations of valley floors in the Basin and Range Province reflect stages of continental rifting. a) Snake Valley, adjacent to Great Basin National Park in Nevada, is 5,500 feet (1,700 meters) above sea level, a high elevation characteristic of early rifting. **b)** Tucson Valley, between the Tucson Mountains and Rincon Mountain units of Saguaro National Park, Arizona, lies at 2,400 feet (730 meters), a lower elevation characteristic of crustal thinning during a later stage of rifting. **c)** Death Valley in Death Valley National Park, California, extends below sea level, and has very thin crust consistent with advanced rifting.

new oceanic crust between the Baja Peninsula and mainland Mexico is so thin that the Gulf of California is below sea level.

Death Valley National Park contains the lowest point in North America, 282 feet (86 meters) below sea level. Rift valley floors can drop down very fast. When that happens, deep water commonly fills the valleys, producing the deepest lakes in the world (see the "Deep Lakes" box). Lake Tahoe, the world's eighth deepest at 1,685 feet (514 meters), lies in a rift valley on the California–Nevada border. The floor of Death Valley is below sea level for two reasons. First, it lies on thin crust of the advanced stage of continental rifting. Second, the Mojave Desert region is very dry, so that water evaporates before it can accumulate in the valley—the floor of Death Valley has dropped below sea level without a lake filling it in.

FAULT-BLOCK MOUNTAINS. The long mountain ranges and parallel valleys in the Basin and Range Province and Rio Grande Rift are classic examples of "horst and graben" structure (Fig. 3.7). The phrase includes the German words **horst**, meaning hill, and **graben**, referring to a depression or ditch, such as the moat around a castle. Basins (rift valleys) are down-dropped grabens, while ranges are uplifted horst blocks. Extensional forces associated with continental rifting break the brittle upper crust into horst and graben blocks separated by normal faults.

The edges of some mountain ranges are steep, nearly planar fault surfaces, known as **escarpments**. A spectacular example of a fault escarpment can be seen along Gillem's Bluff in Lava Beds National Monument (Fig. 3.1). Fault escarpments rise as much as a mile (1½

kilometers) from the valley floors. But escarpments are only the top portions of normal faults that extend a few miles through the brittle upper crust. The rift valleys generally have an asymmetric structure, with a major fault zone bounding one side of the valley. As crustal blocks move along a fault, the land surface tilts the sedimentary and volcanic strata, sometimes ponding lakes against the escarpment.

Lake sediments cover the valley floors in many national parks and monuments. During recent ice ages, the climate of the Basin and Range Province was cooler and wetter. The water level in some rift valleys was as much as 300 feet (100 meters) higher than it is today.

FIGURE 3.7 Idealized cross section of continental rift valleys in the Basin and Range Province. See Figs. 3.1 and 3.8 for examples of rift valleys, lakes, and fault escarpments in national parks. A basin may be referred to as a "rift valley" or "graben," while the term "horst" describes a mountain range. The sedimentary and volcanic strata tilt toward escarpments representing zones of normal faulting that bound rift valleys. The land surface also tilts, sometimes ponding lakes against the fault escarpments.

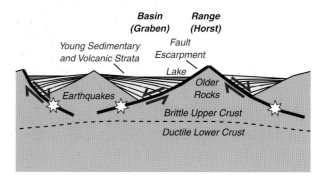

a   b   c

FIGURE 3.8 Continental rift valleys in and near national parks. a) Jackson Hole in Grand Teton National Park, Wyoming. The Teton Range front is the exposed portion of a normal fault along which the valley floor has moved downward a few miles (Fig. 3.9). **b)** Klamath Basin viewed from South Rim Drive in Crater Lake National Park, Oregon. The steep, bare slope on the east (left) side of the basin is the surface expression of an active fault escarpment at Modoc Point. **c)** Basin and Range normal faulting on the east side of the Sierra Nevada Mountains in California. The fault escarpment separates high-mountain terrain in Yosemite, Sequoia, and Kings Canyon national parks from the floor of Long Valley.

When the climate is warm, water cannot accumulate to great depth because it evaporates quickly. Tule Lake in northern California is the shallow vestige of the water that once filled much of the valley floor at Lava Beds National Monument (Fig. 3.1). We can find evidence of this effect on the valley slopes, where notches were carved by erosion along former shorelines.

Fur trappers in the 1800s referred to high valleys surrounded by mountains as "holes." Jackson Hole, Wyoming, which lies partly in Grand Teton National Park, is a rift valley at the northeastern corner of the Basin and Range Province (Fig. 3.8a). The Teton Fault forms a prominent 7,000-foot (2,000-meter) escarpment, separating the uplifted block of the Teton Range from the down-dropped valley (Fig. 3.9). The valley has filled with a thick accumulation of sedimentary and volcanic strata, suggesting that old rocks beneath Jackson Hole have moved downward at least 30,000 feet (9,000 meters) along the Teton Fault. Like many other continental rift valleys, Jackson Hole has prominent lakes (Jackson Lake, Jenny Lake) nestled against the active fault escarpment.

A beautiful view of an active rift valley, at the northwest corner of the Basin and Range Province, is seen looking south from the East Rim Drive of Crater Lake National Park (Fig. 3.8b). The valley floor is flat, covered by young sediments and the Klamath lakes. On the east side, a steep, fresh-looking escarpment (Modoc Point) illustrates that the faulting is active. Two moderate-sized earthquakes that occurred in 1993 indicate that faulting is also active on the west side, beneath the Mountain Lakes National Wilderness Area.

Continental rifting of the Basin and Range Province has been eating away at the Sierra Nevada

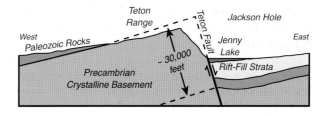

FIGURE 3.9 Diagrammatic cross section of the upper part of the crust in Grand Teton National Park. The rift valley floor (Jackson Hole) lies about 7,000 feet (2,000 meters) lower than the flanking Teton Range. The total vertical offset along the Teton Fault is even more than that, on the order of 30,000 feet (5½ miles, or 9 kilometers; the height of Mount Everest!).

DEEP LAKES

Most of the deep lakes in the world formed in continental rift valleys: Lake Baikal in Siberia (world's deepest; 5,369 feet, 1,637 meters); Lake Tanganyika in East Africa (second deepest; 4,708 feet, 1,435 meters); Lake Malawi in East Africa (fourth deepest; 2,316 feet, 706 meters); Issyk Kul in Central Asia (fifth deepest; 2,297 feet, 700 meters); Lake Tahoe in the Basin and Range Province of the United States (eighth deepest; 1,685 feet, 514 meters).

Badwater

a

Badwater
Escarpment
Smaller
Fault Scarps
Alluvial Fan

b

Turtleback

c

Zabriskie
Point

d

FIGURE 3.10 **Advanced-stage continental rifting in Death Valley National Park. a)** Badwater is the lowest point in North America, 282 feet (86 meters) below sea level. **b) Alluvial fan** deposits at the base of a steep escarpment on the east side of Death Valley. Note smaller fault scarps where the Black Mountains fault zone has displaced very young alluvial fan deposits. **c)** A **turtleback** is a surface of a mountain front that is convex upward (like a turtle shell) and composed of old crystalline (igneous and metamorphic) rocks and mantled in places by younger metamorphic rocks (mostly marble). Turtlebacks are thought to be fault surfaces that have been gently folded, uplifted, and exposed as the rocks of the overlying fault block eroded away. The advanced degree of metamorphism suggests that the rocks once lay at least 9 miles (15 kilometers) below the surface. **d) Badlands topography** eroded into lake deposits at Zabriskie Point.

mountain range in California. Yosemite, Sequoia, and Kings Canyon national parks are bounded on their eastern sides by a spectacular normal fault. The fault escarpment represents more than 7,000 feet (2,000 meters) of elevation change between the High Sierra and the floor of Long Valley (Fig. 3.8c).

Volcanism

The drop in pressure on hot asthenosphere rising at divergent boundaries causes partial melting of mantle material (Fig. 2.15a). High-silica minerals commonly melt at lower temperatures than low-silica minerals do (Table 2.6); the material that melts off the peridotite of the mantle (~40% silica) is a basalt/gabbro composition of somewhat higher silica concentration (~50%). Depending on the composition and volume of rock that it has to melt its way through, the magma may retain its basaltic composition or become enriched in silica as it makes its way to the surface.

At a continental rift zone, the basaltic magma must ascend through continental crust on its journey upward (Fig. 3.11). At an early stage, the ascending magma may melt a lot of continental crust in its path, producing

---

DEATH VALLEY NATIONAL PARK—A LAND JUST BEFORE THE OCEAN MOVES IN

Death Valley National Park displays an advanced stage of continental rifting (Fig. 3.10). Crossing the park from west to east, there are several mountain ranges and intervening rift valleys. Basin and Range normal faulting began about 16 million years ago. The mountain ranges consist of much older Precambrian, Paleozoic, and Mesozoic rocks. The oldest are 1.7-billion-year-old metamorphic rocks, and there are igneous rocks formed 1.2 to 0.8 billion years ago. Those rocks are overlain by Paleozoic sedimentary layers deposited along an ancient passive continental margin of western North America (Chap. 4). Young sedimentary and volcanic rocks, resting on sedimentary rocks about 16 to 35 million years old, partially fill the valleys.

Extreme topographic relief in the park, from the top of Telescope Peak at 11,049 feet (3,268 meters) to the bottom of Death Valley (Badwater Basin) at 282 feet (86 meters) below sea level, is a consequence of offset along normal faults. At least 2 miles (3 kilometers) of young rift strata cover Death Valley, so that vertical fault displacement is at least 4 miles (6 kilometers).

(a) Initial-Stage (Basalt and Rhyolite) Volcanism

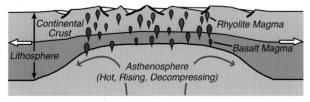

(b) Advanced-Stage (Basalt) Volcanism

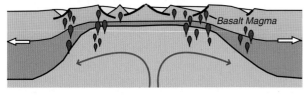

FIGURE 3.11 Two stages of volcanism at a continental rift zone. Magma with basalt composition (dark pattern) forms because of decompression when hot mantle (asthenosphere) rises. **a)** Initial-stage (both high- and low-silica) volcanism. The basalt magma melts some continental crust on its way to the surface, enriching the silica content in places and forming *dacite* to *rhyolite* magma (light pattern). **b)** Advanced-stage (low-silica) volcanism. Through time, the continental crust thins, its silica content is depleted, and more direct pathways for magma migration are established, so that mostly *basalt* erupts.

high-silica (rhyolite) volcanism (~70% silica). As the crust becomes depleted in components that form rhyolite, mostly basalt erupts. The later stage may evolve to a mid-ocean ridge, producing oceanic crust of pure basalt/gabbro composition.

Recent volcanic activity occurs primarily along the edges of the Basin and Range Province and appears to be working its way into adjoining regions. Volcanism in southern Oregon and northern California shows an age progression toward the western edge of the province. At the Steens Mountains in the southeastern corner of Oregon, the initial lava flows are 10 million years old. They become younger and younger westward, to Newberry National Volcanic Monument, south of Bend, Oregon, and Medicine Lake Volcano, just south of Lava Beds National Monument in California. A similar pattern occurs in the San

Francisco Volcanic Field of northern Arizona, where volcanism appears to be migrating eastward and encroaching on the edge of the Colorado Plateau. Bill Williams Mountain, near Williams, Arizona, is a 6-million-year-old lava dome. The volcanism progressed 60 miles (100 kilometers) eastward, to the 1-million-year-old San Francisco Peaks composite volcano and the active cinder cones and lava flows around Sunset Crater Volcano National Monument.

Volcanism also occurs around the edges of the Rio Grande Rift and is "encroaching" on surrounding regions, namely the Colorado Plateau in the west and the North American Craton to the east. About 9 million years ago, volcanic activity began in northern New Mexico near Raton Pass and has progressed eastward to the region of Capulin Volcano National Monument. Similar to the areas fringing the Basin and Range Province in southern Oregon and northern Arizona, the volcanism has both high- and low-silica components.

HIGH-SILICA LAVAS. Both high- and low-silica volcanism occurs in the vicinity of Sunset Crater Volcano National Monument, consistent with the interpretation that Basin and Range volcanism is advancing across the edge of the Colorado Plateau in northern Arizona. O'Leary Peak is a lava dome with huge blocks of rock peeling off its surface, a style of erosion characteristic of high-silica rhyolite and dacite lava (Fig. 3.12a). From the upper portion of the park's Lava Flow Trail, there is a fabulous view of the eastern side of San Francisco Peaks, the remnants of what may have been a large composite volcano (Fig. 3.12b). The strata of San Francisco Peaks include pyroclastic deposits (cinders, bombs, ash, and pumice), as well as lava flows and mudflows (Table 2.7). Rocks are commonly of andesite-to-dacite composition, but with some rhyolite, basaltic andesite, and basalt. Eruptions forming the mountain lasted from about 2.8 million years ago, until a huge blast apparently destroyed the mountain about 300,000 years ago. Using a little imagination, we can envision the slopes of the mountain extending upward to a point at about 16,000 feet (5,000 meters) elevation. One can make analogy with Mount St. Helens, within the Cascadia Subduction Zone in Washington State (Chap. 5). In 1980, high-silica magma migrated

> **RUSTED ROCK**
>
> Volcanic cinders are black or red, sand-to-gravel-sized particles that rain down during an eruption. They are black because the lava was high in iron, or red because some of the iron combines with oxygen to form rust (iron oxide). One can see that rusting is prominent near the tops of cinder cones, where hot gasses and fluids were present during the eruption. After the eruption, interaction is commonly with cold fluids and often in a dry climate; there is not much further oxidation, so most of the cinders on cones remain black. (see Fig. 3.14)

(a) O'Leary Peak

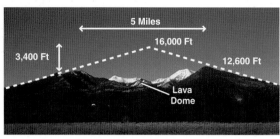

(b) San Francisco Peaks

(c) Mount St. Helens

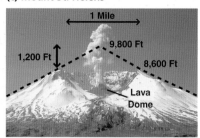

FIGURE 3.12 High-silica volcanism on the fringes of the Basin and Range Province (a and b), compared to recent subduction-zone volcanism in the Cascade Mountains (c). a) O'Leary Peak, adjacent to Sunset Crater Volcano National Monument, Arizona, is the remains of a lava dome that formed between 170,000 and 40,000 years ago. **b)** San Francisco Peaks as viewed from Bonito Park at the monument's western entrance. **c)** Mount St. Helens in Mount St. Helens National Volcanic Monument, Washington, after its 1980 eruption and explosion (Chap. 5). Mount St. Helens lost about 1,200 feet (380 meters) of its height, compared to over 3,000 feet (1,000 meters) that San Francisco Peaks may have lost during a similar explosion. The breached crater (Inner Basin) on San Francisco Peaks is also much wider (≈5 miles; 8 kilometers) than the one on Mount St. Helens (≈1½ miles; 2½ kilometers). San Francisco Peaks has a lava dome (Sugarloaf) similar to the one seen smoking in the Mount St. Helens photo.

upward, causing a bulge on the north side of Mount St. Helens. Gasses trapped under high pressure suddenly exploded, blasting away 1,200 feet (380 meters) from the top of the mountain and forming a breached crater about 1½ miles (2½ kilometers) wide (Fig. 3.12c). A lava dome of sticky dacite has since formed inside the crater. The lateral blast on the east flank of San Francisco Peaks may have been even bigger. The mountain lost 3,000 to 4,000 feet (1,000 to 1,200 meters) of its height, and the breached crater (the Inner Basin) is about 5 miles (8 kilometers) across. A dacite lava dome, Sugarloaf, then developed within the basin.

Newberry Volcano in Oregon is a large shield volcano less than one million years old. But within Newberry National Volcanic Monument there are flows containing obsidian and pumice—both indicators of high-silica volcanism accompanying the initial phase of continental rifting (Fig. 3.13). Pumice is light-colored because of its high silica content. It is also lightweight because, like the foam on top of beer, it has lots of gas bubbles. If it retains enough bubbles after it cools and solidifies, its light weight allows it to float on water! Obsidian is glass because the lava cooled so quickly it did not have time for crystal structure to develop. Even though it is very high in silica (commonly > 70%), obsidian is often black because of slight impurities and the way the glass transmits light.

LOW-SILICA LAVAS. Eruptions of low-silica magma have produced basalt lava flows, cinder cones, shield volcanoes, and spatter cones on and near national park lands in the Basin and Range Province and Rio Grande Rift.

Cinder Cones. A cinder cone generally develops during one eruptive episode, perhaps lasting a month

FIGURE 3.13 High-silica volcanic material in Newberry National Volcanic Monument, Oregon. a) The Big Obsidian Flow, formed 1,300 years ago, contains pumice and obsidian. **b)** Pumice is light colored and floats on water because it has lots of preserved gas bubbles. **c)** Obsidian is glass formed as the lava cooled very quickly.

a

b

c

FIGURE 3.14 Examples of cinder cones on national park lands along the edges of the Basin and Range Province and Rio Grande Rift. a) Sunset Crater Volcano in Sunset Crater Volcano National Monument, Arizona. **b)** Capulin Volcano in Capulin Volcano National Monument, New Mexico. **c)** Schonchin Butte in Lava Beds National Monument, California.

to a few years. Cinder cones form when low-silica magma contains a lot of gas. Rather than flow out as liquid on the surface, the expanding gasses cause the fluid lava to spray out as fountains through long fissures or central vents. As the lava falls through the air, it cools and rains down as solid particles, ranging from watermelon-sized bombs to sand-sized cinders. If the wind is not too strong, the eruptive spray falls straight down, building a symmetrical cinder cone (Fig. 3.14).

ANGLE OF REPOSE

The concept of "angle of repose" may have some economic and social utility. 1) If the stock market rises too fast, beyond its "angle of repose," it collapses back to a less precarious slope. 2) A Pulitzer Prize–winning novel by Wallace Stegner titled *Angle of Repose* delves into the characters' physical and emotional struggles, as they tumble through life to find their own "angle of repose."

"Like sands through the hourglass, so are the days of our lives," and so are the cones of cinders that form around eruption vents (Fig. 3.15). A small cone of sand begins to form when an hourglass is turned over. The cone gradually steepens until it reaches a certain angle, known as the **angle of repose**. A pile of sand grains has internal friction. The friction holds the grains together, but when the surface slope is greater than the angle of repose, the grains slip. The surface slope remains at the angle of repose as more sand falls and the cone grows.

Lava often flows as "tongues" from the base of a cinder cone because of a combination of factors (Fig. 3.16a). First, the pile of cinders is porous and is therefore lightweight (low density). Second, the cinders are not very well compacted. Third, low-silica (basaltic) lava is heavy (high density); if it doesn't have enough gas to propel it, it has difficulty getting all the way up

FIGURE 3.15 Formation of a cinder cone. a) Sand falling in an hourglass builds a cone that steepens until it reaches the angle of repose (α). The cone grows, but the slope remains at that angle. **b)** A cinder cone is similar, reaching the angle of repose, then simply getting bigger as it maintains the same slope. **c)** Paricutín Volcano in Mexico is a cinder cone formed during an eruption in the 1940s. **d)** The road to the top of Capulin Volcano in Capulin Volcano National Monument, New Mexico, reveals layers of cinders dipping at an angle of repose of about 30°.

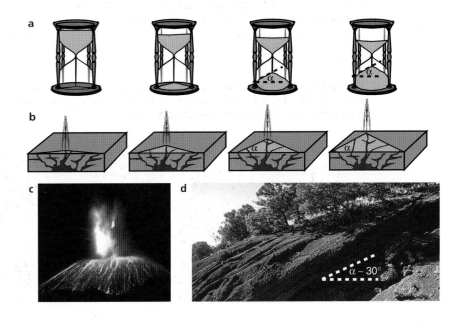

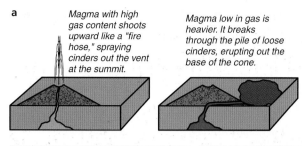

a Magma with high gas content shoots upward like a "fire hose," spraying cinders out the vent at the summit.

Magma low in gas is heavier. It breaks through the pile of loose cinders, erupting out the base of the cone.

b Sunset Crater Volcano — Tongue of Lava

c Lava Butte — Tongue of Lava

FIGURE 3.16 "Tongues" of lava extruding from the bases of cinder cones. A cinder cone is weak because it's a loose pile of material, and low density because there's a lot of pore space between the cinder particles. **a)** Dense, basaltic lava eventually flows out of the base of the cone, rather than fighting gravity to the crater at the summit. **b)** Bonita Lava Flow extruding from the base of Sunset Crater Volcano in Sunset Crater Volcano National Monument, Arizona. **c)** A tongue of lava has extruded from the base of Lava Butte in Newberry National Volcanic Monument, Oregon.

the throat of the cone. The combination of low density and poor compaction makes the pile of cinders weak. And finally, the basaltic lava is fluid; when it rises to the base of the weak cinders, it breaks through, oozing outward. Extensive flows from craters at the tops of cinder cones are rare. Instead it is common to see black "tongues" of basaltic lava extruding from the bases of cinder cones, such as Sunset Crater Volcano and Lava Butte (Fig. 3.16b,c; see also Wizard Island in Fig. 2.19b).

Shield Volcanoes. Low-silica magmas result in fissure eruptions and fluid lava flows. Shield volcanoes develop where there is a large enough volume of magma to build the layers of basalt and cinders into a sizable pile. There are places in the Basin and Range Province and Rio Grande Rift where enough basaltic lava accumulated to build shield volcanoes. Newberry Volcano, the centerpiece of Newberry National Volcanic Monument in Oregon, is the largest active volcano in the continental United States (Fig. 3.17). Another

FIGURE 3.17 Newberry National Volcanic Monument showcases a broad shield volcano in the northwestern corner of the Basin and Range Province in Oregon. a) Newberry Volcano is about 30 miles (50 kilometers) across and is covered with scores of cinder cones. **b)** The top of the volcano is a large crater (collapse caldera) now filled with two lakes. The Big Obsidian Flow and Central Pumice Cone are high-silica components of this predominately low-silica (basaltic) volcano.

(a) Broad Shield Volcano

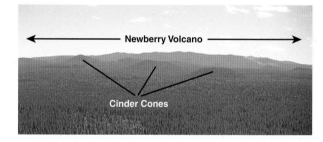

Newberry Volcano — Cinder Cones

(b) Summit Caldera

Central Pumice Cone — Big Obsidian Flow — East Lake — Paulina Lake

FIGURE 3.18 Shield volcano on the fringes of the Rio Grande Rift. Sierra Grande, viewed from the road to the top of Capulin Volcano National Monument, New Mexico.

a

b

FIGURE 3.19 Spatter cones on national park lands in the Basin and Range Province. a) Black Crater in Lava Beds National Monument, California. **b)** Spatter cone, referred to as a "hornito," along Lava Flow Trail in Sunset Crater Volcano National Monument, Arizona.

prominent shield volcano, Sierra Grande, can be observed from the road up Capulin Volcano National Monument in northeastern New Mexico (Fig. 3.18).

Spatter Cones. Anyone who has cooked spaghetti sauce understands how a **spatter cone** forms. Air bubbles pop at the surface, spattering bright, orange-colored sauce on your white stovetop. Similarly, gasses within lava can cause bubbles that pop, spattering hot magma. Spatter cones are prominent components of continental rift volcanism in many parks in the Basin and Range Province and Rio Grande Rift (Fig. 3.19). They commonly occur in a line at the top of basaltic flows, as at Black Crater in Lava Beds National Monu-

THE HEXAGON—NATURE'S IDEAL FORM

Nature loves the hexagon. It is both efficient and sturdy. The efficiency is apparent not only when lava cools and hardens, but also when mud dries and cracks, and even when bees build their honeycombs! As lava cools or mud dries out, the volume decreases, subjecting the material to tensional forces that pull at the material, forming cracks. If the lava or mud is homogeneous and the cooling or drying uniform, then the tension is equal in all directions. The material will try to crack in perfect circles (Fig. 3.22a). But circles have triangular gaps in between. The same is true of other geometric shapes with many sides, such as the eight-sided octagon (Fig. 3.22b). But nature would prefer not to have such gaps, so it chooses the figure that most approximates a circle, but with sides that fit together perfectly, which is the hexa-

gon (Fig. 3.22c). Bees have come to recognize this efficiency—they can store the most honey for their wax by building hexagonal honeycombs. The sturdiness of the hexagonal form is also evidenced in the honeycomb, as well as in human-made objects such as the internal construction of an airplane wing.

Hexagons are apparent in other situations where Earth materials are subjected to tension. This phenomenon occurs on all scales. Compare the near-perfect 120° angles of the columns at Devil's Postpile National Monument (Fig. 3.21b) with the intersection of the Red Sea, East African Rift, and Gulf of Aden (Fig. 3.23), where plates are being pulled apart. Strange how bees, aviators, and geologists all seem to be on the same page!

ment. In Sunset Crater Volcano National Monument, you can see a line of spatter cones formed along the backbone of the Bonita Lava Flow. Such features are also called **hornitos**, from the Spanish word for "small oven."

Lava Tubes. Fluid, basaltic lava doesn't flow only on the surface. As a flow cools and crusts over, the lava may continue to flow underneath. A series of underground channels can form. Some of the passages may eventually drain, leaving a hollow pipe, or **lava tube**. Lava tubes are commonly long and straight, with arched roofs and flat floors, like subway tunnels. In some instances, small amounts of lava can continue to flow along the edges of the floors, leaving long grooves; in places, lava tubes look like underground bowling alleys. A single tube can carry lava for months or even years, as the solidified rock above and below acts as an insulator. The lava can remain hot and fluid and travel for several miles, as occurs today on Kīlauea Volcano in Hawai`i Volcanoes National Park (Chap. 8). In the Basin and Range Province, there are extensive systems of lava tubes in Newberry National Volcanic Monument and Lava Beds National Monument (Fig. 3.20).

Columnar Jointing. Spectacular examples of **columnar jointing** are found in many national park areas, including Devil's Postpile National Monument in California (Fig. 3.21), Devil's Tower National Monument in Wyoming, Mount Rainier National Park in Washington, John Day Fossil Beds National Monument in Oregon, and Shenandoah National Park in Virginia. Other prominent examples are the Palisades of New Jersey (just across the Hudson River from New York

FIGURE 3.20 **Lava tubes form in fluid, basaltic lava.** Valentine Cave in Lava Beds National Monument, California.

a

b

FIGURE 3.21 Columnar jointing in Devil's Postpile National Monument, California. a) The top portion of the lava flow exhibits spectacular columns, each about 60 feet (20 meters) high and 1½ feet (½ meter) across. **b)** The top of the flow shows that individual columns approach the ideal, six-sided hexagon, with cracks (joints) intersecting at 120° angles (Fig. 3.22).

City) and the Giant's Causeway in Northern Ireland. Such columns form within the top portions of lava flows. What causes these spectacular features, and why do they often form such perfect hexagons?

The secret is that the flow must have a homogeneous composition, and that the cooling is uniform. As lava cools, it shrinks and cracks, much as cracks form when mud dries out. When a substance shrinks uniformly, it is under tensional forces that are equal in all directions. The cooling lava will try to crack in circles (Fig. 3.22). But there's a problem, because triangular gaps form between the circles. So the lava will form cracks of the geometric shape with the most sides that will fit together without gaps. That form is six-sided, or hexagonal.

(a) Circles   **(b) Octagons**

(c) Hexagons

(d) Hexagon at Devil's Postpile

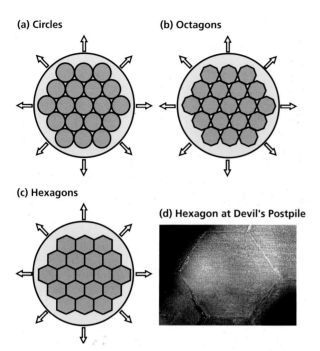

FIGURE 3.22 Formation of columnar jointing in lava flow.
Diagrams represent the top view of a flow that is shrinking as it cools. **a)** If the lava is homogeneous and cools uniformly, the shrinkage causes tensional forces that are equal in all directions. It will try to crack in perfect circles, but there are triangular gaps between the circles. **b)** A figure with eight sides (octagon) will also have gaps. **c)** The figure with the most sides that fits together without gaps is a six-sided hexagon. The cracks (joints) between the hexagons intersect at 120° angles. **d)** Top of lava flow at Devil's Postpile National Monument, California, showing hexagonal joint pattern.

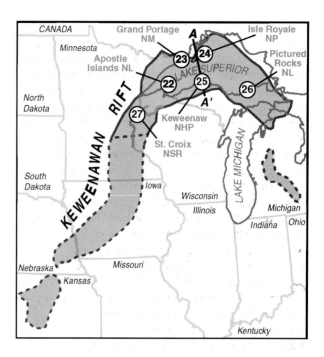

FIGURE 3.24 Map of the mid-continent region of the United States, showing extent of Keweenawan Rift system.
Evidence of ancient continental rifting, including intrusive dikes of gabbro, basalt lava flows, and river and lake deposits, are exposed in Isle Royale National Park as well as other National Park Service sites along Lake Superior and the St. Croix River. The rift can be traced southward beneath younger strata to northeastern Kansas. (Keweenawan Rift rocks: solid outline = exposed; dashed outline = covered by younger layers.) A-A' is the line of cross section in Fig. 3.26. Numbers correspond to ancient continental rift parks in Table 3.1.

FIGURE 3.23 Intersection of the Red Sea, East African Rift, and the Gulf of Aden. Plate divergence pulls the region apart, forming nearly-perfect 120° angles.

ANCIENT CONTINENTAL RIFTING

Several years ago, I was driving cross-country and camped at Interstate State Park, on the St. Croix River between Minnesota and Wisconsin. In the morning, I was amazed to see layers of black lava flows along the bluff above the river. In that part of the world, one would expect to see very old metamorphic and intrusive igneous rocks, but not black basalt flows!

That experience was an introduction to the **Keweenawan Rift**, a very ancient split in the North American continent (Fig. 3.24). About 1.1 billion (1,100 million) years ago, the continent started to rip apart, but it stopped about 40 million years later.

(a) Isle Royale National Park   **(b) Pictured Rocks National Lakeshore**

FIGURE 3.25 Photographic evidence of ancient continental rifting in the Midwest. a) Layers of 1.1-billion-year-old basalt in Isle Royale National Park are part of the Keweenawan Rift system. **b)** Hard sandstone layers in Pictured Rocks National Lakeshore include 1.1-billion-year-old rift strata, as well as late-Cambrian-to-early-Ordovician age (~520- to 480-million-year-old) strata exposed on the northern flank of the Michigan Basin.

Continental rifting does not always progress to the stage where a wide ocean basin develops (Fig. 1.11). Plate divergence may stop before the continent completely rips apart, but after rift valleys, fault-block mountains, igneous intrusions, and volcanic and sedimentary strata form (Fig. 3.1). We can imagine what would result if East Africa or the Basin and Range Province stopped rifting apart. The high mountains would erode away. The cooling and contraction of hot asthenosphere would cause the region of uplifted crust to subside. Shallow seas might periodically flood the area, partially or wholly covering the failed rift with thin sedimentary layers.

Keweenawan Rift

The Keweenawan Rift gets its name because igneous and sedimentary rocks characteristic of continental rifting are found on the Keweenaw Peninsula in Upper Michigan (note Keweenaw National Historical Park in Fig. 3.24). Ancient layers of basalt, lake, and river deposits, along with intrusions of gabbro, are similar to those found in modern rifts, such as East Africa and the Basin and Range Province. The deposits have been traced southwestward, not only to the outcroppings along the St. Croix River, but also in oil-exploration wells in Iowa, Nebraska, and Kansas. The high-density basalt and gabbro result in a remarkable observation, a zone of slightly increased pull in Earth's gravity field that outlines the ancient rift zone from Lake Superior to Kansas. The large volume of basalt lava and gabbro intrusions suggests that the Keweenawan Rift may have evolved very near to the stage of a small ocean basin, similar to the Red Sea or Gulf of California (Fig. 3.5c).

National park lands preserve evidence of the ancient continental rifting in the Lake Superior region. The northern shoreline and interior of Isle Royale National Park contain 1.1-billion-year-old basaltic lava flows, known as the Portage Lake Volcanics (Fig. 3.25). In the southern part of the island, slightly younger sedimentary layers, the Copper Harbor Conglomerate, represent layers deposited after rifting ceased. The region subsided and was later compressed into a large syncline bounded by reverse faults (Fig. 3.26). Across Lake Superior, on the Keweenaw Peninsula of Upper Michigan, the volcanic and sedimentary layers are exposed in Keweenaw National Historical Park. In more recent geologic time, glaciers have scoured out the softer sedimentary layers at the center of the syncline, enhancing the depression that now holds Lake Superior, at 1,333 feet (406 meters) the eleventh-deepest freshwater lake in the world.

FIGURE 3.26 Cross section of the upper crust in the Lake Superior region. The overall structure is a syncline (Fig. 2.8b), with younger sedimentary layers on the inside, and older igneous rocks exposed on the flanks. Volcanic and sedimentary layers in Isle Royale National Park and Keweenaw National Historical Park formed during continental rifting that nearly split the continent apart 1.1 billion years ago. Isle Royale and the Keweenaw Peninsula were the sites of intense copper mining activity from the mid- to late 1800s. The copper occurs within basalt lava flows.

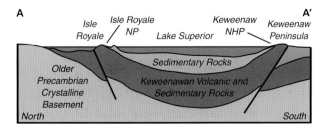

ABOUT MID-OCEAN RIDGES

The Mid-Atlantic Ridge is the active plate boundary where new oceanic crust forms as the passive continental margins of North America and Europe and Africa drift apart (Fig. 3.27). Although no national park lands are on active mid-ocean ridges (Iceland is not part of the United States!), some ancient mid-ocean ridge rocks are preserved in national parks. Those rocks are now high and dry on the continents because they were either: 1) scraped off the top of a subducting plate, then accreted onto the overriding plate (Chap. 5); or 2) pushed over the continental edge when ocean basins closed and continents collided (Chap. 6).

The volcanism at mid-ocean ridges is similar to the late-stage volcanism that occurs in continental rift zones (Fig. 3.11b). The partially-melted asthenosphere produces low-silica magma, resulting in a characteristic sequence of igneous rocks (Fig. 3.28). Lava that pours out onto the ocean floor cools rapidly, forming basalt (50% silica). Lava extruding into water often forms globular features, known as **pillow lavas**; such features in Olympic, Denali, and other national parks identify crust as having formed beneath the ocean. Magma of the same (~50% silica) composition that does not make it to the surface forms intrusive dikes of gabbro, comprising the lower crust. The high-density material that remains below is peridotite (~40% silica), forming the mantle portion of the new lithospheric plate. Together with the overlying layer of deep-sea sediment, such sequences of rocks comprising oceanic crust and uppermost mantle are called **ophiolites**.

When iron-rich basalt cools and hardens on the ocean floor, it becomes magnetized according to the strength and direction of Earth's magnetic field, which

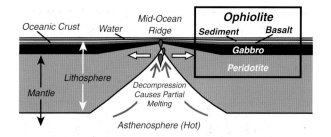

FIGURE 3.28 Formation of an ophiolite. As hot asthenosphere rises at a mid-ocean ridge, the drop in pressure causes it to start melting. The magma cools to form oceanic crust consisting of fine-grained basalt on the seafloor, and coarse-grained gabbro beneath the surface. The heavy, coarse-grained rock left behind forms peridotite of the upper mantle. Deep-ocean sediment layers eventually cover the crust. Although there are no U.S. national parks on mid-ocean ridges, there are rocks in many national parks that formed in such settings but were later thrust onto the North American continent (Chaps. 5, 6, and 11).

flips around about every half million years or so. A compass would sometimes point south instead of north! Geologists can determine the age of oceanic crust worldwide simply by measuring Earth's magnetic field from ships. Maps developed from such measurements show that the newest oceanic crust lies at mid-ocean ridges, and that crust gets progressively older toward the edges of continents (Fig. 3.29). This map pattern is consistent with the longstanding idea of continental drift, and demonstrates that new plate material is created along mid-ocean ridges.

FIGURE 3.27 Effects of opening the Atlantic Ocean. The Mid-Atlantic Ridge is the active boundary between the North American and African plates. The transitions from continental to oceanic crust surrounding the Atlantic Ocean are a long distance from this boundary and are thus passive continental margins.

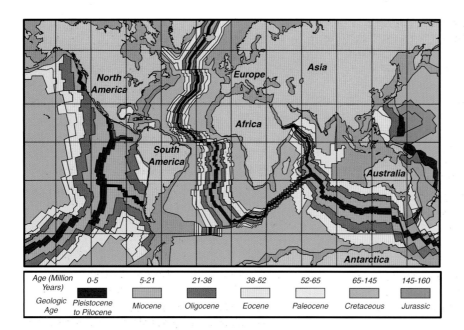

FIGURE 3.29 Age of the ocean floor. From the pattern, it is clear that new crust forms where plates diverge at mid-ocean ridges, then is carried away on top of the plates.

Age (Million Years)	0-5	5-21	21-38	38-52	52-65	65-145	145-160
Geologic Age	Pleistocene to Pilocene	Miocene	Oligocene	Eocene	Paleocene	Cretaceous	Jurassic

FURTHER READING

GENERAL

Collier, M. 1990. *An Introduction to the Geology of Death Valley*. Death Valley, CA: Death Valley Natural History Association. 60 pp.

Duffield, W. A. 1997. *Volcanoes of Northern Arizona*. Grand Canyon, AZ: Grand Canyon Association. 68 pp.

Good, J. M., and K. L. Pierce. 1996. *Interpreting the Landscapes of Grand Teton and Yellowstone National Parks*. Moose, WY: Grand Teton Natural History Assoc. 58 pp.

Lamb, S. 1991. *Lava Beds National Monument*. Tulelake, CA: Lava Beds Natural History Association. 48 pp.

Love, J. D., and J. C. Reed, Jr. 1995. *Creation of the Landscape: The Geologic Story of Grand Teton National Park*. Moose, WY: Grand Teton Natural History Assoc. 120 pp.

McPhee, J. 1980. *Basin and Range*. New York: Noonday Press. 216 pp.

Miller, M. G., and L. A. Wright. 2002. *Geology of Death Valley National Park*. Dubuque, IA: Kendall/Hunt. 72 pp.

TECHNICAL

Anderson, R. E., M. L. Zoback, and G. A. Thompson. 1983. Implications of selected subsurface data on the structural form and evolution of some basins in the northern Basin and Range province, Nevada and Utah. *Geological Society of America Bulletin*, v. 94, p. 1055–1072.

Baker, B. H., P. A. Mohr, and L. A. J. Williams. 1972. *Geology of the Eastern Rift System of Africa*. Boulder, CO: Geological Society of America, Special Paper 136. 66 pp.

Baldridge, W. S., and K. H. Olsen. 1989. The Rio Grande Rift. *American Scientist*, v. 77, p. 240–247.

Bartley, J. M., and B. P. Wernicke. 1984. The Snake Range décollement interpreted as a major extensional shear zone. *Tectonics*, v. 3, p. 647–657.

Cochran, J. R. A model for development of the Red Sea. 1983. *American Association of Petroleum Geologists Bulletin*. v. 67, p. 41–69.

Ewing, J. I., and M. Ewing. 1959. Seismic refraction measurements in the Atlantic Ocean basins, in the Mediterranean Sea, on the mid-Atlantic ridge, and in the Norwegian Sea. *Geological Society of America Bulletin*, v. 70, p. 291–318.

Huber, N. K. 1975. *The Geologic Story of Isle Royale National Park*. U.S. Geological Survey Bulletin 1309. 66 pp.

Keller, G. R., E. G. Lidiak, W. J. Hinze, and L. W. Braile. 1983. The role of rifting in the tectonic development of the midcontinent, U.S.A. *Tectonophysics*, v. 94, p. 391–412.

Kerr, R. A. 1997. Why the west stands tall. *Science*, v. 275, p. 1564–1565.

Love, J. D., J. C. Reed, Jr., R. L. Christianson, and J. R. Stacy. 1973. *Geologic Block Diagram and Tectonic History of the Teton Range, Wyoming—Idaho*. Washington: U.S. Geological Survey, Misc. Geologic Invest., Map I-730.

McCarthy, J., and T. Parsons. 1994. Insights into the kinematic Cenozoic evolution of the Basin and Range-Colorado Plateau transition from coincident seismic refraction and reflection data. *Geological Society of America Bulletin*, v. 106, p. 747–759.

Miller, M. G. 1991. High-angle origin of the currently low-angle Badwater Turtleback fault, Death Valley, California. *Geology*, v. 19, p. 372–375.

Mono Lake Committee Field Guide Series. 2000. *Roadside Geology of the Eastern Sierra Region*. Portland, OR: Geological Society of the Oregon Country, 42 pp.

Sykes, L. R. 1967. Mechanism of earthquakes and nature of faulting on the mid-ocean ridges. *Journal of Geophysical Research*, v. 72, p. 2131–2153.

Vine, F. J. 1966. Spreading of the ocean floor: New evidence. *Science*, v. 154, p. 1405–1415.

Wernicke, B., G. C. Axen, and J. K. Snow. 1988. Basin and Range extensional tectonics at the latitude of Las Vegas, Nevada. *Geological Society of America Bulletin*, v. 100, p. 1738–1757.

Woelk, T. S., and J. Hinze. 1991. Model of the midcontinent rift system in northeastern Kansas. *Geology*, v. 19, p. 277–280.

Zandt, G., S. C. Myers, and T. C. Wallace. 1995. Crust and mantle structure across the Basin and Range-Colorado Plateau boundary at 37°N latitude and implications for Cenozoic extensional mechanism. *Journal of Geophysical Research*, v. 100, p. 10529–10548.

WEBSITES

National Park Service: www.nps.gov

Park Geology Tour:
www2.nature.nps.gov/grd/tour/index.htm

Basin and Range Province:
www2.nature.nps.gov/grd/tour/basin_ra.htm

1. *City of Rocks National Reserve:*
   www2.nature.nps.gov/grd/parks/ciro/index.htm

2. *Death Valley National Park:*
   www2.nature.nps.gov/grd/parks/deva/index.htm

3. *Devil's Postpile National Monument:*
   www2.nature.nps.gov/grd/parks/depo/index.htm

4. *Golden Spike National Historic Site:*
   www.nps.gov/gosp/index.htm

5. *Grand Teton National Park:*
   www2.nature.nps.gov/grd/parks/grte/index.htm

6. *Great Basin National Park:*
   www2.nature.nps.gov/grd/parks/grba/index.htm

7. *Joshua Tree National Park:*
   www2.nature.nps.gov/grd/parks/jotr/index.htm

8. *Lake Mead National Recreation Area:*
   www2.nature.nps.gov/grd/parks/lame/index.htm

9. *Lava Beds National Monument:*
   www2.nature.nps.gov/grd/parks/labe/index.htm

10. *Mojave National Preserve:*
    www2.nature.nps.gov/grd/parks/moja/index.htm

12. *Organ Pipe Cactus National Monument:*
    www2.nature.nps.gov/grd/parks/orpi/index.htm

13. *Saguaro National Park:*
    www2.nature.nps.gov/grd/parks/sagu/index.htm

14. *Sunset Crater Volcano National Park:*
    www2.nature.nps.gov/grd/parks/sucr/index.htm

15. *Timpanogos Cave National Monument:*
    www2.nature.nps.gov/grd/parks/tica/index.htm

Rio Grande Rift:

16. *Bandelier National Monument:*
    www2.nature.nps.gov/grd/parks/band/index.htm

17. *Capulin Volcano National Monument:*
    www2.nature.nps.gov/grd/parks/cavo/index.htm

18. *Carlsbad Caverns National Park:*
    www2.nature.nps.gov/grd/parks/cave/index.htm

19. *Guadalupe Mountains National Park:*
    www2.nature.nps.gov/grd/parks/gumo/index.htm

20. *Petroglyph National Monument:*
    www2.nature.nps.gov/grd/parks/petr/index.htm

21. *White Sands National Monument:*
    www2.nature.nps.gov/grd/parks/whsa/index.htm

Keweenawan Rift:

22. *Apostle Islands National Lakeshore:*
    www2.nature.nps.gov/grd/parks/apis/index.htm

23. *Grand Portage National Monument:*
    www.nps.gov/grpo/index.htm

24. *Isle Royale National Park:*
    www2.nature.nps.gov/grd/parks/isro/index.htm

25. *Keweenaw National Historical Park:*
    www2.nature.nps.gov/grd/parks/kewe/index.htm

26. *Pictured Rocks National Lakeshore:*
    www2.nature.nps.gov/grd/parks/piro/index.htm

27. *St. Croix National Scenic River:*
    www2.nature.nps.gov/grd/parks/sacn/index.htm

Volcanism:
www2.nature.nps.gov/grd/tour/volcano.htm

U.S. Geological survey: www.usgs.gov

Geology in the Parks:
www2.nature.nps.gov/grd/usgsnps/project/home.html

U.S. Forest Service: www.fs.fed.us

Geology:
www.fs.fed.us/geology/mgm_geology.html

11. *Newberry National Volcanic Monument:*
    www.raztrans.com/id138.htm

4

Passive Continental Margins

▲▲▲▲▲▲ *"Most of our national seashores lie along the passive continental margin of the eastern and southern United States. The plate-tectonic history of the region helps us understand the current structure, sedimentation, and geomorphology of the Atlantic and Gulf coasts, as well as the ancient sedimentary layers in Grand Canyon National Park."*

Tranquil beaches and offshore islands reflect the calm tectonic state of coastlines in the eastern United States (Fig. 4.1). But those coastlines belie the forces responsible for their formation, forces strong enough to rip a continent apart, and sustained long enough to open an ocean.

Rachael Carson writes about the evolution of coasts in *The Edge of the Sea,* "Once this rocky coast beneath me was a plain of sand; then the sea rose and found a new shore line. And again in some shadowy future the surf will have ground these rocks to sand and will have returned the coast to its earlier state. And so in my mind's eye these coastal forms merge and blend in a

FIGURE 4.1 Beautiful beaches have developed on the passive continental margins of the Atlantic and Gulf coasts. Gulf Islands National Seashore is the site of dramatic barrier islands, such as Horn Island off the coast of Mississippi.

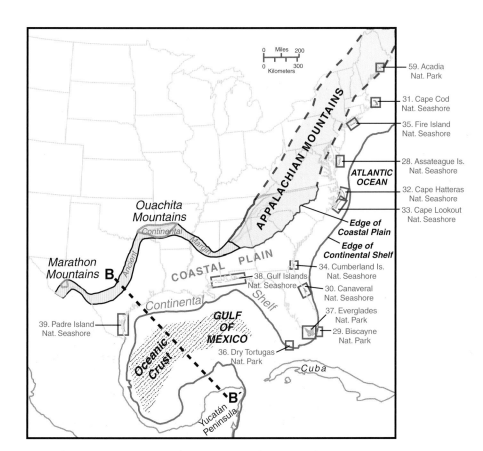

FIGURE 4.2 National parks and seashores along the coasts of the Gulf of Mexico and Atlantic Ocean. The coastal plain and continental shelf mark the current transition from thick crust of North America to thin crust of the Atlantic Ocean and Gulf of Mexico. Oceanic crust lies beneath the central and western Gulf and within the deep Atlantic. An earlier continental margin is now covered by the Appalachian-Ouachita-Marathon mountain chain (Chap. 6). National parks and seashores lie on the gently-sloping surface of the coastal plain and continental shelf, deriving their sedimentary layers from the eroding mountains. Dry Tortugas and Biscayne national parks in Florida lie on the very edge of the continental shelf. Numbers refer to modern passive margin parks in Table 4.1. Acadia National Park is located where the Appalachian Mountains extend to the coast in Maine (Chap. 6). B–B' is the line of the cross section shown in Fig. 4.9.

shifting kaleidoscopic pattern in which there is no finality, no ultimate and fixed reality—Earth becoming fluid as the sea itself."

The eastern and southern coasts of the United States are passive continental margins formed as the Atlantic Ocean and Gulf of Mexico opened about 200 million years ago. Nine national seashores and three national parks lie within the zone of transition from thick continental to thin oceanic crust (Fig. 4.2 and Table 4.1). That change in crustal thickness is responsible for the development of the coastal plain and continental shelf on the eastern seaboard, and ultimately for the wonderful beaches and islands that comprise the national parks and seashores. The western United States is currently the site of earthquakes, volcanism, and mountain building—telltale signs of active plate-boundary processes. But in the past, the western edge of the continent was much like today's East Coast. The spectacular layers of rock in Grand Canyon and other parks in the Colorado Plateau region hint at the sedimentary

TABLE 4.1 National park lands along passive continental margins.

MODERN		ANCIENT
ATLANTIC COAST	**GULF COAST**	**COLORADO PLATEAU**
28. Assateague Island NS, MD	36. Dry Tortugas NP, FL	40. Canyonlands NP, AZ
29. Biscayne NP, FL	37. Everglades NP, FL	41. Cedar Breaks NM, UT
30. Canaveral NS, FL	38. Gulf Island NS, FL/MS	42. Grand Canyon NP, AZ
31. Cape Cod NS, MA	39. Padre Island NS, TX	43. Petrified Forest NP, AZ
32. Cape Hatteras NS, NC		44. Walnut Canyon NM, AZ
33. Cape Lookout NS, NC		
34. Cumberland Island NS, GA		
35. Fire Island NS, NY		

processes that accompanied subsidence of an ancient passive continental margin.

MODERN PASSIVE MARGIN OF NORTH AMERICA

Most of our national seashores lie along the passive continental margin of the eastern and southern United States. The plate-tectonic history of the region helps us understand the current structure, sedimentation, and geomorphology of the Atlantic and Gulf coasts. An earlier eastern edge of North America was some 300 miles (500 kilometers) to the west and north, a margin that now lies roughly beneath the Appalachian and Ouachita mountains (Fig. 4.2). That earlier margin was deformed as Africa and South America came crashing in, forming the mountains and a supercontinent known as Pangea (Fig. 4.3). Later continental rifting left behind rocks representing the closed ocean, as well as pieces of Africa and South America. The modern continental margin developed over the past 200 million years as

Pangea ripped apart, opening the Atlantic Ocean and Gulf of Mexico.

In places, the Atlantic Ocean and Gulf of Mexico opened quite a bit east and south of the mountains that mark the ancient continental margin. Rocks beneath the young sediments of the coastal plain, including most of Florida and regions of offshore Georgia and the Carolinas, are stranded pieces of Africa, left behind when the Atlantic opened. Similarly, as the Gulf of Mexico opened, parts of South America and the Yucatán Peninsula were stranded and now constitute

FIGURE 4.3 Development of the Atlantic and Gulf coasts of North America. a) The ancient Iapetus Ocean begins to close. **b)** The supercontinent of Pangea forms as Africa (along with South America and the rest of Gondwanaland) crashes into North America (Chap. 6). The Appalachian Mountains are part of a much larger zone of continental collision that includes the Marathon and Ouachita mountains in the southeastern United States, the Atlas Mountains in Africa, and the Caledonide Mountains in Greenland, the British Isles, and Scandinavia. Crustal blocks that later became the Yucatán Peninsula and Cuba lie between North and South America. **c)** The modern oceans originated about 200 million years ago when Europe, Africa, and South America ripped away from North America. The Atlantic Ocean continues to widen along the Mid-Atlantic Ridge. The Gulf of Mexico ceased opening about 100 million years ago, when tectonic activity shifted to the Caribbean region, where South America continues to drift southward. National seashores lie on the passive continental margin between the Appalachian-Ouachita-Marathon mountains and the Atlantic Ocean/Gulf of Mexico.

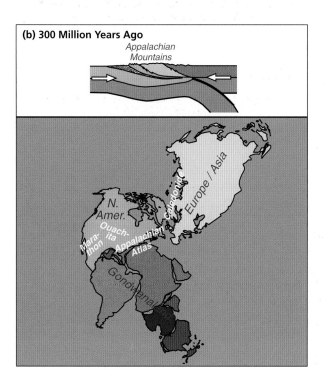

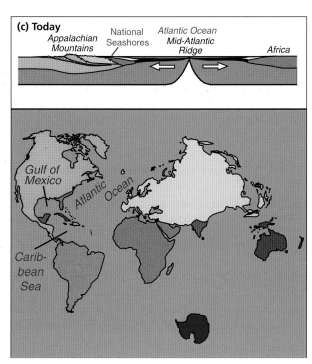

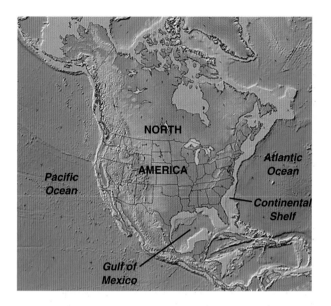

▲ **FIGURE 4.4 Shaded-relief map of North America.** The map reveals information about the thickness of crust and the positions of plate boundaries. Crust of the continents is thick and buoyant, so that its surface rests about 3 miles (5 kilometers) higher than the floor of the surrounding oceans (Fig. 1.6). The extent of the thick crust includes the shallow continental shelves that lie beneath the edges of the Atlantic Ocean and Gulf of Mexico. The eastern part of the continent has relatively smooth topography and broad shelves because it is a passive continental margin, far removed from plate boundaries and their accompanying tectonic activity. The high, rugged topography and narrow shelves of western North America are characteristic of an active continental margin that coincides with plate boundaries.

the southern parts of Alabama, Mississippi, Louisiana, and East Texas.

Structure of Eastern Seaboard

The margins of the North American continent reflect the transition from thick continental to thin oceanic crust (Fig. 4.4). The West Coast is an active continental margin; it coincides with plate boundaries and displays active features like volcanoes, earthquakes, and mountain building (Fig. 4.5). The East Coast is a passive continental margin, lying some 2,000 miles (3,000 kilometers) from the nearest earthquake and volcanic activity along the Mid-Atlantic Ridge.

Continental rifting can progress to a stage where the crust breaks into two separate fragments (Fig. 4.6). Each fragment includes a zone of transition from continental crust, to crust that was stretched and thinned

FIGURE 4.5 Active versus passive continental margins. Continental margins are zones of transition from thick continental to much thinner oceanic crust. **a)** A continental margin is active if it lies at or near an active plate boundary, where earthquakes, volcanoes, and mountain building occur. The Pacific Northwest is an active continental margin because it coincides with subduction at a convergent plate boundary. **b)** A continental margin is passive if it lies a great distance from a plate boundary and its tectonic activity. Passive continental margins developed because of plate divergence that opened the Atlantic Ocean and Gulf of Mexico. **c)** Earthquakes (represented here by red dots) are plentiful along the West Coast of North America. The East Coast lies a long way from the nearest zone of earthquake activity along the Mid-Atlantic Ridge. ▼

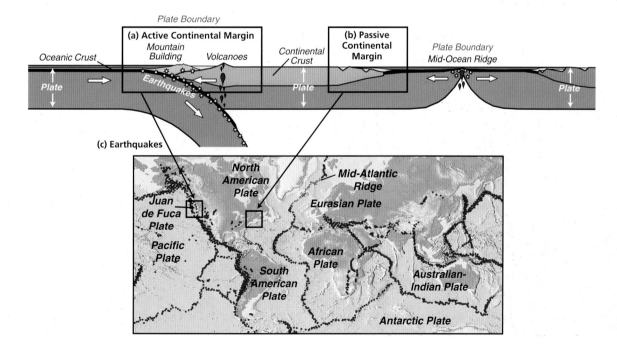

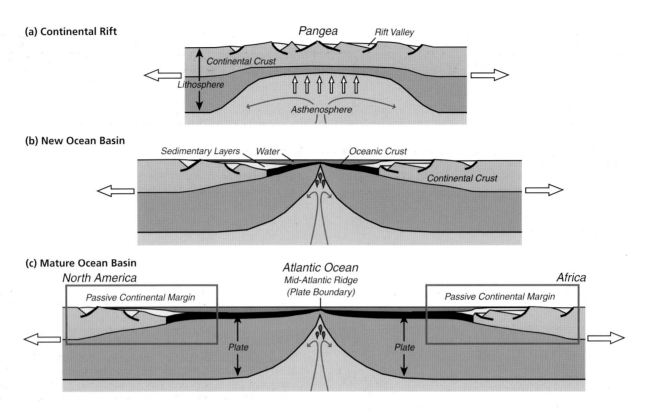

(a) Continental Rift

Pangea   Rift Valley

Continental Crust

Lithosphere

Asthenosphere

(b) New Ocean Basin

Sedimentary Layers   Water   Oceanic Crust

Continental Crust

(c) Mature Ocean Basin

North America

Passive Continental Margin

Atlantic Ocean
Mid-Atlantic Ridge
(Plate Boundary)

Africa

Passive Continental Margin

Plate   Plate

FIGURE 4.6 Plate divergence and the development of passive continental margins. Passive margins abutting the Atlantic Ocean developed as a large continent, **Pangea**, ripped apart about 200 million years ago. Since then, the ocean has continued to widen as Africa drifts away from North America and new ocean floor is created at the Mid-Atlantic Ridge. **a)** Rift valleys form as the crust rips apart and thins. **b)** A narrow basin, with thin oceanic crust, develops as the two continental fragments split apart. **c)** The continents move away from one another as parts of different lithospheric plates. The transitions from thick continental to thin oceanic crust are passive continental margins because they are far away from the plate boundary at the mid-ocean ridge. Thick sedimentary deposits accumulate as the continental margins cool and subside.

during rifting, to thin oceanic crust formed at a mid-ocean ridge. As the ocean expands, the continental fragments move farther and farther from the ridge.

The crust that lies beneath passive continental margins preserves evidence of continental rifting and initial ocean development (Fig. 4.7). Buried beneath coastal plain and continental shelf regions are ancient rift valleys filled with sedimentary and volcanic strata, similar to those forming today in the Basin and Range Province (Fig. 3.8). Farther out to sea, basaltic oceanic crust that formed as the continent drifted away from the mid-ocean ridge lies beneath the continental slope, continental rise, and abyssal plain. National parks and seashores on the Atlantic and Gulf of Mexico coasts lie along the top layer of the thick wedge of sediment deposited as the passive margin slowly subsides.

LAND–SEA TRANSITION. The overall form of a passive continental margin results from the crust thinning from the continent to the ocean. Where crust is thick, its buoyancy makes it stick up far above the denser man-

FIGURE 4.7 Crustal structure of a passive continental margin. Older sedimentary and volcanic layers fill rift valleys formed during the continental rifting (Fig. 3.7). Because the oceanic crust is thinner (and thus less buoyant) than the continental crust, its surface is lower (Fig. 4.8a). As the ocean basin floods with water, a wedge of sedimentary strata (sandstone, shale, and limestone) accumulates along the subsiding margin. The form of the top of the wedge depends on crustal thickness changes, sediment supply, and margin subsidence (Fig. 4.8c). The ocean water is shallow over the thicker continental crust, a region termed the continental shelf. The water level deepens along the continental slope to the continental rise, which overlies the thinner oceanic crust. Beaches and barrier islands on the coastal plain and continental shelf of the Atlantic Ocean and Gulf of Mexico are the sites of several national parks and seashores.

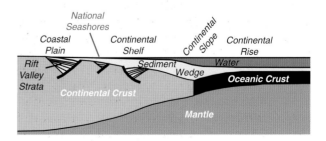

National Seashores

Coastal Plain   Continental Shelf   *Continental Slope*   Continental Rise

Rift Valley Strata   Sediment Wedge   Water

Continental Crust   **Oceanic Crust**

Mantle

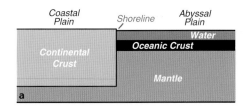 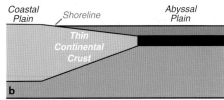 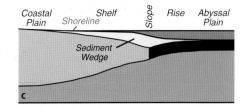

FIGURE 4.8 Factors responsible for the morphology of a passive continental margin. a)
Simplified model with blocks depicting continental and oceanic crust. The greater buoyancy of the thick
continental crust makes it stick up higher than the thin oceanic crust (Fig. 1.6). Only a coastal plain, near
sea level, and an abyssal plain, about 3 miles (5 kilometers) below sea level, would form. **b)** Model including a tapered zone of continental crust that thinned during rifting (Fig. 4.6a). A zone of deepening water
would develop between the coastal and abyssal plains. **c)** Model incorporating a wedge of sediment
draped over the continental margin. A continental shelf develops; beyond its edge is a relatively steep
seafloor, the continental slope, then a gentler continental rise.

tle; conversely, thin oceanic crust sits down much lower
(Fig. 4.8a). The thickness difference results in the top of
oceanic crust sitting down about three miles (5 kilometers) lower than continental crust. A crust of tapering
thickness would cause the gradual change in water
depth depicted in Fig. 4.8b. The situation is complicated
by the fact that, as the margin moves away from the
mid-ocean ridge, it cools, subsides, and is covered by
sediment eroding from the continent (Fig. 4.8c). The
subsidence and sedimentation result in a thin veneer of
sediment lying just above sea level at the **coastal plain**
and a thickening wedge of sediment beneath shallow
water of the **continental shelf.** Farther out to sea,
where thin oceanic crust is encountered, the water deepens abruptly at the **continental slope,** then more gradually along the **continental rise,** to the flat **abyssal
plain**. This morphology is reflected dramatically on a relief map of the Gulf of Mexico, which has developed to
the stage of a small ocean basin (Fig. 4.9).

Shoreline Processes in Our National Parks and Seashores

Picturesque beaches and barrier islands are the hall-
mark of the national parks and seashores on the
Atlantic and Gulf coasts. Those features are part of the
sediment wedges above the transitions from continental to oceanic crust. Coastlines and barrier islands are
features that, geologically speaking, change positions
rapidly, depending on global sea level, sediment supply,
and rate of subsidence of the continental margin. The
national parks and seashores lie along a very gentle
slope that extends hundreds of miles from the landward edge of the coastal plain to the seaward edge of
the continental shelf (Fig. 4.2). Over a distance of
about 300 miles (500 kilometers), elevation changes by

FIGURE 4.9 Relief map and crustal-scale cross section of the Gulf of Mexico.
Note the broad continental shelves off Louisiana, western Florida, and the Yucatán
Peninsula. National seashores and parks include coastlines and barrier islands that lie
on the gentle surface of the coastal plain and continental shelf. Cross section B–B'
shows that only the very central part of the Gulf of Mexico lies on oceanic crust. The
northern Gulf and its coastal plain are underlain by continental crust that was
thinned during rifting, and is covered with up to 10 miles (16 kilometers) of sedimentary strata. The layer of white salt was deposited when the Gulf was narrow
and water circulation restricted; it later formed salt domes that penetrate the overlying sedimentary layers.

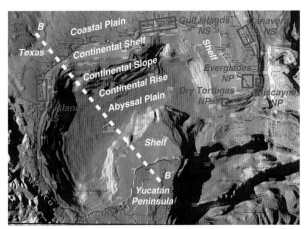

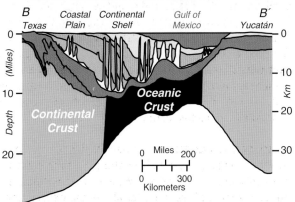

only about 1,000 feet (300 meters), from 400 feet (120 meters) above sea level to 600 feet (200 meters) below. The average slope is only about 0.3°!

SEDIMENTATION. Sedimentary basins are low areas of Earth's surface where eroded rock particles (sediments) accumulate. Most sedimentary basins form because of subsidence at plate boundaries or on passive continental margins. At divergent plate boundaries, sedimentation occurs during continental rifting and continues as the ocean basin opens and the continents drift away (Fig. 4.6). **Rift valley strata** (sedimentary deposits from eroding mountain ranges, along with lava flows) accumulate as a continent rips apart. Passive continental margins subside below sea level as continental fragments drift apart, because lithospheric plates cool as they move away from the mid-ocean ridge. A gradually thickening **sediment wedge** covers the rift valley strata and older continental rocks, as surrounding mountains erode and the passive margin continues to subside. Continental crust is too thick and buoyant to subside much, and cannot develop deep basins. The sedimentary layers along a passive continental margin can reach enormous thickness because they are deposited on continental crust that was thinned during the rifting stage, or on even thinner oceanic crust. The sedimentary wedge underlying the continental shelves of the Atlantic Ocean and Gulf of Mexico is as much as 10 miles (16 kilometers) thick (Fig. 4.9).

The beautiful sand beaches in national seashores are the products of sedimentary deposition. Those sediments can be quite different, depending on their origins.

THE OPENING OF AN OCEAN

As you stand on the flowing sands of Cape Hatteras or of other national seashores, close your eyes and imagine you're in the interior of a giant continent, called Pangea, 250 million years ago. The surface rises upward, as the land splits apart. With your friends the dinosaurs, you watch as volcanic eruptions, along with layers of sand and mud, fill subsiding rift valleys. After several million years, the continent completely rips apart, and you watch Africa slowly drift away from North America. More sand, mud, and limestone layers are deposited along the subsiding edge of the continent. Now open your eyes and gaze upon the mighty Atlantic Ocean; feel its waves continue to swirl the sands around your toes.

During the last ice age, ending about 15,000 years ago, a great continental ice sheet extended down from Canada and covered all of New England (Chap. 10). At its terminus, it dumped large amounts of sedimentary debris into a **terminal moraine**, forming Long Island, New York, just east of Manhattan, as well as the islands of Nantucket and Martha's Vineyard off the coast of Massachusetts. **Recessional moraines** formed as the climate warmed and the ice melted and receded northward. Lakes developed between the receding ice and moraines, including one very large one where mud, silt, and sand were deposited in what is now Cape Cod in Massachusetts. As the climate further warmed, the ice sheet disappeared from the region and the sea level rose. The beaches of Fire Island and Cape Cod national seashores are glacial moraine and lake deposits that have been reworked by wave action since the last ice age, forming eroded cliffs and long ridges of sand, including **spits** that are connected to the mainland, and **barrier islands** that lack a connection (Fig. 4.10a).

Glacial ice did not extend as far as the Atlantic and Gulf coasts of the southeastern United States. Much of the sedimentary deposits in those areas derive from erosion of the nearby North American continent, particularly the Appalachian-Ouachita-Marathon mountain range (Fig. 4.2). Most of the sedimentary material making up Assateague Island, Cape Hatteras, Cape Lookout, Cumberland Island, and Canaveral national seashores on the Atlantic Coast come from erosion of the Appalachian Mountains (Fig. 4.10b). The southern portion of Florida is some distance from the mouths of large rivers that would carry a lot of clastic (eroded) sedimentary grains. Instead, the beaches in those areas contain shell fragments made mostly of calcium carbonate, material much softer than the sand and silt found on beaches farther north. The warm waters produce fabulous coral reefs around Biscayne, Everglades, and Dry Tortugas national parks (Fig. 4.10c).

At Gulf Islands National Seashore along the Florida Panhandle and Mississippi coast, the incoming sediment is primarily from the Apalachicola River a bit farther east. Those sediments derive from erosion of a small part of the southern Appalachian Mountains and contain a lot of the mineral quartz. Quartz is a hard substance that does not dissolve in cold water, while other minerals may dissolve or are broken into fine grains during transport. Upon entering the Gulf of Mexico, the sediment is carried westward by water action known as **longshore current**. The beaches of Gulf Islands National Seashore are so spectacularly white because they are composed of more than 99 percent clear quartz grains (Fig. 4.10d). The finer material is

(a) Fire Island NS

(b) Cape Hatteras NS

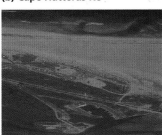

(c) Dry Tortugas NP

(d) Gulf Islands NS

FIGURE 4.10 Beaches in national seashores are made of sediments derived from different sources. a) Fire Island National Seashore in New York contains glacial moraine deposits that were reworked by wave action. **b)** Cape Hatteras National Seashore in North Carolina contains mixed sedimentary grains transported by rivers from the Appalachian Mountains. **c)** The beaches of Dry Tortugas National Park at the western end of the Florida Keys are fragments of shells broken up by wave action, and coral reefs are developed in the shallow, warm waters. **d)** The pure white sands of Gulf Islands National Seashore in Florida and Mississippi are grains of quartz derived from the southern Appalachian Mountains.

winnowed out by the current and carried farther west, where it mixes with mud carried from the interior of the continent by the Mississippi River and deposited on the Louisiana and Texas coasts. It is not until the region of Padre Island National Seashore that the mud content lessens and sandy beaches once again appear.

MIGRATING SHORELINES. Shorelines are not static features. They move in and out as sea level rises and falls. Beaches and barrier islands also migrate parallel to the coastline because of erosion and deposition by longshore currents. In national seashores, we can witness a natural flow and ebb, not only of water and tides, but of the land surface itself.

Sea Level Changes. As Earth's climate fluctuates, sea level drops when water is concentrated in polar ice caps, and rises as the ice melts. The shoreline migrates back and forth, from near the edge of the continental shelf during the peak of ice ages, to somewhere landward of its current position when climate warms (Fig. 4.11). During the peak of the last ice age, for example, when sea level was about 400 feet (120 meters) lower, the Atlantic and Gulf coastlines were about 100 miles (160 kilometers) to the east and south, respectively (Fig. 4.12). Lands that now make up our national seashores were high and dry at that time, developing their current morphology as the Earth warmed and the coast migrated to its current position over the past 18,000 years. At other times, when the climate was warmer, the coastlines were 70 miles (100 kilometers) farther inland. Nowadays the concern is that people are forcing the issue by burning hydrocarbons and spew-

FIGURE 4.11 Migration of the shoreline as sea level changes along a passive continental margin. The surface of the coastal plain and continental shelf tilts seaward at only about 3 feet per mile (0.6 meter per kilometer), or 0.3°. **a)** A thin layer of water covers the continental shelf, so that national seashores generally lie about 100 miles (160 kilometers) landward of the shelf edge. **b)** When Earth's atmosphere is cold, a lot of water is concentrated in the polar ice caps, causing sea level to drop. The beaches and barrier islands along the Atlantic Ocean and Gulf of Mexico migrate to near the edge of the continental shelf. **c)** Warmer temperatures cause melting of the ice caps and a rise in sea level. Shorelines migrate landward as coastal plains flood.

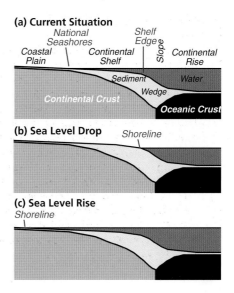

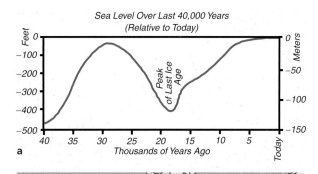

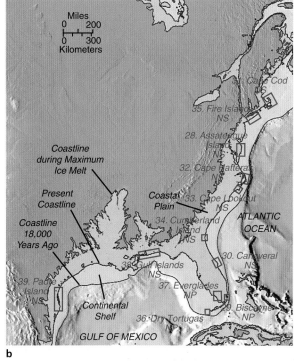

FIGURE 4.12 The position of the coastline is sensitive to changes in sea level. a) Sea level has been more than 300 feet (100 meters) lower than its present position twice in the past 40,000 years. **b)** During the peak of the last ice age, about 18,000 years ago, the coastline was near the edge of the continental shelf. It would have been possible to walk all the way out to Dry Tortugas National Park! At other times during the past 5 million years, sea level was higher than today, so that areas of the coastal plain were flooded. National parks and seashores lie in the zone between the maximum and minimum levels. Numbers refer to modern passive margin parks in Table 4.1.

FIGURE 4.13 Migration of a barrier island. At West Ship Island in Gulf Islands National Seashore, Mississippi, the longshore current erodes sand from the "upstream" end of the island and deposits it on the other tip. Through the years, the island gradually moves "downstream."

ing other materials into the atmosphere, perhaps causing Earth's atmosphere to warm through a "greenhouse effect." If that is the case, then within only a few centuries, water could flood the coastal plains and overrun huge population centers.

Longshore Currents. On a shorter time scale, narrow spits and offshore barrier islands migrate in directions parallel to the shoreline. Natural movements of water, called longshore currents, can transport large volumes of sand and mud. The current erodes material from the upstream side of an island, then carries and deposits it on the downstream side. Over a period of just a few decades, barrier islands, such as those in Gulf Islands National Seashore, can migrate several miles (Figs. 4.1 and 4.13). National seashores are places where we can see geologic changes in our own lifetimes—and observe, as Rachel Carson so vividly revealed, "Earth becoming fluid as the sea itself."

ANCIENT PASSIVE MARGIN OF NORTH AMERICA

The sandy beaches of our national seashores are just the top portions of thick sedimentary wedges that drape the subsiding edge of the eastern part of North America. Imagine what would happen if the whole East Coast were lifted upward, so that deep canyons carved through those layers. But we needn't just imagine—we can go to places

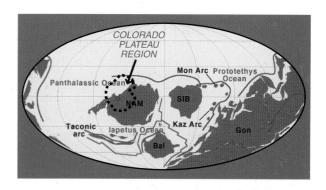

FIGURE 4.14 Continental configuration near the beginning of the Paleozoic Era. Five hundred million years ago, western North America was smaller than it is today, and near the equator. Note approximate position of the Colorado Plateau region, along a passive continental margin. The red star represents the area of Grand Canyon National Park. NAM = North America; SIB = Siberian craton; Bal = Baltica; Gon = Gondwanaland.

such as Grand Canyon National Park and see layers similar to those that lie beneath the beaches of Cape Hatteras, Padre Island, and other national seashores.

The shapes of continents, and their positions relative to one another, are images that are deeply embedded in our minds. But continents are constantly evolving through time. The western portion of North America was not even there 500 million years ago (Fig. 4.14). Instead, a passive continental margin, much like the East Coast and Gulf Coast

---

OUR WORLD TURNED UPSIDE DOWN

The image below (Fig. 4.15) just doesn't seem right. But why? There's no rational reason why north should be up and south down—it's simply that the first map-

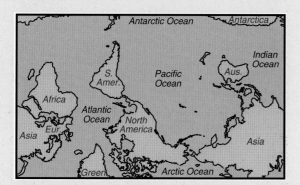

FIGURE 4.15 Upside-down map of the world.

makers were from the Northern Hemisphere and portrayed their part of the world on the top of maps. But suppose that South Americans or Australians had made the first maps, with the southern continents on the top? Would such a map induce vertigo for us today? So much of what we perceive as "normal" is really what we grew up with; we have become used to it. Similarly, the slow processes of geology might lull us into believing that our physical world is firmly fixed, but that's only because our time of observation is so brief. The positions and shapes of continents are a case in point. In fact, a bit of artistic license was used in drawing the maps in Fig. 4.3. Although the relative positions of continents are fairly accurate, for clarity their shapes and sizes are drawn as they appear today. The maps in Figs. 4.14 and 4.17 are more realistic, reflecting the fact that much of western North America was simply not there 500 million years ago!

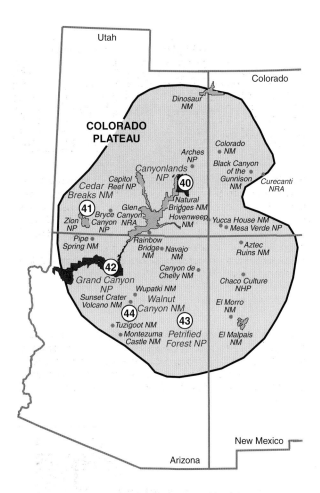

FIGURE 4.16 National park lands on the Colorado Plateau. The region was part of a passive continental margin on the western edge of the North American continent during the Paleozoic Era, and remained so until around the middle part of the Jurassic Period of the Mesozoic Era (Fig. 4.24). The sedimentary layering of the parks shown in red, including Grand Canyon National Park, Petrified Forest National Park, and Walnut Canyon National Monument, as well as the lower part of the layering at Canyonlands National Park and Cedar Breaks National Monument, were deposited along the ancient passive margin (the numbers reflect those in Table 4.1). Early-to-middle-Jurassic strata of some other parks (Zion, Capitol Reef, Navajo, Arches, Grand Staircase-Escalante) might also be argued as passive margin layering. But from the late-Jurassic Period onward, a convergent plate boundary was active along the western margin of the continent, so that the strata in other parks were no longer deposited in a passive margin setting. Those strata were deposited in foreland basins associated with the convergent tectonic activity, or by shallow seas that lapped up on the continental platform (Chap. 10).

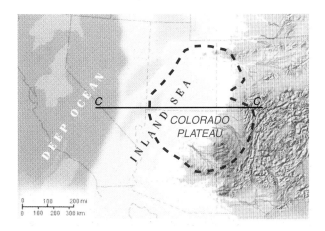

FIGURE 4.17 Map of the southwestern United States during the early Paleozoic Era, about 500 million years ago. The Colorado Plateau region straddled a shallow inland sea that periodically rose and fell, lapping over the edge of the continent. In a plate-tectonic framework, the area spanned the continental shelf and coastal plain of a passive continental margin. (Line C–C' is the cross section in Fig. 4.18).

today, was subsiding and being covered by sedimentary layers. North America was also much farther south than it is today (Chap. 10). The deposits along the continental shelf were similar to those forming today around the Florida Keys and Bahama Islands—limestone layers intermingled with sandstone and shale, and including fossils of tropical plants and animals.

National Parks on the Colorado Plateau

The Colorado Plateau is a region of high elevation in the southwestern part of the United States (Fig. 4.16). The uplift that elevated the plateau is a relatively recent event, occurring within the past 70 million years. During much of the Paleozoic and Mesozoic eras, from about 540 to 170 million years ago, the region was part of the ancient passive continental margin (Figs. 4.17 and 4.18). Since then, the continent has grown westward as oceanic material and blocks of thicker crust have been added to its edge during plate convergence (Chap. 11). Sedimentary layers of the passive margin were left stranded within the interior of the continent. As the Colorado Plateau rose upward, rivers carved downward, exposing those layers on the walls of canyons (Fig. 4.19).

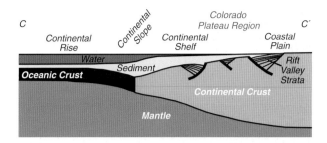

FIGURE 4.18 Schematic cross section of the passive continental margin of western North America during the early Paleozoic Era. (Line C–C' in Fig. 4.17.) The Colorado Plateau region included the coastal plain as well as shallow water of the continental shelf. Rift valley strata from an earlier breakup of the continent are called the Grand Canyon Supergroup, while the igneous and metamorphic rocks of the inner portions of Grand Canyon are part of the underlying, older continental crust (Fig. 4.21).

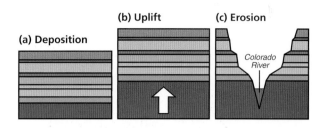

FIGURE 4.19 The rock layers of the Grand Canyon region are much older than the canyon itself. a) Sedimentary layers deposited in shallow seas that periodically lapped over the continental shelf and coastal plain during the Paleozoic Era from 540 to 250 million years ago. **b)** Uplift of the Colorado Plateau did not begin until about 70 million years ago. **c)** Most of the erosion and carving of the Grand Canyon by the Colorado River occurred during the past 6 million years.

The rock layers exposed in canyons in parks on the Colorado Plateau are like the pages of a giant history book (Fig. 4.20). At Grand Canyon National Park, the initial chapters, now seen deep down in the canyon along the Colorado River, tell us about igneous and metamorphic processes that accompanied mountain building more than a billion years ago (Fig. 4.21). Those rocks became part of the hard crust of an ancient continent. As that continent rifted apart, other layers filled rift valleys; those tilted layers can be seen farther up the canyon. Strata in the upper portion of the canyon were deposited on the passive continental margin that developed as a broad ocean opened.

SEAS COME AND GO. The passive margin strata in the Grand Canyon, as well as layers in some other parks on the Colorado Plateau, are mostly sandstone, limestone, and shale (Fig. 4.22). Those layers were deposited as shallow seas periodically lapped back and forth across the ancient continental shelf and coastal plain of western North America. Along shorelines, where currents and wave action were vigorous, coarse-grained sand was deposited. Farther offshore, in calmer waters, finer-grained silt and mud accumulated. Still farther out, beyond the reach of sand, silt, and mud, dissolved organic material settled out on the ocean floor as fine lime mud. The typical rocks that formed were sandstone in the nearshore and coastal plain environment, then progressively shale and limestone farther out on the continental shelf.

FIGURE 4.20 Rock layers of the Grand Canyon. The names and ages of formations are those encountered on a hike down the South Kaibab Trail in Grand Canyon National Park. From top to bottom, the layers are passive margin sediment, rift valley strata, and older continental crust (Fig. 4.18). Fm = Formation; Ss = Sandstone.

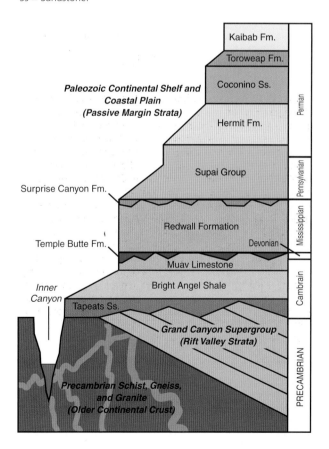

(a) Upper Canyon

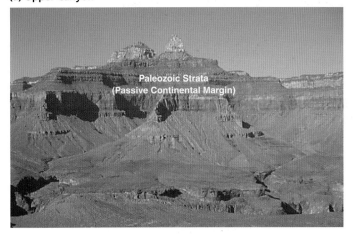

Paleozoic Strata
(Passive Continental Margin)

(b)Middle Canyon

Paleozoic Strata
(Passive Continental Margin)

Grand Canyon Supergroup
(Rift Valley Strata)

(c) Lower Canyon

Metamorphic
Rocks

Granite
Dikes

FIGURE 4.21 The Colorado River has eroded through Paleozoic passive continental margin strata and older rocks, forming the Grand Canyon. a) The upper portion of the canyon reveals shallow continental shelf and coastal plain sedimentary layers. **b)** The Grand Canyon Supergroup consists of continental rift valley strata preserved beneath the passive margin strata. **c)** In the deeper portions of the canyon, the river has cut down to the older metamorphic and igneous rocks of the underlying continental crust.

When the land surface drops down slightly, or sea level rises, the ocean advances over the continental edge (Fig. 4.23). Such a **transgression** results in layers of sandstone, covered by shale, and then by limestone as the water gradually deepens. During a **regression**, as the sea gradually moves out, the sequence is reversed, with limestone covered by shale and then by sandstone. Numerous advances and retreats of the sea led to the layered sequences of rock at Grand Canyon and other parks on the Colorado Plateau. In today's dry climate of the Southwest, sandstone and limestone layers erode very slowly, forming steep cliffs, while more easily eroded shale layers result in gentle slopes. The overall effect is the "stairstep" appearance seen in the layering at Grand Canyon and other parks (Figs. 4.20 and 4.22).

The strata observed in parks on the Colorado Plateau span a range of tectonic settings. Canyons have cut down deeply enough to expose the passive continental margin strata in only a handful of the parks (Fig. 4.24). The layers exposed in most of the parks on the Colorado Plateau relate to a more recent tectonic episode, when plate convergence resulted in the formation of mountains that shed sediment into adjacent, low-lying regions known as **foreland basins** (Chap. 10).

a

b

c

FIGURE 4.22 Ancient passive continental margin layers in parks on the Colorado Plateau. a) Canyonlands National Park, Utah. **b)** Petrified Forest National Park, Arizona. **c)** Walnut Canyon National Monument, Arizona.

(a) Transgression

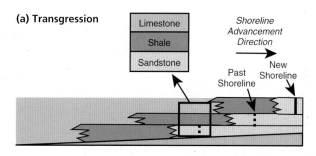

(b) Regression

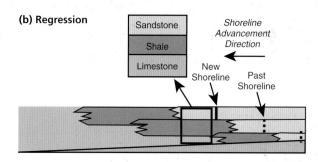

FIGURE 4.23 Development of sedimentary rock layers in the Colorado Plateau region. a) During a transgression, the sea advances over the continental shelf and coastal plain of the passive continental margin. Sandstone is initially deposited along the advancing shoreline, and then shale and limestone as the water deepens. **b)** As the sea moves out during a regression, the deepwater limestone is covered by mud and silt that turns into shale, and then into sandstone along the retreating shoreline.

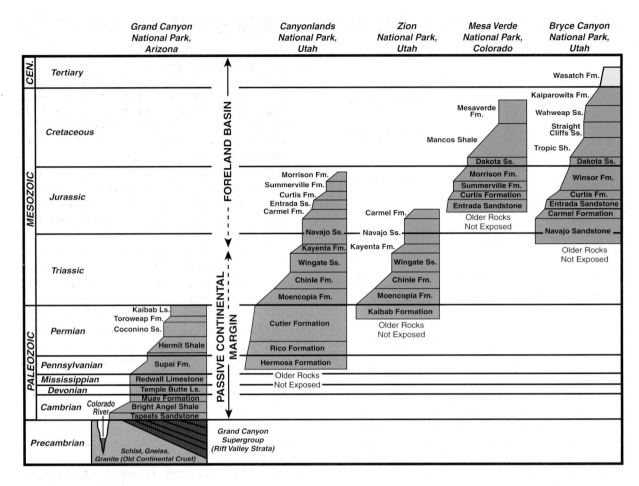

FIGURE 4.24 Stratigraphic sections for selected parks on the Colorado Plateau. Rock layers in the parks portray a variety of tectonic episodes and settings. The Precambrian rocks exposed in the lower part of Grand Canyon represent ancient mountain building, with accompanying igneous and metamorphic processes, and a continental rifting stage. The Paleozoic rocks higher up in Grand Canyon, as well as the lower layers seen in Canyonlands National Park, reveal an ancient passive continental margin of North America (Figs. 4.17 and 4.18). In the Mesozoic Era, plate convergence was the dominant tectonic force in western North America. By the Jurassic Period, deposition in the Colorado Plateau region consisted predominantly of sediment eroding from rising mountains. Such foreland basin strata comprise the upper layers exposed in Canyonlands and Zion national parks, as well as the strata seen in Mesa Verda and Bryce Canyon national parks (Chap. 10).

FURTHER READING

GENERAL

Beus, S. S., and M. Morales (editors). 2003. *Grand Canyon Geology*, 2nd Ed. New York: Oxford University Press. 432 pp.

Carson, R. 1955. *The Edge of the Sea*. Boston: Houghton Mifflin. 304 pp.

Chronic, H. 1988. *Pages of Stone, Geology of Western National Parks and Monuments: Grand Canyon and the Plateau Country*. Seattle: Mountaineers. 158 pp.

Fletcher, Colin. 1967. *The Man Who Walked through Time*. New York: Random House. 247 pp.

Hamblin, W. K., and J. R. Murphy. 1969. *Grand Canyon Perspectives: A Guide to the Canyon Scenery by Means of Interpretive Panoramas*. Provo, UT: H & M Distributors. 48 pp.

Leatherman, S. P. (editor). 1979. *Barrier Islands: From the Gulf of St. Lawrence to the Gulf of Mexico*. New York: Academic Press. 325 pp.

Leatherman, S. P. 1982. *Barrier Islands Handbook*, 2nd Ed. College Park, MD: University of Maryland Press. 109 pp.

Powell, J. W. 1961. *Exploration of the Colorado River and Its Canyons*. New York: Dover Publications. 397 pp.

Price, L. G. 1999. *An Introduction to Grand Canyon Geology*. Grand Canyon, AZ: Grand Canyon Association. 63 pp.

Sprinkel, D. A., T. C. Chidsey, and P. B. Anderson (editors). 2000. *Geology of Utah's Parks and Monuments*. Salt Lake City: Utah Geological Association. 644 pp.

Wagner, S. S. 2002. A geology training manual for Grand Canyon National Park. M.S. thesis, Oregon State University. 184 pp.

TECHNICAL

Boillot, B. 1981. *Geology of the Continental Margins*. New York: Longman. 115 pp.

Dobson, L. M., and R. T. Buffler. 1997. Seismic stratigraphy and geologic history of Jurassic rocks, northeastern Gulf of Mexico. *American Association of Petroleum Geologists Bulletin*, v. 81, p. 100–120.

Elston, D. P., G. H. Billingsley, and R. A. Young (editors). 1989. *Geology of the Grand Canyon, Northern Arizona (with Colorado River Guides)* (28th International Geological Congress Field Trip Guidebook T15/315). Washington: American Geophysical Union. 239 pp.

Ewing, J. I., and M. Ewing. 1959. Seismic refraction measurements in the Atlantic Ocean basins, in the Mediterranean Sea, on the mid-Atlantic ridge, and in the Norwegian Sea. *Geological Society of America Bulletin*, v. 70, p. 291–318.

Gardner, J. V., M. E. Field, and D. C. Twichell. 1996. *Geology of the United States' Seafloor: The View from Gloria*. New York: Cambridge University Press. 364 pp.

Grow, J., R. Mattlick, and J. Schlee. 1979. Multichannel seismic depth sections and interval velocities over outer continental slope between Cape Hatteras and Cape Cod, in *Geological and Geophysical Investigations of Continental Margins*, edited by J. S. Watkins, L. Montadert, and P. W. Dickinson. Tulsa: American Association of Petroleum Geologists, Memoir 29. p. 65–83.

Hall, D. J. 1983. The rotational origin of the Gulf of Mexico based on regional gravity data, in *Studies in Continental Margin Geology*, edited by J. S. Watkins and C. L. Drake. Tulsa: American Association of Petroleum Geologists, Memoir 34, p. 115–126.

Jenney, J.P., and S. J. Reynolds (editors). 1989. *Geologic Evolution of Arizona*. Tucson: Arizona Geological Society. 886 pp.

Le Pichon, X. 1968. Sea-floor spreading and continental drift. *Journal of Geophysical Research*, v. 73, p. 3661.

McClay, K. R. 1989. Physical models of structural styles during extension, in *Extensional Tectonics and Stratigraphy of the North Atlantic Margins*, edited by A. J. Tankard and H. R. Balkwill. Tulsa: American Association of Petroleum Geologists, Memoir 46, p. 95–110.

McQuarie, N., and C. G. Chase. 2000. Raising the Colorado Plateau, *Geology*, v. 28, p. 91–94.

Morgan, P., and C. A. Swanberg. 1985. On the Cenozoic uplift and tectonic stability of the Colorado Plateau. *Journal of Geodynamics*, v. 3, p. 39–63.

Pilger, R. H. 1978. A closed Gulf of Mexico, Pre-Atlantic Ocean plate reconstruction and the early rift history of the Gulf and North Atlantic. *Transactions, Gulf Coast Association of Geological Societies*, v. 22, p. 385–393.

Pratson, L. F., and W. F. Haxby. 1996. What is the slope of the U.S. continental slope? *Geology*, v. 24, p. 3–6.

Salvador, A. 1987. Late Triassic-Jurassic paleography and origin of Gulf of Mexico basin. *American Association of Petroleum Geologists Bulletin*, v. 71, p. 419–451.

Schouten, H., and K. D. Klitgord. 1994. Mechanistic solutions to the opening of the Gulf of Mexico. *Geology*, v. 22, p. 507–509.

Trehu, A. M., K. D. Klitgord, D. S. Sawyer, and R. T. Buffler. 1989. Atlantic and Gulf of Mexico continental margins, in *Geophysical Framework of the Continental United States*, edited by L. C. Pakiser and W. D. Mooney. Boulder: Geological Society of America, Memoir 172, p. 349–382.

Walper, J. L., and C. L. Rowett. 1972. Plate tectonics and the origin of the Caribbean Sea and the Gulf of Mexico. *Transactions, Gulf Coast Association of Geological Societies*, v. 22, p. 105–116.

Walper, J. L., F. H. Henk, E. J. Loudon, and S. N. Raschilla. 1979. Sedimentation on a trailing plate margin: The

Northern Gulf of Mexico. *Transactions, Gulf Coast Association of Geological Societies*, v. 29, p. 188–201.

Wilhelm, O., and M. Ewing. 1972. Geology and history of the Gulf of Mexico. *American Association of Petroleum Geologists Bulletin*, v. 56, p. 575–599.

Winkler, C. D., and R. T. Buffler. 1988. Paleogeographic evolution of early deep-water Gulf of Mexico and margins, Jurassic to Middle Cretaceous (Comanchean). *American Association of Petroleum Geologists Bulletin*, v. 72, p. 318–346.

WEBSITES

National Park Service: http://www.nps.gov

Park Geology Tour:
www2.nature.nps.gov/grd/tour/index.htm

Shoreline Geology:
www2.nature.nps.gov/grd/tour/coastal.htm

Altlantic Coast:

28. Assateague Island National Seashore:
www2.nature.nps.gov/grd/parks/asis/index.htm

29. Biscayne National Park:
www2.nature.nps.gov/grd/parks/bisc/index.htm

30. Canaveral National Seashore:
www2.nature.nps.gov/grd/parks/cana/index.htm

31. Cape Cod National Seashore:
www2.nature.nps.gov/grd/parks/caco/index.htm

32. Cape Hatteras National Seashore:
www2.nature.nps.gov/grd/parks/caha/index.htm

33. Cape Lookout National Seashore:
www2.nature.nps.gov/grd/parks/calo/index.htm

34. Cumberland Island National Seashore:
www2.nature.nps.gov/grd/parks/cuis/index.htm

35. Fire Island National Seashore:
www2.nature.nps.gov/grd/parks/fiis/index.htm

Gulf Coast:

36. Dry Tortugas National Park:
www2.nature.nps.gov/grd/parks/drto/index.htm

37. Everglades National Park:
www2.nature.nps.gov/grd/parks/ever/index.htm

38. Gulf Islands National Seashore:
www2.nature.nps.gov/grd/parks/guis/index.htm

39. Padre Island National Seashore:
www2.nature.nps.gov/grd/parks/pais/index.htm

Colorado Plateau:
www2.nature.nps.gov/grd/tour/cplateau.htm

40. Canyonlands National Park:
www2.nature.nps.gov/grd/parks/cany/index.htm

41. Cedar Breaks National Monument:
www.nps.gov/cebr/index.htm

42. Grand Canyon National Park:
www2.nature.nps.gov/grd/parks/grca/index.htm

43. Petrified Forest National Park:
www2.nature.nps.gov/grd/parks/pefo/index.htm

44. Walnut Canyon National Monument:
www.nps.gov/waca/index.htm

U.S. Geological Survey: http://www.usgs.gov

Geology in the Parks:
www2.nature.nps.gov/grd/usgsnps/project/home.html

Stratigraphic Sections for Parks on Colorado Plateau:
pubs.usgs.gov/gip/geotime/section.html

PART III

Convergent Plate Boundaries

Spectacular national park landscapes result from two plates moving toward one another. Chapter 5 presents the mountain ranges and volcanic features that form where one plate subducts beneath the other. The dramatic effects of an ocean closing and continents colliding are examined in Chapter 6.

Many national parks are in regions where plates are converging. Visitors can witness mountains as they are forming, and sometimes experience the accompanying earthquake and volcanic activity. Geologic structures and various types of rocks in other parks show where convergent boundaries existed long ago.

Where plates converge, the one capped by thinner oceanic crust descends beneath the one with thicker continental crust. Two parallel mountain ranges develop above such a **subduction zone**—a coastal range consisting of sedimentary strata and hard rock lifted out of the sea (accretionary wedge), and a volcanic range farther inland (volcanic arc). Sometimes an entire ocean closes during subduction, causing blocks of continental crust to collide; a **collisional mountain range** forms as the crust is compressed, crumpled, and thickened.

The shaded-relief map shows how national park lands relate to the topography formed at both active and ancient convergent plate boundaries. At the **Cascadia Subduction Zone**, the Coast Range is an uplifted mountain chain forming parallel to the active volcanoes of the Cascade Range. The low region between the two ranges is the Puget Sound area of Washington and the Willamette Valley in Oregon. Similar topography hints at past subduction in California: the Coast

Subduction zones and collisional mountain ranges form where plates converge. a) A plate with thinner (less buoyant) oceanic crust normally descends (subducts) beneath a plate with thicker (more buoyant) continental crust. An accretionary wedge develops where oceanic material is scraped off the descending plate, while a volcanic arc forms above the region where the plate gets hot enough to "sweat" fluids. Magma chamber rocks can be exposed if subduction stops and the volcanoes erode away. **b)** A collisional mountain range develops where blocks of thick continental crust crash together, uplifting the region to great elevation as the crust thickens.

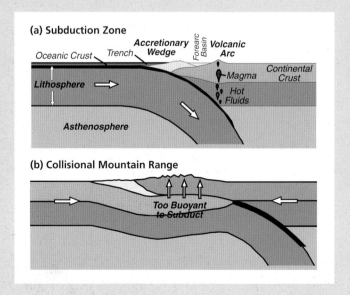

Range is the uplifted mountain range; the **Sierra Nevada** is the eroded volcanic range; and the Great Valley is the intervening depression. National park lands are also found where the Pacific Plate is subducting beneath **southern Alaska**.

The **Appalachian Mountains** in the eastern United States and Canada represent the collision of continents 500 to 300 million years ago. In their prime, those mountains probably had peaks as high as those in the modern zone of continental collision stretching from the Himalayas to the Alps. But after 300 million years, the Appalachians have eroded to more modest heights. The ancient zone of continental collision continues southwestward, but young sediments of the Gulf coastal plain cover most of it. It surfaces as the **Ouachita Mountains** of western Arkansas and southeastern Oklahoma, and the **Marathon Mountains** in the Big Bend area of West Texas. Another, younger zone of continental collision is the **Brooks Range**, stretching west to east across northern Alaska.

TEN QUESTIONS: Convergent Plate Boundaries

1. Why does an oceanic plate subduct beneath a continental plate?
2. Why are the Olympic and Cascade mountain ranges so different?
3. Are the Olympic Mountains getting higher or lower?
4. Why are subduction zone volcanoes so steep?
5. Why do subduction zone volcanoes sometimes explode?
6. Why is a devastating earthquake expected in the Pacific Northwest?
7. Why is the rock in parks in the Sierra Nevada Mountains so good for climbing?
8. How did the Appalachian Mountains form?
9. How high were the Appalachians at the time of collision?
10. How deep were the metamorphic rocks that are now exposed at the surface in the Appalachians?

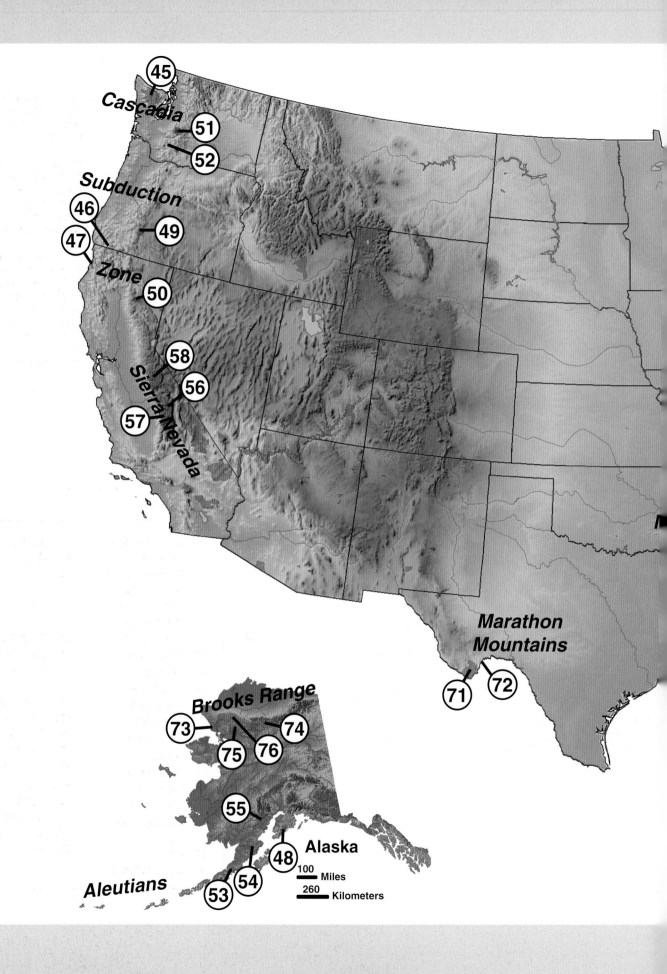

Cascadia

45

51

52

Subduction

46

49

47

Zone

50

Sierra Nevada

58

56

57

Marathon
Mountains

71

72

Brooks Range

73

74

75

76

55

48

Alaska

54

53

Aleutians

100 Miles

260 Kilometers

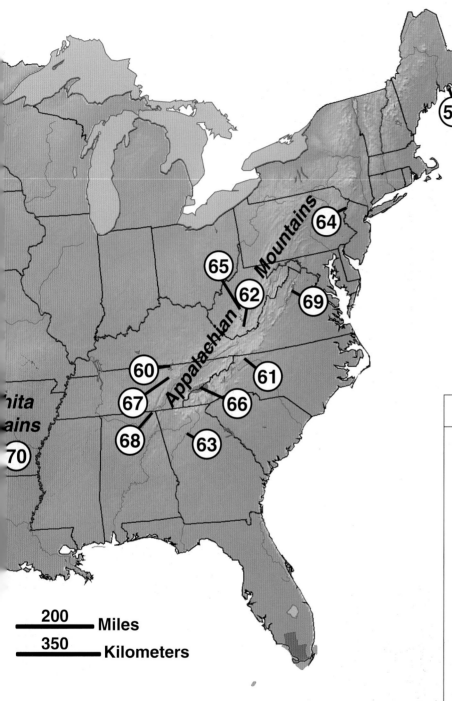

National park lands lie in a variety of convergent plate boundary settings. The Cascadia Subduction Zone has two distinct mountain ranges: a coastal range just above where the Juan de Fuca Plate begins to subduct, and the volcanic Cascade Range farther inland, where the top of the plate is deeper. The Sierra Nevada are the eroded roots of a volcanic range that formed when the subduction zone extended farther south. Another subduction zone is found where the Pacific Plate extends beneath southern Alaska and the Aleutian Islands. Parks in the eastern United States show effects of the continental collision that formed the Appalachian Mountains and their extension into the Ouachita and Marathon mountains. The Brooks Range in northern Alaska is a younger collisional mountain range.

200 Miles
350 Kilometers

Map Key

Cascadia Subduction Zone:
45. Olympic NP
46. Oregon Caves NM
47. Redwood N&SP
49. Crater Lake NP
50. Lassen Volcanic NP
51. Mt Rainer NP
52. Mt. St. Helens NVM

Southern Alaska:
48. Kenai Fjords NP
53. Aniakchak NM
54. Katmai NP
55. Lake Clark NP

Sierra Nevada:
56. Kings Canyon NP
57. Sequoia NP
58. Yosemite NP

Appalachian/Ouachita/Marathon:
59. Acadia NP
60. Big South Fork NR

61. Blue Ridge Parkway
62. Bluestone NSR&RA
63. Chattahoochee River NRA
64. Delaware Water Gap NRA
65. Gauley River NRA
66. Great Smoky Mountains NP
67. Obed WSR
68. Russell Cave NM
69. Shenandoah NP
70. Hot Springs NP
71. Big Bend NP
72. Rio Grande WSR

Brooks Range:
73. Cape Krusenstern NP
74. Gates of the Arctic NP
75. Kobuk Valley NP
76. Noatak NPres

5

Subduction Zones

Inspiring mountain peaks; explosive volcanoes; violent earthquakes. National parks present impressive evidence of the dynamic processes occurring where one plate subducts beneath another (Fig. 5.1). The parks demonstrate both the variety of features that result from subduction, and the systematic way that those features form (Table 5.1). In the Pacific Northwest, an offshore plate capped by thin oceanic crust subducts beneath North America. Many features in Pacific Northwest parks are comparable to those in parks in southern Alaska where subduction is occurring. The massive granite peaks of parks in the Sierra Nevada Mountains of California are the eroded remnants of magma chambers that formed beneath ancient subduction-zone volcanoes.

Continental crust is thicker, and therefore more buoyant, than oceanic crust. A plate with oceanic crust will subduct beneath one capped by continental crust (Fig. 5.2).[1] Two mountain chains, one uplifted and one volcanic, form parallel to the **deep-sea trench** at the surface juncture of the plates. Just landward of the trench, where the top of the subducting plate is shallow and cold, some of the ocean sediment and underlying rock are scraped off and deformed into a wedge shape. Those materials attach (or accrete) to the overriding plate. Portions of the resulting **accretionary wedge** rise above sea level as the sediment and rock are compressed, folded, and faulted, forming long ridges and valleys.

Farther from the trench, the top of the subducting plate can reach depths of 50 to 100 miles (80 to 160 kilometers), where it is so hot that fluids are driven from its crust. The fluids rise, melting silicate minerals from the mantle and crust of the overriding plate. Magma that makes it to the surface erupts as a curving chain of volcanoes, the **volcanic arc**. The low area between the two mountain ranges lies in front of the volcanic arc and is called a **forearc basin**.

FIGURE 5.1 Crater Lake National Park in winter.

[1]Two plates that are both capped by oceanic crust sometimes converge. The older plate, which is colder and more dense, generally subducts beneath the younger one, forming a volcanic island arc. An example is the western Aleutian Islands in Alaska. Subduction-zone park lands in the United States all involve a plate capped by oceanic crust subducting beneath one with continental crust, so the discussion here is limited to that situation.

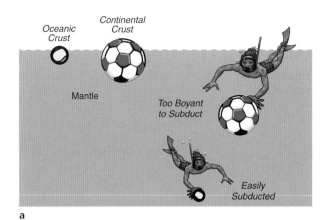

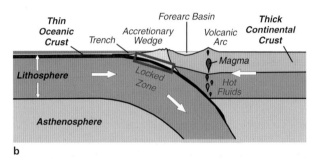

FIGURE 5.2 Ocean–continent subduction zone at a convergent plate boundary. a) A swimmer can easily "subduct" a tennis ball to the bottom of a pool but has a more difficult time with the larger (more buoyant) soccer ball. **b)** Where plates converge, the one with thinner oceanic crust subducts beneath the one capped by thicker (more buoyant) continental crust. The ocean floor is depressed at a deep-sea trench, the surface expression of the plate boundary. Two parallel mountain ranges form on the overriding plate: an accretionary wedge where material scrapes off the oceanic floor and a volcanic arc above the region where the plate gets hot enough to "sweat" fluids and trigger melting. The sediment-filled depression between the two ranges is a forearc basin. The locked zone is where the two plates stick together for centuries, only to let go suddenly as a devastating earthquake.

CASCADIA SUBDUCTION ZONE

The Pacific Northwest is an exciting place to observe geologic processes in action (Fig. 5.3). Parks in the coastal ranges of Washington, Oregon, and northern California contain rugged mountains of rocks that were manufactured in the ocean, then scraped off the plate and lifted out of the sea (Fig. 5.4). But parks in the Cascade Mountains—within the same subduction zone—are dramatically different. They contain explosive volcanoes formed as fluids rise from the top of the subducting plate and generate magma as they melt their way to the surface. And as time ticks on, the region awaits sudden release of energy locked between the converging plates as a devastating earthquake.

ACTIVE		ANCIENT
ACCRETIONARY WEDGE	**VOLCANIC ARC**	**VOLCANIC ARC**
Coast Ranges	*Cascades*	*Sierra Nevada*
45. Olympic NP, WA	49. Crater Lake NP, OR	56. Kings Canyon NP, CA
46. Oregon Caves NM, OR	50. Lassen Volcanic NP, CA	57. Sequoia NP, CA
47. Redwood N&SP, CA	51. Mount Rainier NP, WA	58. Yosemite NP, CA
	52. Mount St. Helens NVM, WA	
Southern Alaska	*Southern Alaska*	
48. Kenai Fjords NP, AK	53. Aniakchak NM, AK	
	54. Katmai NP, AK	
	55. Lake Clark NP, AK	

TABLE 5.1 National parks above subduction zones.

FIGURE 5.3 Three-dimensional view of the tectonics of the Pacific Northwest. The Juan de Fuca Plate forms and moves away from the Pacific Plate along the Juan de Fuca Ridge. The Coast Range comprises rock and sediment uplifted from the ocean. The Cascades are volcanoes above the zone where the top of the plate gets so hot it sweats water and other fluids. Puget Sound and the Willamette Valley remain near sea level as the two surrounding mountain ranges develop. The landscapes of Olympic and Mount Rainier national parks illustrate the contrasting appearance of mountains in the coastal and volcanic ranges.

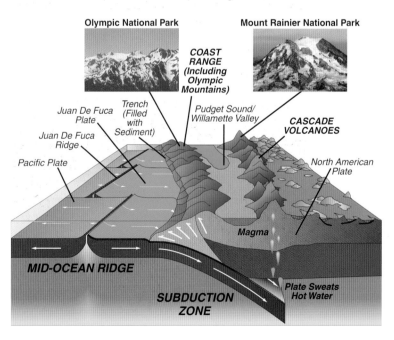

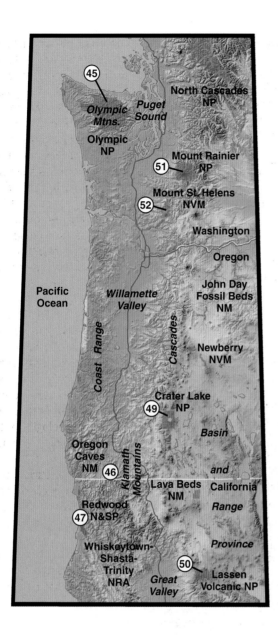

FIGURE 5.4 Subduction-zone parks in the Pacific Northwest.
Olympic National Park, Oregon Caves National Monument, and
Redwood National and State Parks are within the Coast Range
(an accretionary wedge). Whiskeytown-Shasta-Trinity National
Recreation Area is on the edge of the coastal ranges and the Great
Valley (a forearc basin). Mount Rainier National Park, Mount St.
Helens National Volcanic Monument, Crater Lake National Park,
and the western part of Lassen Volcanic National Park are in the
Cascade Mountains (a volcanic arc). The eastern part of Lassen, as
well as Lava Beds National Monument and Newberry National
Volcanic Monument, are on the western edge of the Basin and
Range Province (Chap. 3, "Continental Rifts"). North Cascades
National Park is made of older rocks that were added to North
America before the modern Cascades volcanic arc formed (Chap.
11, "Accreted Terranes"). John Day Fossil Beds National
Monument lies within the Columbia Plateau (Chap. 9, "Hotspots
beneath Continental Crust"). Numbers identify active subduction-
zone parks in Table 5.1.

Tectonics of the Pacific Northwest

Two hundred miles (350 kilometers) off the coast of
Washington, Oregon, and northern California, the
Pacific Ocean floor is ripping apart (Fig. 5.5). Hot mantle rises and begins to melt, pumping molten magma
that hardens along the edges of the Pacific and Juan de
Fuca plates. As the Juan de Fuca plate moves eastward,
it carries hard crust and sediment layers toward North
America, as if on a giant conveyer belt. At the edge of
the continent, the plate plunges downward and some
of the layers are scraped off the top and squeezed upward as the Olympic and other coastal mountains.

Farther east, the top of the Juan de Fuca Plate descends deeper and deeper. The sedimentary layers and
hard crust are metamorphosed because of the great temperatures and pressures at such depths. A by-product of
the metamorphism is the release of hot fluids, especially
water and carbon dioxide. Being lighter than the surrounding rock, those fluids rise and trigger the melting
of rock in their path. The melting produces silica-rich
magma that is so thick and pasty that it sometimes traps
gas under high pressure. The magma erupts through the
Cascade Mountain volcanoes in a variety of ways.
Occasionally it explodes if pressure on the trapped gas is
released suddenly, as occurred at Mount St. Helens in
1980. More often, it oozes out on the surface and hardens to gray-to-pink rocks called andesite, dacite, and
rhyolite (Table 2.6). The high gas content can cause
frothy particles of pumice to spew forth, or fine ash to
rise high in the air. Hot magma can release water or melt
large amounts of glacial ice; when steep parts of a volcano collapse as landslides, mudflows can roar down
valleys, then quickly solidify into layers as hard as concrete. Hot clouds of gas can roar down the mountain

FUN FACTOIDS!

- The Juan de Fuca Plate moves northeastward,
  toward and under the North American Plate, at
  about 1.2 inches (3 centimeters) per year. Your
  fingernails grow at about the same rate!
- Even though it moves slowly, given enough
  time the plate covers some distance. Every
  1 million years, the plate moves about 20 miles
  (30 kilometers)!
- In about 12 million years, crust of the Juan de
  Fuca Plate makes the 200 mile (350 kilometer)
  journey from where it is manufactured at the
  Juan de Fuca Ridge to where it plunges downward at the Cascadia Subduction Zone.

slopes, devastating everything in their paths. Layer upon layer of different materials have built the Cascade Mountains into "strato" or "composite" volcanoes reaching elevations over 14,000 feet (4,300 meters). Their steep profiles are due to the high-silica lava, which tends to stick to the sides rather than flowing long distances.

Coastal Ranges (Accretionary Wedge)

As the Juan de Fuca Plate begins its downward plunge about 50 miles (80 kilometers) offshore, it creates a depression on the seafloor. In other areas where an ocean plate subducts, such as off the west coast of South America, the deep-sea trench is evidenced by a depression on the seafloor that is 5 miles (8 kilometers) deep. In the Pacific Northwest there is no such depres-

VOLCANOES ALL IN A LINE

It may seem curious that Cascade volcanoes lie within a fairly narrow zone extending northward from Lassen Peak in California to Mount Garibaldi in British Columbia. But the reason is clear when you realize that the line of volcanoes is directly above the magic depth of about 50 miles (80 kilometers), where the top of the subducting Juan de Fuca Plate becomes so hot it begins to "sweat" (Fig. 5.3). The plate does not reach that depth until about 150 miles (250 kilometers) inland from the coast.

sion, because of three factors. First, the subducting Juan de Fuca Plate is young, and therefore hot and buoyant. Second, the plate converges with North America at a relatively slow rate of only about 1.2 inches (3 centimeters) per year. Third, because of the wet climate, an enormous amount of sediment is deposited off the coast, so that the depression fills in as fast as it forms. The thick accumulation of sediment supplies new material that is added to the base of the coastal mountains and maintains their topography.

OLYMPIC MOUNTAINS. The Olympic Mountains are a giant recycling machine (Fig. 5.6). Over a period of about 20 million years, a sand grain might be eroded, carried to the sea, and buried; the grain might then be moved eastward back beneath the mountains and shoved upward, to be eroded again and carried back to the sea. Such recycling highlights the renewable aspects of the geologic landscape, which is the very foundation of biological and ecological systems.

The geologic map of Olympic National Park indicates where various types and ages of rocks are found on the surface (Fig. 5.8). The pattern reveals a core area of 18- to 45-million-year-old sedimentary and metamorphic strata surrounded by 45- to 57-million-year-old basalt known as the Crescent Terrane. The core area strata are mainly sandstone and shale layers that were subjected to various degrees of metamorphism during subduction.

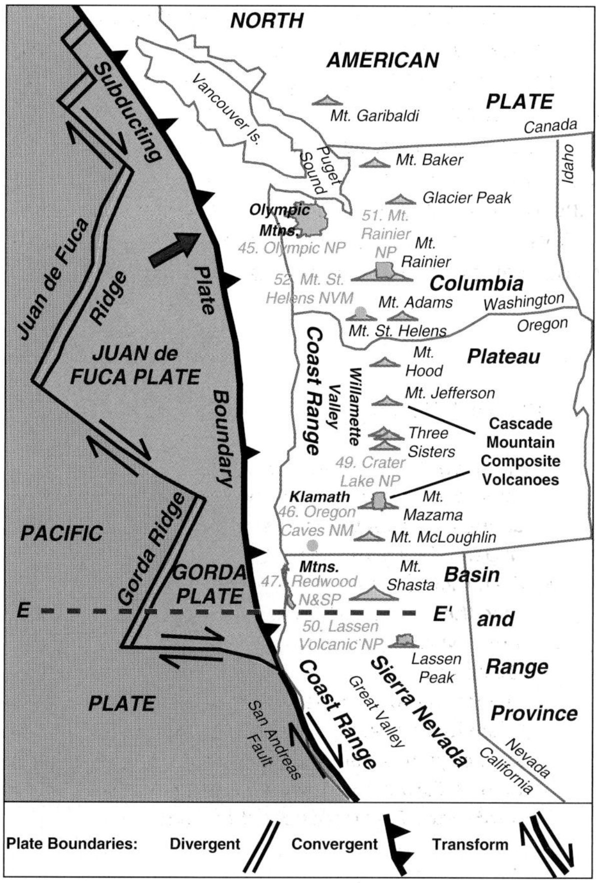

◀ **FIGURE 5.5 The landscapes of the Pacific Northwest form by interactions along three types of plate boundaries.**
1. Divergent Plate Boundary: The Juan de Fuca Plate (including its southern segment, the Gorda Plate) is manufactured where it diverges from the Pacific Plate along the Juan de Fuca/Gorda Ridge system. 2. Convergent Plate Boundary: The Juan de Fuca plate subducts northeastward beneath the North American Plate, forming the Coast Range (including the Olympic and Klamath mountains) and the Cascade volcanoes. Puget Sound, the Willamette Valley, and the Great Valley are low areas between the two ranges.
3. Transform Plate Boundary: The San Andreas Fault is the boundary where the Pacific Plate is sliding northwestward past the North American Plate. Other tectonic features include the Basin and Range Province (a continental rift zone where the North American plate is ripping apart), the Columbia Plateau (perhaps the place where the Yellowstone hotspot originally surfaced), and the Sierra Nevada (the eroded remnants of an extension of the Cascade volcanic arc). E–E' is the line of the cross section in Fig. 5.16.

(a) Erosion and Deposition

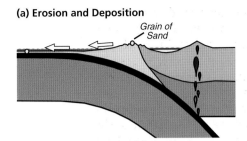

(b) Subduction

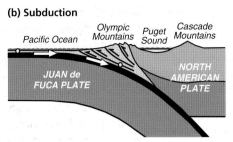

(c) Uplift and Erosion

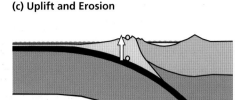

FIGURE 5.6 Recycling of the Olympic Mountains. a) A grain of sand eroded from high in the Olympic Mountains is transported by streams out into the Pacific Ocean, where it settles on the seafloor. **b)** The grain, now embedded in a layer of sandstone, is carried eastward on top of the Juan de Fuca Plate. In 5 million years it lies 10 miles (15 kilometers) beneath the Olympics. **c)** The sandstone layer is scraped off the top of the plate and squeezed upward. Fifteen million years later, the sand grain once again erodes from the surface of an Olympic mountain peak.

A visit to Olympic National Park is an opportunity to witness features of a developing accretionary wedge. The convergence between the Juan de Fuca and North American plates results in compression. Reverse faulting is common, especially the low-angle, thrust-fault variety (Fig. 2.7b). In fact, the Cascadia Subduction Zone can be thought of as a gigantic thrust fault, where the North American Plate is the "hanging wall" that is being shoved over the Juan de Fuca "foot wall"

OLYMPIC-SIZED RECYCLING MACHINE

A grain of sand erodes from a lofty peak in the Olympic Mountains. It's carried by a stream, perhaps one similar to today's Hoh River, to the shores of the Pacific Ocean. There it's swept 100 miles out to sea and settles with other particles of sand, silt, and mud onto the ocean floor. Gradually a mile or more of younger sedimentary layers covers our sand grain, until it is no longer free. Through time it's cemented to other grains and is now part of a layer of sandstone, and the surrounding mud hardens to shale.

Those layers are not standing still. They're carried on top of the Juan de Fuca Plate, back toward the North American continent, as if on a giant conveyor belt. Moving about 1.2 inches per year, they return to near the coast in about 3 million years, where the plate begins a downward plunge. As the plate relentlessly grinds away, the layers are transported still farther eastward and deeper and deeper beneath the edge of North America. After another 2 million years, the layers lie 10 miles beneath the central part of the Olympics. There the temperature and pressure are so great that the sandstone and shale layers harden into metamorphic rocks called metasandstone and slate.

Now something different happens. Instead of continuing their journey eastward and deeper into the Earth, the layers are scraped off the top of the Juan de Fuca Plate and rise upward. The scraping is accompanied by huge earthquakes that rattle the surface above. At the same time, the actions of wind, water, and ice wear away the overlying rock. In another 15 million years, our grain of sand finds itself once again a mile above the sea. Only briefly it experiences the sunlight and stars, wind and snow and rain, before being once again eroded and carried to the sea.

(a) Uplift and Erosion

(b) Transportation

(c) Deposition

(d) Uplift and Erosion

FIGURE 5.7 The path of a recycling grain of sand from the Olympic Mountains.

THE SURFACE WATER RECYCLING MACHINE

Like our grain of sand in Fig. 5.6, a drop of water goes through a similar cycle, from high in the mountains, down a stream, and out into the ocean. The difference is that, instead of being shoved a few miles down into the Earth, the water droplet evaporates and rises a few miles upward into the air, is carried eastward by wind, and falls back on the Olympics as mist, rain, or snow. And the water cycle might encompass a few weeks or months rather than 20 million years!

basalt of the Crescent Terrane has eroded away, so that the underlying sedimentary and metamorphic rocks now form the core area of the Olympics. The basalt that remains forms a large "horseshoe" pattern around the north, east, and south portions of the park. The rocks were also deformed along numerous smaller faults and folds as they were caught in the vise between the converging North American and Juan de Fuca plates. The rocks were tilted so much that in places, layers have nearly vertical orientations (Fig. 5.10).

Igneous, sedimentary, and metamorphic rocks are all found in abundance in Olympic National Park (Fig. 2.10). Most of the igneous rocks were manufactured in the ocean and are basalt. The spectacular **pillow lavas**

(Fig. 5.3). One large thrust fault in the park is the zone where the older Crescent Terrane rocks have overridden younger core area strata (Fig. 5.8). This fault zone has different names along its length, including the Calawah and Hurricane Ridge faults on the north. Numerous other reverse faults are found throughout the park, some so small that they offset layers just a few inches, while others displace strata several miles.

The overall structure of the Olympic Range can be viewed as the development of two large structures (Fig. 5.9). First, the older rocks of the Crescent Terrane were pushed over the younger layers along a gigantic thrust fault. Then, as the layers uplifted, they were folded into a large, plunging anticline (Fig. 2.8d). Much of the

FIGURE 5.8 The Olympic Mountains contain rocks and structures of a developing accretionary wedge. a) Schematic cross section along line D–D' on the map. Basalt rocks of the Crescent Terrane are thrust over younger sedimentary and metamorphic rocks exposed in the core area. Numerous thrust faults cut core area rocks as the rocks are squeezed upward out of the sea. **b)** Geologic map of Olympic National Park. Older layers of basalt (Crescent Terrane) form a "horseshoe" around younger layers of sandstone, shale, and slate in the core area of the Olympics. The hard rocks are covered by very young layers of gravel, sand, and silt that were deposited by continental glaciers in the northern and northwestern parts of the Olympic Peninsula, and by streams and mountain glaciers in other areas.

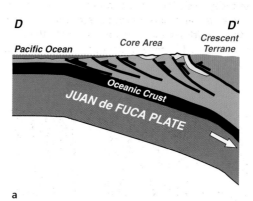
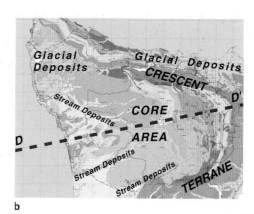

ROCK CONTORTION

Although the broad structure of Olympic National Park is an upfold, like an anticline, there is an interesting twist. For a normal stratigraphic sequence, with younger layers deposited on top of older layers, anticlinal folding and erosion produce a structure with the oldest strata in the center (Fig. 2.8a). But the sequence of rock layers in Olympic National Park is not normal; older basalt rocks of the Crescent Terrane were thrust over the younger sedimentary layers (Fig. 5.9a). When the strata were deformed into a broad anticline and tilted eastward, and the older rocks eroded from the top, a "window" into the younger rocks was established (Fig. 5.9b). So the Olympic Mountains are an unusual type of anticline that exposes the younger layers in the middle.

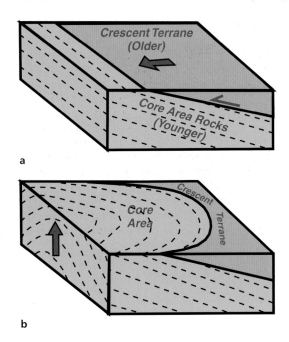

a

b

FIGURE 5.9 Structural development of the Olympic Mountains. **a)** Thrust faulting. The Crescent Terrane basalt was thrust over younger sedimentary and metamorphic layers that now form the core area of the Olympics. (Compare the frontal view with the cross section in Fig. 5.8a.) **b)** Eastward-plunging anticline. The region was folded upward and tilted eastward. The underlying, younger core area rocks were exposed as some of the Crescent Terrane was eroded from the top of the anticline. (Compare the eroded surface with the map in Fig. 5.8b.)

FIGURE 5.10 Sedimentary rocks found along the Hurricane Hill Trail in the core area of Olympic National Park have been tilted so much that the layers are vertical.

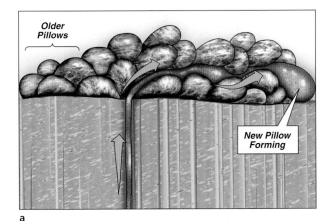

a

b

FIGURE 5.11 Basalt "pillows" are evidence that rocks of the Olympic Mountains formed beneath the ocean. **a)** As a pulse of lava pours out into the cold seawater, it cools quickly and hardens into a pillow form that settles on the seafloor. Other pulses created pillows that piled up on, and filled the gaps between, previously formed pillows. **b)** Close-up of pillows on Hurricane Ridge Road, just beyond the tunnels in Olympic National Park.

seen on the road up Hurricane Ridge indicate that at least some of the 50-million-year-old Crescent Formation formed on the seafloor and was later uplifted (Fig. 5.11). Basalt lava is thin and runny and can flow for miles and miles. But if the lava spills out from beneath the ocean, it encounters cold water and hardens quickly. A pulse of lava assumes a globular shape that flattens into a pillow form as it flops onto the sea bottom. Successive pulses deposit more pillows between the gaps and on top of earlier pillows.

Most of the sedimentary rocks in Olympic National Park result from sand and mud deposited in huge, fan-shaped patterns beneath the ocean. Such layers of thick sandstone and thin shale are called **turbidites** because they were deposited by fast-moving undersea flows of water known as **turbidity flows** (Fig. 5.12). Turbidity flows are often triggered by earthquakes.

High volumes of sediment make the flows dense, so they move quickly down the edge of the continental shelf. As a turbidity flow reaches the flat bottom of the deep ocean, it begins to slow down, dropping first the coarse, heavy sand, then gradually the finer silt. The very fine particles continue farther out to sea as mud. During the time between turbidity flows, the ocean water is quiet, so that a thin layer of mud is deposited on top of the sand and silt. A turbidite sequence commonly has thick layers of sandstone (the sand de-

posited quickly by a turbidity flow) interspersed with thin shale layers (the mud deposited slowly during the long time between turbidity flows).

The degree to which a rock was metamorphosed provides a clue to how far the rock was pushed below the surface, and for how long (Table 2.5). Some of the rocks in Olympic National Park, especially along the coast, are still sandstone and shale, indicating that those rocks were never more than about 5 miles (8 kilometers) below the surface (Fig. 5.13). But in other places, particularly in the

FIGURE 5.12 Turbidite sequence on Beach 3 north of Kalaloch, Olympic National Park. a) The thick sandstones and thin shales were originally deposited as horizontal layers. **b)** Today the rock layers are nearly vertical. Part *(a)* shows the same photograph as part *(b)*, but rotated to illustrate the orientation of layers before tilting.

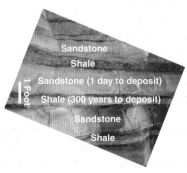
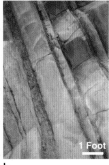

a   b

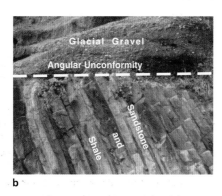

b

a

FIGURE 5.13 Evidence for uplift and deformation of rocks in Olympic National Park (Beach 3 north of Kalaloch). a) Layers rotated from original horizontal orientation. Sandstone and shale layers were lifted out of the sea and tilted to a nearly vertical position. **b)** Angular unconformity. The tilted sandstone and shale layers below the unconformity are about 20 million years old, whereas the glacial gravel layers above are less than 1 million years old.

core area of the mountains, rocks were pushed down 7 to 10 miles (11 to 16 kilometers) deep. Shales developed to slate and phyllite, but never to higher-grade schist (Fig. 5.14). The high pressure caused some of the metamorphic rocks to develop preferred directions of cracking, known as **cleavage**. Slate sometimes shatters along the perpendicular cleavage planes into long, skinny blocks, an effect known as **pencil cleavage**.

Olympic National Park has mountains nearly 8,000 feet (2,500 meters) in elevation. Those heights are maintained because the Olympic Mountains are in a **topographic steady state**. The amount of uplift, caused by layers added to the base of the mountains by the Juan de Fuca Plate as it subducts, is roughly balanced by erosion due to wind, water, and ice. As more and more layers are added from below, the range continually rises. But erosion on the surface keeps the mountains at about the same maximum elevation of 1 to 1½ miles (1.5 to 2.5 kilometers). The overall height

FIGURE 5.14 Metamorphic rocks in Olympic National Park. a) Sandstone and shale layers on Hurricane Hill Trail have changed to metasandstone (light color) and slate. **b)** The slate on Obstruction Point Road has well-developed planes of weakness, such that the rock shatters into long "pencils."

(a) Metamorphosed Turbidite Sequence

(b) Pencil Cleavage in Slate

THE ORIGIN OF DEEP CANYONS

Ice and water have carved canyons deep into the interior of the Olympic Mountains. But why are the Olympic canyons so deep? Why is any canyon deep? Consider the Grand Canyon in Arizona (Chap. 4). The Colorado Plateau, 7,000 feet (2,000 meters) above sea level, is being eaten away by the Colorado River a mile below the canyon rim. You might think of the river as a knife carving downward into a wedding cake. But there's

another way to look at it. Hold the knife steady and move the cake upward. The Colorado River has been holding steady at 2,000 feet elevation while the Colorado Plateau moves slowly upward. Likewise, rivers such as the Elwha in Olympic National Park strive to remain near sea level, yet the Olympic "plateau" continues to move upward as material is added from below.

FUN FACTOID!

Material added to the base of the Olympic Mountains uplifts the range at a rate of about $\frac{4}{100}$ of an inch per year (1 millimeter per year). From a depth of 10 miles (16 kilometers), it takes about 15 million years to elevate the material back to sea level. If it were possible for uplift to continue at that rate with no accompanying erosion, then in 15 million years the Olympics would have peaks twice as high as Mount Everest!

PONDERANCES ON TIME

Olympic High School
The youngest hard sedimentary layers at Olympic National Park are 18 million years old. This is rather curious, considering what we learned about the Olympic-sized recycling machine—that the entire cycle covers a similar span of time, about 20 million years. So where are the younger layers? The answer is that they are going through earlier parts of the cycle. They are either still on the Juan de Fuca Plate and headed toward North America, or they are beneath the Olympics and haven't yet made it back up to the surface.

So the youngest (18-million-year-old) rocks exposed on Olympic beaches are like the (18-year-old) students about to graduate from "Olympic High School." We know there are younger students, but we don't see them in the graduation ceremonies because they are in the lower grades and haven't yet risen to the highest level where they are about to enter the real world.

of the mountains is maintained as the Olympic Mountains recycle themselves!

It is no coincidence that uplift and erosion are in balance in the Olympic Mountains. This is often the case where mountains form, because erosion increases as topography rises higher and higher. When mountains begin to form, the erosion rate is slow, so that uplift exceeds erosion (Fig. 5.15a). But if the mountains were to rise really high, then the rate of erosion would exceed uplift (Fig. 5.15b). A natural topographic height is established where rates of uplift and erosion are equal (Fig. 5.15c).

ACCRETIONARY WEDGE IN OREGON AND CALIFORNIA.
The landscape of the Pacific Northwest consists of a coastal and a volcanic mountain range, with a forearc basin in between (Fig. 5.3). That typical subduction-zone morphology is disrupted in southern Oregon and north-

ern California, where there is no low area (forearc basin) equivalent to Puget Sound or the Willamette Valley. Instead, the Klamath Mountains extend from near the Pacific Ocean all the way to the Cascades (Fig. 5.16). The Klamath Mountains are made up of "exotic" or accreted terranes because their hard crust and sedimentary layers

FIGURE 5.15 A developing mountain range commonly reaches a steady topographic state. a) Coastal mountain range begins to form because of plate subduction. When elevations are low, the erosion rate might be 0.01 inch per year, while the uplift rate is 0.04 inch per year. The net effect is that the topography rises by 0.03 inch per year (0.04 – 0.01 inch per year). **b)** If the mountains were to grow to great height, the erosion might accelerate to 0.06 inch per year, so that the net effect would be to decrease topography by 0.02 inch per year (0.04 – 0.06 inch per year). **c)** The mountains find a natural height where the rate of erosion equals the uplift rate. The uplift/erosion rate of 0.04 inch (1 millimeter) per year in the Olympic Mountains results in topography that is about 5,000 to 8,000 feet (1,500 to 2,500 meters) above sea level.

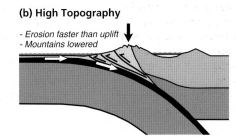

(a) Low Topography
- Erosion slower than uplift
- Mountains rise

(b) High Topography
- Erosion faster than uplift
- Mountains lowered

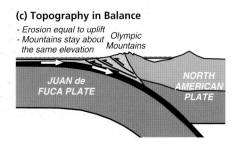

(c) Topography in Balance
- Erosion equal to uplift
- Mountains stay about the same elevation

Olympic Mountains

JUAN de FUCA PLATE *NORTH AMERICAN PLATE*

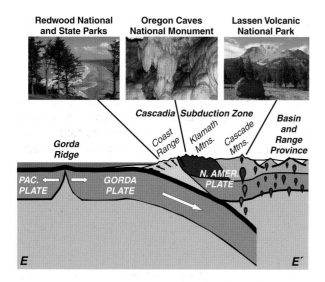

Redwood National and State Parks   **Oregon Caves National Monument**   **Lassen Volcanic National Park**

FIGURE 5.16 Formation of mountain ranges in southern Oregon and northern California. (Cross section E–E' in Fig. 5.5.) The Gorda Plate, the southern portion of the Juan de Fuca Plate, rips apart from the Pacific Plate at the Gorda Ridge. A narrow accretionary wedge (Coast Range) forms where oceanic rocks are scraped off the Gorda Plate as it subducts beneath the North American Plate. The Klamath Mountains are older oceanic and subduction-zone rocks that slammed into North America during the early part of the Cretaceous Period (about 100 million years ago). A prominent forearc basin, such as Puget Sound or the Willamette Valley, is lacking in southern Oregon and northern California because of the accreted Klamath Mountain terranes. Farther east, dehydration of the subducting Gorda Plate leads to the formation of Lassen Peak and other Cascade volcanoes. Redwood National and State Parks lie in the young accretionary wedge (Coast Range). Oregon Caves National Monument is in the Klamath Mountains. The western portion of Lassen Volcanic National Park is in the Cascades, while features on the eastern side of the park relate to continental rifting in the Basin and Range Province.

FIGURE 5.17 Oregon Caves National Monument.
a) The marble formations are part of a block of thicker crust that slammed into North America, forming the Klamath Mountains.
b) Groundwater dissolves some of the calcite (calcium carbonate) of the marble, forming flowstone structures, such the Banana Grove Drapery on the park's cave tour.

a   b

formed some distance from the Pacific Northwest (Chap. 11). The blocks of crust were brought in by transform and convergent plate motion. As they came crashing in, the blocks were uplifted, deformed, and accreted to the edge of the continent.

The rocks of the Klamath Mountains are distinct from other rocks in the Pacific Northwest. They include uplifted oceanic plate material known as ophiolite (Fig. 3.28). Peridotite, the dark-green, heavy igneous rock of Earth's mantle, is abundant, as are basalt and gabbro comprising oceanic crust. Intrusive diorite and its volcanic equivalent, andesite, are also found in the region, revealing an ancient subduction zone. In addition to sandstone and shale, the sedimentary layering includes limestone, suggesting that rocks of the Klamath Mountains were deposited in tropical waters, far south of their present location.

The Klamath Mountains are a complex mass of rocks that were folded and faulted, first when they were part of an ancient subduction zone about 150 million years ago, and then more recently when they slammed into the Pacific Northwest. Rocks were pushed down to significant depths, where high temperature and pressure metamorphosed the ophiolite to rocks called **greenschist** and **serpentinite**, and the limestone to **marble**.

Many caves have formed in the volcanic landscapes of the Pacific Northwest. Those caves are typically long, smooth-walled lava tubes, such as Ape Cave in Mount St. Helens National Volcanic Monument, and the numerous examples at Newberry National Volcanic Monument and Lava Beds National Monument (Fig. 3.20). The caverns at Oregon Caves National Monument are different (Fig. 5.17). They are caused by groundwater dissolving out calcium carbonate from the marble found in the Klamath Mountains. Oregon Caves contain **stalactites** dripping from the cave ceilings and **stalagmites** forming on cave floors, similar to the spectacular features found in the limestone caverns at Carlsbad Caverns National Park in New Mexico and Mammoth Cave National Park in Kentucky (Chap. 10).

Redwood National and State Parks are a collection of federal and state lands in the northwestern corner of California. The region includes some of the old rocks of the Klamath Mountains, as well as the modern accretionary wedge (Fig. 5.18). Much of the rock exposed in the park is **mélange**, a contorted mixture of rock-layer fragments of different sizes. Rocks of the accretionary wedge were subjected to high pressures, yet relatively low temperatures, as they were dragged down into the Earth at the subduction zone. The situation is analogous to pushing an ice cube into a cup of hot coffee.

The pressure on the cube increases right away, but it takes time for the ice to heat up and melt. In northern California, the Gorda Plate (the southern portion of the Juan de Fuca Plate) is like the ice cube. As it subducts to 5 to 15 miles (8 to 25 kilometers) beneath the Coast Range, the overlying rock instantly subjects it to high pressure, but the plate does not heat up nearly so fast. The sedimentary layers on top of the plate (sandstone, siltstone, conglomerate, and shale) were metamorphosed under such conditions to quartzite, phyllite, and schist (Table 2.5). Basalt that formed through seafloor volcanism, as well as some of the underlying peridotite of Earth's mantle, metamorphosed to a greenish rock called serpentinite. The ocean sediments and underlying hard crust that were metamorphosed, folded, faulted, and uplifted as the mélange of the California Coast Range are known as the **Franciscan Group** (discussed further in Chap. 7).

Evidence of the ongoing tectonic activity at Redwood National and State Parks includes the rugged coastline, as well as frequent landslides and earthquakes. As at Olympic National Park, erosion rates are a response of a system trying to maintain a steady topographic state (Fig. 5.15). But the California coastal mountains have been eroding at an accelerated rate because of intense mining and logging activities over the past 150 years. The result is the buildup of sand

FIGURE 5.18 Geologic processes at Redwood National and State Parks. a) Ongoing subduction and uplift of oceanic rocks lead to a steep and rapidly-eroding coastline. **b)** As rocks were dragged down the subduction zone, they were metamorphosed and contorted into a chaotic mixture known as mélange.
c) Rapid uplift of the Coast Range and Klamath Mountains, along with mining, logging, and other human activities, results in erosion and transport of a large volume of sediment to the sea by streams such as the Klamath River.

and mud deposits near the mouths of rivers, causing coastlines to build outward at unnatural rates.

Cascade Mountains (Volcanic Arc)

Volcanoes in national parks in the Cascade Mountains have eruptive "personalities" that reflect the complex magma systems associated with subduction zones. Mount Rainier National Park showcases a 14,411-foot (4,393-meter) composite volcano. The mountain consists mainly of andesite lava flows and mudflows covered by numerous glaciers. The volcano in Mount St. Helens National Volcanic Monument has been known to explode, as it did so dramatically in 1980, reducing its elevation from 9,677 to 8,365 feet (2,950 to 2,500 meters). Crater Lake National Park has the seventh deepest lake in the world. The lake partially fills a collapsed crater (caldera) that formed when a 12,000-foot (3,700-meter) composite volcano, Mount Mazama, erupted 7,700 years ago. Lassen Peak in Lassen Volcanic National Park is an amalgamation of dacite-to-rhyolite lava domes.

PARKS IN THE WASHINGTON CASCADES. National park lands in the Cascade Mountains of Washington State include Mount Rainier National Park, a unit of the National Park Service, and Mount St. Helens National Volcanic Monument, managed by the U.S. Forest Service. The two volcanoes showcased in those parks illustrate two quite different types of volcanic hazards.

Many people have an indelible image of the 1980 eruption of Mount St. Helens, involving an explosion that destroyed the summit and north flank of the mountain (Fig. 5.19). That eruption was a consequence of high-silica, pasty magma that accumulated beneath the mountain (Fig. 5.20). As the magma rose in early

---

CASCADE SAUNA

Why the Juan de Fuca Plate Sweats
Metamorphism of rocks on top of the Juan de Fuca Plate is ultimately responsible for Cascade volcanism. The crust of the ocean is mainly basaltic (the igneous rocks basalt and gabbro). As the plate descends to greater and greater depths and encounters higher temperature and pressure, the crustal rocks metamorphose to eclogite. Eclogite is more dense than basalt, formed as the rock is compressed and fluids (mostly water) are driven off. That hot water rises and triggers melting within the overriding North American plate, forming magma.

FIGURE 5.19 Mount St. Helens National Volcanic Monument. a) The mountain before the 1980 eruption. As magma rose, the north flank began to bulge outward. **b)** At 8:32 A.M. on May 18, 1980, an earthquake beneath the mountain caused the oversteepened north flank to fail as a giant landslide. **c)** The sudden pressure release was like opening a shaken bottle of coke, allowing gasses to expand outward. **d)** The resulting lateral blast expelled material at speeds up to 600 miles per hour (1,000 kilometers per hour). **e)** The blast toppled giant Douglas Fir trees as if they were matchsticks. **f)** A lava dome grew for six years inside the crater formed by the lateral blast.

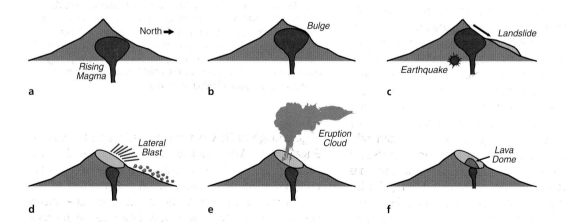

FIGURE 5.20 Eruption of Mount St. Helens in 1980.
a) Magma moves to a shallower position beneath the north flank of the mountain. **b)** The rising magma causes the north side of the mountain to bulge. Gasses remain trapped under high pressure within the pasty magma. **c)** An earthquake beneath the mountain on the morning of May 18 causes the steep bulge to move downhill as a giant landslide. **d)** The gasses expand suddenly, causing a lateral blast out of the north side of the mountain. The mountain loses about 1,200 feet (380 meters) of its height. **e)** A huge cloud of volcanic ash rises upward from the mountain and is carried northeastward by the prevailing winds. **f)** Smaller eruptions of sticky lava over the next six years build a lava dome within the explosion crater. Renewed activity in 2004 has produced steam eruptions and continued lava-dome growth.

POACHED ROCK

Sunset Amphitheater on the west flank of Mount Rainier appears distinctly different in color than much of the remainder of the mountain. The rock has been modified and its color and texture changed. Rising sulfur gasses, combined with meltwater, produce a weakly acidic solution that chemically changes and weakens the rock. Rock that was once solid andesite becomes a crumbly, orange-yellow mass of clay. This process, known as **hydrothermal alteration**, has weakened the fractured rock along an east-west trending zone across Mount Rainier's summit. The weakened rock is responsible for many small landslides and at least one large event, the Electron Mudflow, that swept the west side of Mount Rainier 600 years ago.

a

b

FIGURE 5.21 Fire and ice of Mount Rainier National Park. a) The high elevation and wet climate lead to a great amount of snow that compacts and forms the 25 glaciers on the slopes of Mount Rainier. **b)** The walls of glacial valleys reveal why Mount Rainier is called a "strato" or "composite" volcano. Its layering consists not only of lava flows (steep cliffs), but also of volcanic mudflows, pumice, and other materials.

1980, a bulge developed on the north side of the mountain. Because the magma was so pasty, gas was trapped within it under high pressure. On the morning of May 18, 1980, a moderate (magnitude 5.1) earthquake occurred beneath the mountain. The shaking caused the oversteepened north flank to collapse as a giant landslide. The effect was much like opening a shaken soda pop or uncorking a bottle of champagne. Gas under high pressure had an outlet, so the mountain exploded northward, devastating an area of 250 square miles (650 square kilometers) and killing 57 people. Ash rose 60,000 feet (18 kilometers) into the air and drifted northwestward, blanketing communities throughout eastern Washington and parts of Idaho and Montana.

Mount Rainier is also dangerous. It has steep slopes and volcanic layers that, in places, have been chemically altered by hot water and gasses into weak clay minerals. As a composite volcano, Mount Rainier comprises a complex array of eruptive materials. Liquid lava has poured out mostly as andesite, but the layering also includes ash, pumice, and other material (Fig. 5.21). Although Mount Rainier could potentially explode, more immediate hazards involve huge volumes of mud flowing through valleys on the flanks of the volcano (Fig. 5.22). Such features form when materials on the steep sides of a volcano are saturated with water. Deadly mixtures of mud, sand, rock, and water, known by the Indonesian word *lahar*, are also called *mudflows* or *debris flows*. Lahars on all scales are common hazards on Mount Rainier. The Osceola Mudflow 5,600 years ago and Electron Mudflow just 600 years ago were both

large enough to flow down river valleys all the way to Puget Sound. Many people live in the valleys that will carry future mudflows; in fact, some towns are built directly on top of very young mudflow deposits.

Cascade volcanoes come and go over time, much like bubbles popping in spaghetti sauce. The high peaks in the chain represent only the very latest glimpse in time. Mount Rainier began to grow about 1 million years ago. Its magma extruded through and covered much older rock (Fig. 5.23). Much of the material at the base of Mount Rainier—where not covered by volcanic, glacial, or river deposits—is intrusive rock that formed about 17 million years ago. The coarse-grained, high-silica rock is common in the Tatoosh Mountains and is called the Tatoosh Granodiorite (Fig. 5.24). It

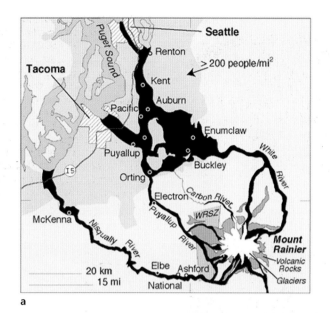

a

b   c

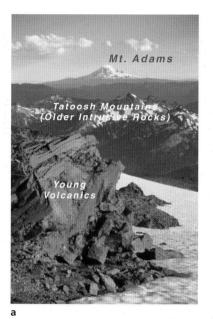

a

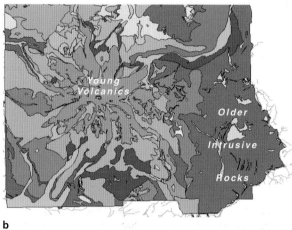

b

FIGURE 5.22 Downstream effects from volcanic hazards at Mount Rainier. a) The orange shows very young lava flows, which are mostly within the boundaries of Mount Rainier National Park. The black shows where lahars (fast-moving mudflows) and associated floods have inundated stream valleys and other low-lying areas in the past 6,000 years. Note the positions of towns (represented by dots) and the city limits of Seattle and Tacoma (shown as diagonal lines). The gray area labeled WRSZ is the Western Rainier Seismic Zone. **b)** The brown color of the Nisqually River in the southern part of Mount Rainier National Park is due to a large amount of sediment from melting glaciers. During a lahar event, the swollen river flows much faster and has larger amounts of mud and other debris. **c)** Lahar deposits near the Nisqually River, just west of the park.

FIGURE 5.23 Mount Rainier National Park is built on a base of older intrusive rocks. a) View to southeast from below Camp Muir. Mount Adams is a 12,276-foot (3,742-meter) composite volcano. The Tatoosh Mountains are high-silica (granodiorite) intrusives that comprised the magma chambers of volcanoes formed 17 million years ago. The young volcanic layers on Mount Rainier are mostly andesite lava flows and mudflows. **b)** Geologic map of the park. The base of Mount Rainier consists mostly of the older igneous intrusive rocks. The younger volcanic and glacial materials on the mountain's flanks and summit were deposited since Mount Rainier began to form about a million years ago.

(a) Fine-Grained Andesite

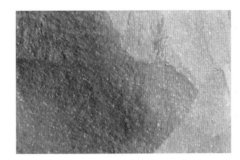

(b) Coarse-Grained Granodiorite

FIGURE 5.24 Texture of igneous rocks at Mount Rainier National Park. a) Andesite on Skyline Trail near Panorama Point. The rock is a grayish color because of its intermediate (~60%) silica composition (Tables 2.3 and 2.6). Close-up on right shows that the rock is fine-grained because the magma flowed out on the surface and cooled quickly. **b)** Tatoosh Granodiorite at Christine Falls turnout on the road between Longmire and Paradise. The light color is due to a high-silica composition between that of granite (~70% silica) and diorite (~60% silica). Close-up on the right shows coarse crystal grains that formed as the magma cooled slowly within the Earth.

IS THIS AN ACTIVE VOLCANO?

Visitors to national parks in the Cascade Mountains are often perplexed to learn that many of the peaks are active volcanoes. They wonder why, for example, Mount Rainier is considered active, even though they see no lava or clouds of steam spewing out of the mountain. What makes a volcano "active"?

"Volcanically active" can be thought of in the same way as "physically active." A person does not have to jog or bicycle 24 hours a day, 365 days a year to be considered physically active—a couple of times a week would probably suffice! Likewise, a mountain is volcanically active if it has been erupting every few decades or centuries, or perhaps even a few times during the past several thousand years.

The forms of High Cascade peaks can hint at whether the volcanoes are active or extinct (Fig. 5.25). A volcano that has not erupted in a long time shows the jagged effects of glacial erosion, while one that has been covered frequently by volcanic material since the last ice age has a smoother profile.

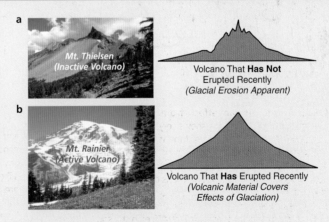

FIGURE 5.25 The profiles of Cascade peaks may reveal if the volcanoes are active or not. a) Mount Thielsen can be viewed from the North Rim Drive of Crater Lake National Park, Oregon. The volcano has been extinct for half a million years, so glaciers have eroded it heavily. **b)** Mount Rainier in Mount Rainier National Park, Washington. Even though glaciers are carving it up, recent eruptions have provided lava, mudflows, and other materials that maintain the smoother profile of the mountain.

Oregon Cascades (View from East)

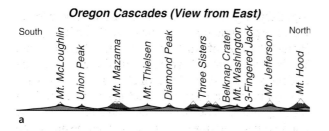

b

FIGURE 5.26 Cascade Mountains in Oregon. a) The high-silica, composite peaks (purple) are built on a base of interlocking shield volcanoes (gray). Active volcanoes, such as Mount Hood, Mount Jefferson, and the Three Sisters, maintain smooth profiles. The jagged shapes of Union Peak, Mount Thielsen, Mount Washington, and Three-Fingered Jack are due to glacial erosion of extinct volcanoes. Crater Lake National Park is on Mount Mazama, a volcano that exploded and collapsed. **b)** The Cascade Mountains are volcanoes formed above the line where the top of the Juan de Fuca Plate gets so hot that it starts to sweat (Fig. 5.3).

formed deep beneath ancient Cascade Mountain volcanoes that have since eroded away, exposing the magma chamber rock. Similar coarse-grained rock is common in southern California, illustrating that the Sierra Nevada Mountains are an eroded volcanic arc.

PARKS IN THE SOUTHERN CASCADES. The High Cascade volcanoes in Washington State poke up through old igneous and metamorphic rocks. But the southern Cascades of Oregon and northern California are built on top of a series of young, interlocking shield volcanoes (Fig. 5.26). This backbone of basaltic lava flows is covered in places by higher-silica, andesite-to-rhyolite lavas, forming steep-sided composite volcanoes such as Mount Hood and the Three Sisters. Smaller cinder cones form on the slopes of both the shields and composite volcanoes.

Crater Lake National Park offers an exciting opportunity for visitors not only to enjoy incredible scenery, but also to contemplate the effects of a large volcanic eruption that led to the collapse of an entire mountain. At the park, you can see many volcanoes from a distance and walk on the flanks of one. You can take a boat ride inside a recently active volcano, and explore its internal architecture.

Crater Lake partially fills the hole caused by the eruption and collapse of a large volcano, Mount Mazama, that occurred 7,700 years ago (Fig. 5.27). Similar to other volcanoes in the southern Cascades, Mount Mazama erupted basaltic lava flows in its early stages, about 400,000 years ago. The base of the mountain has thin black layers characteristic of a shield volcano. In its upper parts, the mountain is more than just lava flows and is composed of many different materials, making it a **composite volcano**. The materials have a distinct layering, or stratification, so the term **stratovolcano** is also applicable.

The giant, steep-sided crater that holds Crater Lake is called a **caldera**, from the Spanish word for "boiling pot" or "caldron." The word is especially applicable because for some time after the big eruption and collapse of Mount Mazama, more lava and other materials poured out onto the crater floor (Fig. 5.28). The original caldera depth was about 4,000 feet (1,200 meters); the subsequent eruptions filled the hole with about 1,000 feet (300 meters) of volcanic material. The activity was

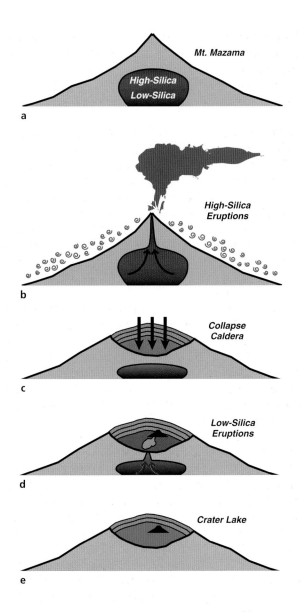

FIGURE 5.27 Formation of Crater Lake. a) Mount Mazama de-velops into a 12,000-foot (3,500-meter) composite volcano through numerous eruptions spanning more than 400,000 years. The magma chamber beneath the mountain settles out with lighter (high-silica) material on the top, and heavier (low-silica) material on the bottom. **b)** High-silica eruptions of volcanic ash, pumice, and dacite lava 7,700 years ago empty out the upper part of the magma chamber. **c)** The top of the mountain collapses as rubble into the emptied-out magma chamber, forming a caldera 5 miles (8 kilometers) across and 4,000 feet (1,200 meters) deep. Some dense magma remains at the bottom of the magma cham-ber. **d)** Intermittent eruptions of low-silica lava from the lower part of the magma chamber coat the bottom of the caldera and create the cinder cone that will become Wizard Island. **e)** Rainfall and snowmelt partially fill the caldera, forming Crater Lake. The amount of water added each year by precipitation is exactly bal-anced by evaporation and seepage through porous layers on the caldera walls.

especially vigorous on the west side of the caldera, where large outpourings of fluid (basaltic-andesite) lava formed the base of what would become Wizard Island. The late-stage volcanic activity sealed the floor of the caldera, so that rain and snowmelt could no longer es-cape (much as a coating of sealant prevents water from seeping into a wood deck). The water rose for a few hundred years, until evaporation and seepage out of the porous strata on the sides of the caldera exactly equaled input from rain and snow.

The eruption that resulted in the formation of the Crater Lake caldera 7,700 years ago was a far greater event than the 1980 eruption of Mount St. Helens, ex-pelling 50 to 100 times as much material (Fig. 5.29). The mountains are similar in that they are steep-sided, composite volcanoes. As at Mount St. Helens, magma rose to a shallow position prior to the big eruption at Mount Mazama. Trapped gas could not escape from the sticky magma, so that an eruption occurred as pressure was released and the gas expanded rapidly. The moun-tains are different in that Mount St. Helens blew out to the north, forming an asymmetric, breached crater; there are no confining walls to trap a "crater lake." The crater on top of Mount St. Helens is about 1 mile (1½ kilometers) across, whereas the Mount Mazama caldera spans 5 miles (8 kilometers). Mount St. Helens lost about 1,200 feet (350 meters) of its height, whereas Mount Mazama lost about 5,000 feet (1,600 meters).

Layers on the walls of the Crater Lake caldera ap-pear horizontal, but mostly they tilt outward, away from the crater rim (Fig. 5.30). The layers are lava flows, ash, pumice, mudflows, and other material that flowed down the sides of the erupting volcano. If you project the layers and the surface slopes upward to above the center of the lake, you can envision a moun-tain about 12,000 feet (3,500 meters) high.

Magma doesn't just flow out of the top of a vol-cano; it often erupts out of the sides (Fig. 5.31). Sheets of magma work their way through existing cracks, make their own cracks, or melt their way through the volcanic layers. Some of the magma flows out on the surface as lava, but some remains underground, cools, and solidi-fies as a **dike** (from the word used for an artificial sea-wall). Dikes appear on the Crater Lake caldera walls today because the summit area collapsed like a piston, slicing away and exposing the layers. The dikes are ob-vious for two reasons. First, they cut across horizontal strata on the sides of the volcano. Second, igneous rocks that cooled below Earth's surface are commonly harder, and more resistant to erosion, than the horizon-tal strata they intrude (in this case, soft layers such as volcanic ash and mudflows). Devil's Backbone is a

FIGURE 5.28 Collapse caldera features on subduction-zone volcanoes. a) Lava flows can be seen on the floor of the caldera at Aniakchak National Monument and Preserve in Alaska. The caldera is breached on the far side, so that it is not filled with a lake. **b)** Digital, shaded-relief map of the floor of Crater Lake, looking north. As at Aniakchak, eruptions of fluid, basaltic lava sealed the bottom of the caldera. Wizard Island erupted on a plat-form of basaltic andesite lava that built up on the west side of the caldera. Merriam Cone is a submerged cinder cone that ex-tends to within 100 feet (30 meters) of the surface in the northeast part of the lake. **c)** Crater Lake hides features on the caldera floor. Note the tongue of lava that flowed from the base of Wizard Island (Fig. 3.16). **d)** Shaded-relief view of the west side of the caldera. The circular fea-ture just off the east (left) side of Wizard Island is a submerged, dacite lava dome. Pillow lavas form only underwater (see Fig. 5.11), indicating that the lake had already risen to a significant level before eruptions on the floor of the caldera ceased.

(a) Aniakchak NM

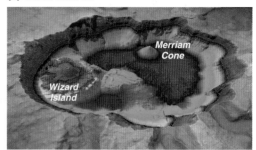

(b) Crater Lake NP

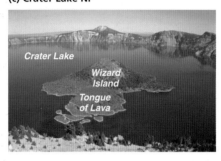

(c) Crater Lake NP

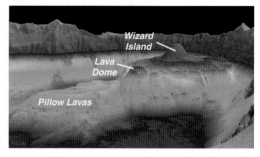

(d) Crater Lake NP

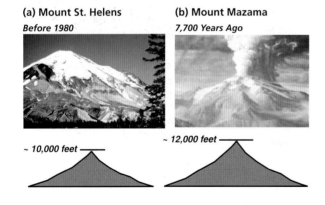

(a) Mount St. Helens

Before 1980

~ 10,000 feet

(b) Mount Mazama

7,700 Years Ago

~ 12,000 feet

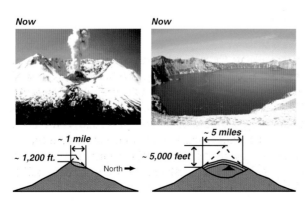

Now

~ 1 mile

~ 1,200 ft.

North ➡

Now

~ 5 miles

~ 5,000 feet

FIGURE 5.29 Comparison of craters formed on Mount St. Helens and Mount Mazama. a) Mount St. Helens National Volcanic Monument, Washington. The 1980 eruption formed a crater about 1 mile (1½ kilometers) wide and breached on its north side; it cannot hold a lake. The mountain lost about 1,200 feet (350 meters) of its height. **b)** Crater Lake National Park, Oregon. Mount Mazama lost about 1 mile (1,600 meters) of its height when it erupted and collapsed 7,700 years ago. The resulting crater was about 4,000 feet (1,200 meters) deep and 5 miles (8 kilometers) across.

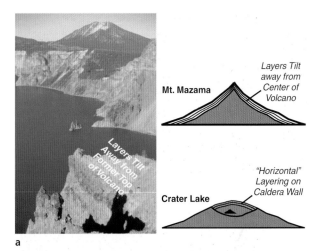

FIGURE 5.30 Crater Lake National Park provides an opportunity to see the internal layering of a composite volcano.
a) Overlooking Phantom Ship (the island in the lake) and Mount Scott (beyond the caldera wall) from Garfield Peak Trail. Note that the layers in the foreground are tilted away from the now void region that was once the upper portions of Mount Mazama.
b) Looking over the edge of Wizard Island at layers on the northwest caldera wall. The layers appear horizontal, but they are on the flank of the volcano and tilt away from the observer.

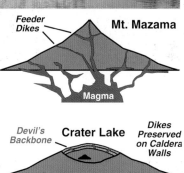

FIGURE 5.31 The interior plumbing system of a volcano can be seen from Crater Lake. The Devil's Backbone is a prominent dike that was exposed on the caldera walls when the summit of the mountain collapsed.

prominent dike on the northwest wall of the caldera. It is nearly vertical, about 50 feet (15 meters) thick, and can be traced upward more than 1,500 feet (500 meters) from the water to the crater rim. The dike is so resistant to erosion that it sticks out several feet compared to the horizontal strata; hence the name "Backbone."

When magma erupts violently, the explosion can leave a crater on the top or flanks of a volcano. High-silica magma, of dacite-to-rhyolite composition, often pours out into the crater. The magma stays together as a **lava dome** because it is sticky. Such a dome has formed in recent years in the explosion crater on Mount St. Helens (Figs. 2.20d and 5.19f). At Crater Lake, there are two examples of explosion craters that formed on the flanks of Mount Mazama some time before its ultimate eruption: Llao Rock and Redcloud Cliff, on the north and east sides of the caldera, respectively (Fig. 5.32). The craters were filled with sticky dacite lava. The lava

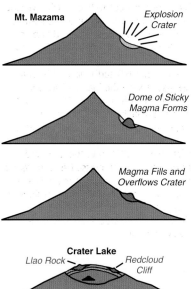

FIGURE 5.32 Dissected lava domes on the caldera walls of Crater Lake. Llao Rock formed on the north flank of Mount Mazama when a small explosion crater filled with sticky, high-silica dacite. Redcloud Cliff is a similar dome on the east side of the lake.

FIGURE 5.33 Lassen Volcanic National Park. Bumpass Hell displays geothermal features that rival those of Yellowstone National Park, including hot springs and mudpots.

remained thick and covered only a small area because of its resistance to flow. Later, when Mount Mazama erupted and collapsed, portions of the explosion craters and their domes fell into the caldera. The cross-sectional views we now see on the caldera wall resemble birds—the main part of the domes are the bodies, the flows that spilled out of the craters are the wings.

Lassen Volcanic National Park lies above the southern portion of the subducting Juan de Fuca Plate, known as the Gorda Plate (Fig. 5.5). Outside of Alaska and Hawai`i, Mount St. Helens and Lassen Peak were the only two U.S. volcanoes to erupt in the twentieth century. The Lassen eruptions began in 1914 and continued vigorously for three years, and then sporadi-

FIGURE 5.34 Contrasting volcanic activity at Lassen Volcanic National Park. West Side (a, b): High-silica volcanic features are due to subduction of the Gorda Plate. **a)** Brokeoff Mountain, at 9,235 feet (2,815 meters) elevation, is part of the remnants of the ancient composite volcano known as Mount Tehama. **b)** Chaos Crags comprise a ridge line of overlapping, dacite lava domes formed about 1,000 years ago. East Side (c, d): Low-silica volcanism relates to continental rifting of the Basin and Range Province. **c)** The 6,907-foot (2,105-meter) hill in the northeast corner of the park, appropriately named Cinder Cone, formed as fluid, basaltic magma erupted about 350 years ago. The summit region contains two concentric craters. **d)** The dark-colored tongue of basalt lava that oozed out of the base of Cinder Cone reveals the fluid nature of the low-silica magma.

a

b

c

d

cally until 1921. There is evidence that magma remains at shallow depth beneath the mountain. Bumpass Hell in the southwest part of the park (Fig. 5.33) has active hot springs and mudpots—the most extensive geothermal features in the United States outside Yellowstone National Park (Chap. 9). Although Lassen Peak rises to 10,457 feet (3,187 meters) elevation, it is the remnant of a much taller and broader volcano known as Mount Tehama. Similar to Mount Mazama's eruption and collapse 7,700 years ago that ultimately resulted in Crater Lake, a catastrophic event about 27,000 years ago led to the destruction of Mount Tehama.

Volcanic features reveal that Lassen Volcanic National Park overlaps two tectonic provinces (Figs. 5.4 and 5.16). In the western part of the park, Lassen Peak and Brokeoff Mountain are remnants of the ancient composite volcano, Mount Tehama, and Chaos Crags is a ridge of intermingled lava domes made of dacite (Fig. 5.34a,b). The high-silica lavas that comprise those features are part of the subduction-zone volcanism forming the Cascade Mountains. The eastern portions of the park display quite different features (Fig. 5.34c,d). Dark basaltic lava flows and cinder cones represent low-silica volcanism caused by continental rifting on the western edge of the Basin and Range Province, similar to features in Newberry National Volcanic Monument and nearby Lava Beds National Monument (Chap. 3).

Pacific Northwest Earthquakes

Earthquakes in the Pacific Northwest are of four varieties, each in some way related to the convergence of the Juan de Fuca and North American plates. Visitors to national parks in the region can gaze over the landscape and imagine their position relative to the subducting plate. As the cold Juan de Fuca Plate descends, its brittle upper portion contorts and cracks as earthquakes (Fig. 5.35a). The largest and most devastating earthquakes occur where the two plates lock together for centuries, then suddenly let go (Fig. 5.35b). Olympic National Park, Oregon Caves National Monument, and Redwood National and State Parks lie directly above such a "locked zone" in the Pacific Northwest. Smaller events occur within the overriding North American Plate, where plate convergence causes compression and breaking along shallow fault lines (Fig. 5.35c), or where magma rises beneath the Cascade Mountains (Fig. 5.35d). Earthquakes beneath Mount St. Helens National Volcanic Monument, as well as Mount Rainier, Crater Lake, and Lassen Volcanic national parks, are of the latter two types.

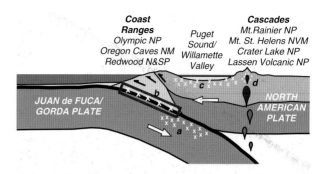

FIGURE 5.35 Types of earthquakes in the Pacific Northwest. a) Within the Juan de Fuca Plate. The top portion of the Juan de Fuca Plate stays cold and brittle for some time as it subducts, so that "subduction slab" or "Benioff Zone" earthquakes occur to depths of about 30 miles (50 kilometers) beneath the Olympic/-Puget Sound region. The magnitude 6.8 earthquake of February 28, 2001, was of this type. **b)** Between the Juan de Fuca and North American plates. Beneath the offshore and Coast Range/Olympic Mountains regions, the plate boundary is less than 20 miles (30 kilometers) deep, where rocks are still cold and brittle. A devastating earthquake can occur if the plates lock together for centuries, then suddenly let go (Fig. 5.36). **c)** Within the North American Plate. The overriding plate is compressed as it converges with the subducting plate. The Seattle area and Western Rainier Seismic Zone (WRSZ in Fig. 5.22a) are two of the many regions of shallow earthquakes that result. **d)** Rising magma beneath the Cascades. The colder, brittle rock above cracks as small earthquakes and earthquake swarms occur.

EVIDENCE FOR LARGE PACIFIC NORTHWEST EARTHQUAKES. The boundary between the Juan de Fuca and North American plates is similar to that of other regions where a plate with oceanic crust descends beneath one with continental crust, such as western South America and southern Alaska. Like other ocean/continent subduction zones, the morphology of the Cascadia Subduction Zone consists of a coastal mountain range, a volcanic arc, and a depression in between. But unlike other such regions, there is no written record of a large earthquake (magnitude ≥8) occurring in the Pacific Northwest, even though written records have been kept for the past 200 years. Geologic evidence, however, indicates that large earthquakes accompanied by subsidence of coastal areas and flooding by **tsunami** (giant sea surges often mistakenly referred to as "tidal waves") have indeed occurred. The last four big earthquakes were about 300, 1,100, 1,300, and 1,700 years ago. An explanation is that stress builds as the Juan de Fuca and North American plates lock together along their boundary (Fig. 5.36). After 200 to 800 years of continued convergence, the stress becomes so great that the plates suddenly unlock, sending off shock

(a) Before Earthquake

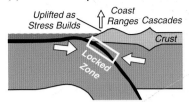

(b) During Earthquake

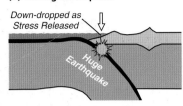

FIGURE 5.36 **Uplift and down-drop of the coastal region due to a great subduction-zone earthquake. a)** Plates lock together. The region above the locked zone is contorted and slowly uplifts in some places and subsides in others, because the plates continue to converge but have no place to go. Where uplift occurs, shallow bays rise a few feet out of the sea and develop into marshes. **b)** Plates suddenly unlock. After a few centuries, stress builds to a level where the plates snap apart and the energy is released as a giant earthquake. The uplifted areas drop down abruptly, again becoming shallow bays.

waves as a giant earthquake. If this interpretation is correct, Northwest residents should take heed, because the last great earthquake apparently occurred 300 years ago!

The last huge earthquake is thought to have struck the Pacific Northwest on January 26, 1700. Without direct observation, how is it possible to date such an event so precisely? Carbon dating of tree stumps in buried marshes suggests the trees died about 300 ± 10 years ago. Written records from Japan indicate that a giant tsunami wave hit that country on January 27, 1700. The pattern of when and where the waves struck suggests that the tsunami was generated in the Pacific Northwest of the United States. Traveling about the speed of a jetliner, it takes about 12 hours for such waves to cross the Pacific Ocean. By combining the carbon age dating and tsunami data, we can determine the date with such precision.

VISUALIZING A GREAT SUBDUCTION-ZONE EARTHQUAKE. The national parks in the Pacific Northwest provide an exceptional setting to appreciate the danger and likelihood of a major earthquake. Visitors can imagine the Juan de Fuca Plate beginning its downward

---

THE FEBRUARY 28, 2001, EARTHQUAKE

The February 28, 2001, earthquake in the Puget Sound region was only a "wake-up call," reminding us to prepare for far greater events. Ground shaking was felt by visitors to Olympic and Mount Rainier national parks, as well as Mount St. Helens National Volcanic Monument, but the earthquake was well west of the Cascades. It was a **subduction slab** earthquake, occurring within the top portion of the Juan de Fuca Plate (Fig. 5.35a). Its epicenter was 11 miles (18 kilometers) northeast of Olympia, Washington, or about 15 miles (24 kilometers) west-southwest of Tacoma and 36 miles (58 kilometers) south-southwest of Seattle. It has been named the **Nisqually Earthquake**, because its epicenter was near the mouth of the Nisqually River. Such quakes are common in subduction zones, sometimes showing that a cold slab extends as deep as 400 miles (700 kilometers; Fig. 1.2). But the Juan de Fuca Plate subducting beneath Washington State is young and relatively hot, having formed at the Juan de Fuca Ridge within the last 10 million years (Figs. 5.3 and 5.5). The region of the plate that is still cold and brittle is shallow, extending only beneath the Olympic Mountain/Puget Sound area (Fig. 5.37).

The magnitude of the Nisqually Earthquake was 6.8. It resulted in intensities only as high as VIII (Table 2.8) because it occurred 32 miles (52 kilometers) below the surface. It caused about $2 billion in property damage but would have been worse had the earthquake been shallower or of larger magnitude. Seismic waves lose energy as they travel to the surface, much as your voice loses volume as it travels to the back rows of an auditorium. Intensities were not as severe as they would have been for a shallower quake.

As destructive as the February 28 earthquake was, two other varieties that occur in the Pacific Northwest can be far more severe. A **locked zone earthquake** (Fig. 5.35b) would occur at shallower depth, and its magnitude (perhaps as high as 9) could release seismic energy equivalent to 1,000 Nisqually earthquakes (Fig. 2.24). A quake within the overriding North American Plate could have magnitude similar to that of the February 28 event, but it would be much shallower (Fig. 5.35c). Many such active faults are in the Seattle/Puget Sound area. The magnitude 6.9 earthquake that struck Kobe, Japan, on January 17, 1995, occurred on a similar type of shallow fault, with a focal depth of only 1 mile (1.6 kilometers), and caused over 5,000 deaths and $200 billion damage.

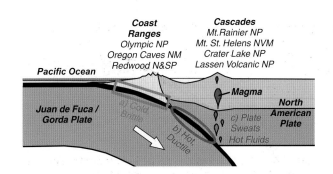

FIGURE 5.37 Behavior of Juan de Fuca–North American plate boundary. a) Where the boundary is shallow, the rocks are relatively cold—they break in a brittle manner. This is the "locked zone" where huge earthquakes occur. **b)** Farther east, the plate boundary is deeper and hotter. Rocks deform in a ductile fashion, flowing instead of snapping as earthquakes. **c)** The Juan de Fuca Plate gets still hotter where it extends deeper. Hot fluids rise, forming the Cascade volcanoes.

plunge about 50 miles (80 kilometers) out in the Pacific Ocean (Fig. 5.37). The plate encounters high temperatures as it extends into the Earth at a shallow angle. Beneath the Cascade Mountains, the top of the plate is about 50 miles (80 kilometers) deep. Very large earthquakes will not occur along the plate boundary beneath Mount St. Helens National Volcanic Monument, or Mount Rainier, Crater Lake, and Lassen Volcanic national parks, because rocks are so hot that they deform in a ductile fashion, like pulling taffy. The taffy will deform but will not snap.

Beneath the Coast Range, however, the situation is quite different. The top of the plate is only 5 to 15 miles (8 to 25 kilometers) deep. Rocks at those depths are still relatively cold, so that they break in a brittle manner. Imagine bending a piece of peanut brittle. When it finally breaks, it gives off a snapping sound. You can imagine what would happen if the subducting

GEOLOGIC EVIDENCE FOR GREAT SUBDUCTION-ZONE EARTHQUAKES

Geologists cite two lines of evidence that great earthquakes struck the Pacific Northwest at intervals ranging from 200 to 800 years, with an average "recurrence interval" of about 300 to 500 years. The first relates to what happens when converging plates lock together. In places, Earth's surface slowly bends upward to absorb the stress (Fig. 5.36). In 300 years, the land can move up about 6 feet (2 meters). If the plates suddenly unlock, the surface abruptly drops down the same amount. In many areas along the Washington, Oregon, and northern California coasts, one can find evidence of marsh plants, including trees, that were suddenly killed and partially buried by deposits of sand. Above the sand are bay muds and then more marsh and sand deposits. This sequence repeats at least seven times over the past 3,000 years. Each sequence may indicate the time between large earthquakes. As the plates lock and the surface rises, shallow bay areas rise up and become marshes. Then, when the earthquake occurs as the plates suddenly unlock, the surface abruptly drops below sea level and is covered by sand from the resulting giant tsunami wave. Over the next few hundred years, the area slowly rises up again until the next big earthquake.

The second line of evidence concerns large turbidity flow deposits (turbidites) off the coast (Fig. 5.12). A big earthquake can rattle the region so much that pieces of the edge of the continental shelf break up and travel down the continental slope as turbidity flows. Drilling and carbon age dating have documented 14 such turbidites for the past 7,700 years, many that correlate with age dates obtained from the buried marshes.

FIGURE 5.38 National park lands plotted on shaded-relief map of Alaska. Parks highlighted in southern Alaska lie on the accretionary wedge and volcanic arc formed by northward subduction of the Pacific Plate. F–F′ is the line of the cross section in Fig. 5.39. Numbers are as in Table 5.1. Wrangell–St. Elias National Park is included with the "accreted terrane" parks presented in Chap. 11.

Juan de Fuca Plate is locked against the North American Plate for centuries. The "snapping" sounds that occur when the plates suddenly let go are seismic waves that travel to the surface and severely shake the region. Visitors to Olympic National Park, Oregon Caves National Monument, or Redwood National and State Parks can look down and imagine the cold, brittle locked zone about 10 miles (16 kilometers) beneath their feet. The locked zone extends from about 30 miles (50 kilometers) offshore, completely beneath the Coast Range to the Puget Sound/Willamette Valley area. The two plates may be locked along the subduction-zone boundary from Vancouver Island to northern California (Fig. 5.5). A sudden unlocking could produce an earthquake every bit as big as the one that occurred in Alaska in 1964, when a similar plate boundary snapped. Even if only $\frac{1}{10}$ of the length of the boundary suddenly lets go, the resulting earthquake could be bigger than the magnitude 7.8 earthquake that devastated San Francisco in 1906 (the largest earthquake to strike the lower 48 states during the twentieth century).

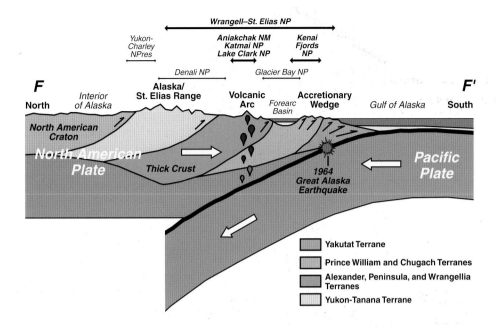

FIGURE 5.39 National parks in southern Alaska relate to subduction of the Pacific Plate beneath North America. Plate convergence during the past 200 million years has attached accreted terranes to the North American Craton (Chap. 11). Kenai Fjords National Park has oceanic sedimentary layers that have been metamorphosed, uplifted, and deformed as part of the accretionary wedge. Aniakchak National Monument, Katmai National Park, and Lake Clark National Park are part of the active volcanic arc built on older accreted terranes. Wrangell–St. Elias National Park spans accreted terranes, active volcanoes, and uplifting accretionary wedge material.

SOUTHERN ALASKA SUBDUCTION ZONE

Five park sites in southern Alaska showcase landscapes formed by the subduction of the northward-moving Pacific Plate beneath North America (Figs. 5.38 and 5.39). One of them, Kenai Fjords National Park, lies within the accretionary wedge of uplifting oceanic sedimentary strata and hard crust. Three others, Aniakchak National Monument, Katmai National Park, and Lake Clark National Park, are farther inland, along the active volcanic arc. Another, Wrangell–St. Elias National Park, is a vast region that extends across the accretionary wedge and volcanic arc and is built of accreted terranes (Chap. 11).

Accretionary Wedge

Visitors to Kenai Fjords National Park witness Earth processes in action, including plate tectonics, glaciation, mountain building, earthquake activity, and changes in global climate and sea level. Like Olympic, Oregon Caves, and Redwood, Kenai Fjords lies within a coastal mountain range formed above an ocean/continent subduction zone. Sandstone and shale layers at Kenai Fjords are commonly metamorphosed and were so deformed during uplift that they are vertical in places. Not all of the rocks are of oceanic origin, as evidenced by blocks of granite.

Much of the northern part of Kenai Fjords National Park is covered by the Harding Icefield, and glaciers run down to the spectacular fjords along the Kenai Peninsula (Fig. 5.40). The fjords are broad valleys that

FIGURE 5.40 Kenai Fjords National Park. Bear Glacier just southwest of Resurrection Bay.

are being invaded by the ocean because of the rise in sea level since the end of the last ice age. Popular ranger activities in the park include interpretive programs presented during boat tours of the fjords; the tours travel in and out of bays, where visitors can witness the glaciers and mountains in the background.

Kenai Fjords National Park lies right above the focus of the Great Alaska Earthquake of 1964, one of the two largest ever recorded (the other, in 1960, occurred along the subduction zone off western South America). Effects of the 1964 earthquake are quite spectacular in the park; the coastline dropped, so much that in places it lies submerged beneath about 8 feet (2½ meters) of water.

Volcanic Arc

More than 40 volcanoes have erupted in Alaska since Russians first arrived in the mid-1700s—a great many of the volcanoes are in areas that are now national parks. They are part of the volcanic arc that extends northeastward from the Aleutian Islands, across the Alaska Peninsula and Aniakchak National Monument, Katmai and Lake Clark national parks, and then bends eastward to Wrangell–St. Elias National Park.

Aniakchak National Monument and Preserve is on the Alaska Peninsula, a part of the thick continental crust of North America. The line of volcanoes continues southwestward as the Aleutian Islands, an island arc

chain developed on thinner oceanic crust. The eruption of Aniakchak about 3,400 years ago rivals the eruption and collapse of Mount Mazama that formed Crater Lake (Fig. 5.27). The original composite volcano was about 7,000 feet (2,000 meters) high, but it lost about 2,500 feet (750 meters) of its height. The 6-mile (10-kilometer) wide caldera is about 2,000 feet (600 meters) deep. Like Mount Mazama, Aniakchak spewed lava across its caldera floor for centuries after its collapse (Fig. 5.41). The volcano is still active, having erupted in 1931. But unlike Mount Mazama, Aniakchak does not hold a large lake because the caldera wall is breached on its east side. Wave-cut shorelines high on the caldera walls suggest that a lake did partially fill the depression for some time. As the lake rose to over 300 feet (100 meters), it topped the lowest spot on the caldera rim and wore away the notch that the Aniakchak River now flows through as it drains the caldera. This opening is known as The Gates. Surprise Lake on the north side of the caldera is a small remnant of the earlier lake.

Katmai National Park and Preserve lies where the world's largest volcanic event of the twentieth century occurred. In 1912, the Katmai region erupted an estimated 7 cubic miles (30 cubic kilometers) of ash and pyroclastic material, roughly 50 times the amount that came out of Mount St. Helens in 1980. The collapse caldera from the Katmai eruption is about 3 miles (5 kilometers) across and 2,000 feet (600 meters) deep (Fig. 5.42a). The caldera has partially filled with

FIGURE 5.41 Aniakchak National Monument and Preserve, Alaska. a) Aniakchak Caldera formed during a large eruption and collapse about 3,400 years ago. **b)** Surprise Lake is the remnant of a larger lake that partially filled the caldera. **c)** The Gates is a notch in the caldera wall that drained the larger lake. **d)** Pyroclastic and other volcanic materials coated the caldera floor after the big eruption and collapse.

a

b

c

d

(a) Katmai Caldera

(b) Trident Volcano

(c) Novarupta Lava Dome

(d) Valley of 10,000 Smokes

FIGURE 5.42 Katmai National Park and Preserve, Alaska. a) Katmai Caldera formed when eruptions drained magma from beneath Katmai Volcano in June 1912. The caldera is about 3 miles (5 kilometers) across. Katmai Volcano was 7,513 feet (2,290 meters) high before the caldera formed. The lake was about 800 feet (250 meters) deep, and still rising, when it was measured in the 1970s. **b)** Trident Volcano is a cluster of andesite and dacite composite cones. The round feature in the center foreground is Novarupta. **c)** Novarupta is a lava dome developed during the 1912 eruption that formed Katmai Caldera. **d)** The Valley of 10,000 Smokes filled with 600 feet (200 meters) of volcanic ash during the 1912 eruption. The "smokes" were from fumaroles expelling gasses trapped beneath ash. Rim of Katmai Caldera is on the upper left.

water—an approximately half-scale version of the better-known Crater Lake in Oregon.

Recent research has revealed a rather curious finding about how Katmai Caldera formed. Katmai Volcano did collapse, but unlike Mount Mazama 7,700 years before, the material expelled from its magma chamber did not spew forth out of the top of the volcano. Rather, it flowed underground to a vent 6 miles (10 kilometers) away! The vent was the source for most of the ash and pyroclastic flows expelled during the 1912 eruptions, and it has since filled with a lava dome known as Novarupta (Fig. 5.42b,c).

During the 1912 eruptions, a thick flow of pyroclastic material filled a valley on the north side of Katmai Volcano. For years after that, steam from numerous fumaroles filled the area, so that it became known as the Valley of 10,000 Smokes (Fig. 5.42d). Geologists later determined that the thick deposits in the valley were not from the summit of Katmai Volcano, but from the Novarupta vent, near a cluster of five volcanoes known as Trident Volcano.

HIGH-TECH VOLCANIC HAZARD

The initial phase of eruption of Mount Redoubt in 1989 illustrates how a volcano can pose an unexpected hazard in our modern age—and why monitoring volcanic activity can be so important. On December 15, a Boeing 747 jet, with 244 passengers and crew, was en route from Amsterdam to Anchorage. Without warning, the plane flew right through the eruption cloud at 25,000 feet (7,500 meters) altitude. Silica-rich ash melted and coated the hot engine turbines with glass, causing all four of them to shut down. After a terrifying free fall that lasted eight minutes, the pilots finally managed to restart the engines at only 6,000 feet (2,000 meters) altitude. Fortunately, no one was injured. But the incident highlighted the importance of monitoring volcanic activity and making the information immediately available to the airline industry and other segments of the public.

a   b

c   d

FIGURE 5.43 Lake Clark National Park and Preserve, Alaska. a) April 21, 1990, eruption of Redoubt Volcano, viewed across Cook Inlet from the Kenai Peninsula. **b)** The 10,197-foot (3,108-meter) high Redoubt Volcano on August 13, 1990. Iliamna Volcano at the left-center is another active volcano in the park. **c)** The Drift River Valley was covered by lahar deposits following the 1989 to 1990 eruptions of Redoubt Volcano. **d)** A lava dome now partially fills the eruption crater of Redoubt Volcano.

Whereas Katmai National Park has a Crater Lake look-alike, Lake Clark National Park and Preserve contains an active composite volcano with a profile and recent history reminiscent of Mount St. Helens'. In 1989 and 1990, Mount Redoubt, on the eastern side of the park, had four explosive eruptions that resulted in large volumes of ash and mudflows. And, as at Mount St. Helens, a lava dome later grew within the breached crater near the mountain's summit (Fig. 5.43).

At 20,625 square miles (53,396 square kilometers), Wrangell–St. Elias National Park and Preserve is the largest U.S. national park—six times the size of Yellowstone. The park is a complex amalgamation of blocks of continental and oceanic crust that have slammed into North America (Chap. 11, "Accreted Terranes"). Poking through the mass of some of the highest mountains in North America are very young volcanoes formed by the ongoing Pacific Plate subduction (Fig. 5.44). Among them are Mount Wrangell, a 14,163-foot (4,317-meter) volcano that last erupted during the early part of the twentieth century. The summit of Mount Wrangell contains a collapse caldera about 2½ miles (4 kilometers) wide and 3½ miles (5½ kilometers) long, thought to have formed about 50,000 years ago. Unlike the water-filled calderas of Mount Katmai and Crater Lake, the caldera on Mount Wrangell is filled with ice. Plumes of steam seeping out along the rim of the caldera reveal that Mount Wrangell is indeed

Mt. Wrangell   Mt. Drum

a

Mt. Sanford   Mt. Wrangell

b

FIGURE 5.44 Wrangell–St. Elias National Park, Alaska. a) Mount Wrangell has erupted in historic time; Mount Drum was active between 700,000 and 240,000 years ago. The Copper River is in the foreground. **b)** Mount Wrangell and Mount Sanford are piles of intermediate-silica (andesite) lava flows that have broad, shield-like profiles.

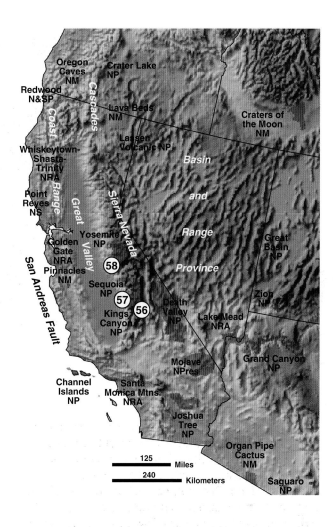

FIGURE 5.45 Shaded-relief map of California and surrounding regions, showing National Park Service lands in red. The Coast Range (an accretionary wedge), Great Valley (a forearc basin), and Sierra Nevada (a volcanic arc) reflect the earlier subduction-zone history of California. Rocks in Yosemite, Kings Canyon, Sequoia, and Joshua Tree national parks contain exumed magma-chamber rock that formed beneath the ancient volcanoes. Numbers indicate the ancient volcanic arc parks listed in Table 5.1.

an active volcano. Mount Sanford (16,237 feet; 4,949 meters) is another potentially active volcano in the park. Both volcanoes are made primarily of andesite lava flows, typical of subduction-zone volcanoes, although each has a broad, shield-like appearance. Yet another volcano in the park, Mount Drum (12,010 feet; 3,661 meters), is considered merely dormant.

ANCIENT SUBDUCTION ZONE

National parks in the Pacific Northwest and southern Alaska offer graphic evidence of the subduction of plates capped by thin oceanic crust beneath the thicker continental crust of the North American Plate. Other parks, including some in California, offer evidence of past subduction-zone activity. Still other parks, in the frontal ranges of the Rocky Mountains, reveal details of how low-angle subduction might have deformed the continental craton farther east (Chap. 10).

Sierra Nevada

A glance at a map of the western United States might suggest that the Sierra Nevada Mountains are a continuation of the Cascade Mountain volcanoes (Fig. 5.45). In the past, that was true. The present continental margin of western North America involves subduction

(a) 40 Million Years Ago

(b) 18 Million Years Ago

(c) Today

FIGURE 5.46 Tectonic evolution of the Sierra Nevada. a) The Farallon Plate subducts beneath North America. A continuous volcanic arc extends all the way from Alaska to Mexico. **b)** A mid-ocean ridge separating the Pacific and Farallon plates reaches the subduction zone, splitting the Farallon Plate into the Juan de Fuca and Cocos plates. Between them, the Pacific and North American plates slide along a transform plate boundary, the volcanic arc shuts off, and the Basin and Range continental rift zone begins to develop. **c)** As the Cocos and Juan de Fuca plates become smaller, the transform plate boundary expressed by the San Andreas and other faults grows longer. The remaining volcanic arc includes the Cascades in the Pacific Northwest. The Sierra Nevada are the eroded roots of the once-extensive volcanic arc in California.

zones off Mexico and the Pacific Northwest, connected by a transform plate boundary in central and southern California. But more than 30 million years ago, the entire West Coast was a subducting plate boundary, with a volcanic arc extending all the way from British Columbia through Mexico, including the region of the Sierra Nevada Mountains (Fig. 5.46). A divergent plate boundary (mid-ocean ridge) separated the subducting Farallon Plate from the Pacific Plate. About 28 million years ago, the mid-ocean ridge first encountered the subduction zone. The Pacific and North American

plates were then in contact, so that transform plate motion was manifest between the two plates along the San Andreas Fault in California (Chap. 7). The volcanoes shut down and have largely eroded away. The rocks of the Sierra Nevada Mountains are not from volcanoes, but rather the hardened remnants of magma chambers that once fed volcanoes. Magma-chamber rock makes up the hardened monoliths seen in Yosemite and other national parks in the Sierra Nevada Mountains (Figs. 5.47 and 5.48). But what happened to the volcanoes? And why, if there has been so

FIGURE 5.47 Yosemite National Park. a) Half Dome is the eroded remnant of cooled magma-chamber rock that formed beneath volcanoes when the region was part of a subduction zone. **b)** Close-up shows that the rock is coarse-grained and high in silica, with a composition between that of granite and diorite (granodiorite; Table 2.6). Compare with the eroded magma-chamber rock exposed at the base of Mount Rainier (Fig. 5.24b).

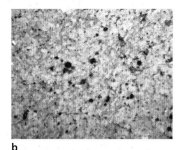

a

b

much erosion, are the Sierra Nevada Mountains so high?

At some volcanic arcs, such as the Andes in western South America, the crust becomes very thick. That may have been the case while the ancient Sierra Nevada volcanoes were developing. The thick crust would have meant that, because of isostatic uplift (Chap. 2), the region would have had high topography, as in the Andes today. When the subduction stopped, the volcanoes started to erode, removing some of the weight that held down the thick, buoyant crust. The crust rebounded upward, raising the topography to almost its initial height. Over time, several miles of rock eroded away, removing the volcanoes and exposing the granodiorite magma-chamber rock at the surface. The uplift has been enhanced by Basin and Range continental rifting encroaching on the east side of the Sierras (Fig. 3.8c). Hot, elevated asthenosphere thus helps maintain elevations well over 10,000 feet (3 kilometers) in Yosemite, Sequoia, and Kings Canyon national parks.

FIGURE 5.48 Sequoia–Kings Canyon National Park.

FURTHER READING

GENERAL

Alaska Division of Geological and Geophysical Surveys. 1998. *Volcanoes of Alaska*. Fairbanks: ADGGS Information Circular 38 (map, photos, and text).

Alt, D. P., and D. W. Hyndman. 1984. *Roadside Geology of Washington*. Missoula, MT: Mountain Press. 288 pp.

Alt, D., and D. W. Hyndman. 1995. *Northwest Exposures: The Geologic Story of the Northwest*. Missoula, MT: Mountain Press. 443 pp.

Chronic, H. 1986. *Pages of Stone: Geology of Western National Parks and Monuments. 2: Sierra Nevada, Cascades, and Pacific Coast*. Seattle: Mountaineers. 170 pp.

Clark, E. E. 1953. *Indian Legends of the Pacific Northwest*. Berkeley: University of California Press. 225 pp.

Decker, R., and B. Decker. 2001. *Volcanoes in America's National Parks*. New York: Norton. 256 pp.

Driedger, C. L. 1986. *A Visitor's Guide to Mount Rainier Glaciers*. Longmire, WA: Pacific Northwest National Parks and Forests Association. 80 pp.

Gardner, J. V., and P. Dartnell. 2001. *2000 Multibeam Sonar Survey of Crater Lake, Oregon: Data, GIS, Images, and Movies*. Washington: U.S. Geological Survey, Digital Data Series DDS-72 (compact disk).

Harris, S. L. 1988. *Fire Mountains of the West: The Cascade and Mona Lake Volcanoes*. Missoula, MT: Mountain Press. 379 pp.

Hoblitt, R. P., J. S. Walder, C. L. Driedger, K. M. Scott, P. T. Pringle, and J. W. Vallance. 1998. *Volcano Hazards from Mount Rainier, Washington*, revised edition: Open-file Report 98-428. 11 pp., 1 pl.

Loomis, B. F. 1926. *Pictorial History of the Lassen Volcano*. Mineral, CA: Loomis Museum Association. 96 pp.

McPhee, J. 1998. Rising from the Plains, in *Annals of the Former World*, Book 3. New York: Farrar, Straus and Giroux. p. 280–425.

Moore, J. G. 2000. *Exploring the Highest Sierra*. Stanford, CA: Stanford University Press. 427 pp.

Orr, E. L., W. N. Orr, and E. M. Baldwin. 1992. *Geology of Oregon*, 4th Ed. Dubuque, IA: Kendall/Hunt. 254 pp.

Plafker, G. 1965. Tectonic deformation associated with the 1964 Alaska earthquake. *Science*, v. 148, p. 1675–1687.

Rau, W. W. 1973. *Geology of the Washington Coast between Point Grenville and the Hoh River*. Olympia: Washington Department of Natural Resources, Geology and Earth Resources Division, Bulletin 66. 58 pp.

Rau, W. W. 1980. *Washington Coastal Geology between the Hoh and Quillayute Rivers*. Olympia: Washington Department of Natural Resources, Geology and Earth Resources Division, Bulletin 72. 57 pp.

Schaffer, J. P. 1999. *Yosemite National Park: A Natural History Guide to Yosemite and Its Trails*, 4th Ed. Berkeley: Wilderness Press. 288 pp.

Scott, K. M., and J. K. Vallance. 1995. *Debris Flow, Debris Avalanche, and Flood Hazards at and Downstream from Mount Rainier*. Washington: U.S. Geological Survey, Hydrologic Investigation Atlas HA-729, 1:100,000. 9 pp., 2 pl.

Scott, K. M., J. W Vallance, and P. T. Pringle. 1995. *Sedimentology, Behavior, and Hazards of Debris Flows at Mount Rainier, Washington*. U.S. Geological Survey, Professional Paper 1547. 56 pp.

Sisson, T. W. 1995. *History and Hazards of Mount Rainier, Washington*. Washington: U.S. Geological Survey Open-file Report 95-642. 2 pp.

Sisson, T. W., J. W. Vallance, and P. T. Pringle. 2001. Progress made in understanding Mount Rainier's hazards. *EOS: Transactions of the American Geophysical Union*, v. 82, p. 113–120.

Sullivan, J. M., and L. C. Sullivan. 2001. *Kids' Guide to National Parks of California and Oregon*. Corvallis, OR: E & S Geographic and Information Services. 116 pp.

Tabor, R. W. 1987. *Geology of Olympic National Park*. Seattle: Northwest Interpretive Association. 144 pp.

Tilling, R. I. 1982. *Eruptions of Mount St. Helens: Past, Present, and Future*. Washington: U.S. Government Printing Office. 46 pp.

Yeats, R. S. 1998. *Living with Earthquakes in the Pacific Northwest*. Corvallis: Oregon State University Press. 309 pp.

TECHNICAL

Atwater, B. F. 1992. Geologic evidence for earthquakes during the past 2000 years along the Copalis River, southern coastal Washington. *Journal of Geophysical Research,* v. 97, p. 1901–1919.

Atwater, B. F., and A. L. Moore. 1992. A tsunami about 1000 years ago in Puget Sound, Washington. *Science*, v. 258, p. 1614–1617.

Babcock, R. S., C. A. Suczek, and D. C. Engebretson. 1994. The Crescent "Terrane," Olympic Peninsula and southern Vancouver Island. In *Regional Geology of Washington State*, edited by R. Lasmanis and E. S. Cheney. Olympia: Washington Division of Geology and Earth Resources, Bulletin 80, p. 141–157.

Benioff, H. 1954. Orogenesis and deep crustal structure: Additional evidence from seismology. *Geological Society of America Bulletin*, v. 65, p. 385–400.

Brandon, M. T., M. Roden-Tice, and J. I. Garver. 1998. Late Cenozoic exhumation of the Cascadia accretionary wedge in the Olympic Mountains, NW Washington state. *Bulletin Geological Society of America*, v. 110, p. 985–1009.

Byrne, D. E., D. M. Davis, and L. R. Sykes. 1988. Loci and maximum size of thrust earthquakes and the mechanics of the shallow region of subduction zones. *Tectonics*, v. 7, p. 833–857.

Couch, R. W., and R. P. Riddihough. 1989. The crustal structure of the western continental margin of North America. 1989. In *Geophysical Framework of the Continental United States*, edited by L. C. Pakiser and W. D. Mooney. Boulder, CO: Geological Society of America, Memoir 172, p. 103–128.

Crandell, D. R. 1971. *Postglacial Lahars from Mount Rainier Volcano, Washington*. Washington: U.S. Geological Survey. Professional Paper 667. 75 pp.

Crandell, D. R., and D. R. Mullineaux. 1967. *Volcanic Hazards at Mount Rainier, Washington*. Washington: U.S. Geological Survey, Bulletin 1238. 26 pp.

Crowley, J. K., and D. R. Zimbelman. 1997. Mapping hydrothermally altered rocks on Mount Rainier, Washington, with airborne visible/infrared imaging spectrometer (AVRIS) data. *Geology*, v. 25, p. 559–562.

Davis, D. M., F. A. Dahlen, and J. Suppe. 1983. Mechanics of fold-and-thrust belts and accretionary wedges. *Journal of Geophysical Research*, v. 88, p. 1153–1172.

Dragovich, J. D., P. T. Pringle, and T. J. Walsh. 1994. Extent and geometry of the Mid-Holocene Osceola Mudflow in the Puget Lowland—implications for Holocene sedimentation and paleogeography. *Washington Geology*, v. 22, no. 3, p. 3–26.

Driedger, C. L. 2000. Surface elevation measurements on Nisqually Glacier, Mount Rainier, WA 1931–1998 (abstract). *Washington Geology*, v. 28, no. 12, p. 24.

Driedger, C. L., and P. M., Kennard. 1986. Ice volumes on Cascade volcanoes: Mount Rainier, Mount Hood, Three Sisters, and Mount Shasta. U.S. Geological Survey, Professional Paper 1365. 28 pp. + plates.

Duncan, R. A. 1982. A captured island arc chain in the Coast Range of Oregon and Washington. *Journal of Geophysical Research*, v. 7, p. 10827–10837.

Dzurisin, D., D. J. Johnson, and R. B. Symonds. 1983. *Dry Tilt Network at Mount Rainier*. Washington: U.S. Geological Survey, Open-file Report 83-277. 18 pp.

Ewert, J. K., S. R. Brantley, and B. Myers. 1994. *The Next Eruption in the Cascades*. Washington: U.S. Geological Survey, Open-file Report 94-585. 5 pp.

Fahnestock, R. K. 1963. *Morphology and Hydrology of a Glacial Stream—White River, Mount Rainier*. Washington: U.S. Geological Survey, Professional Paper 422-A. 70 pp.

Fiske, R. S., C. A. Hopson, and A. C. Waters. 1963. *Geology of Mount Rainier National Park*. Washington: U.S. Geological Survey Professional Paper 444. 93 pp.

Grow, J. A. 1973. Crustal and upper mantle structure of the central Aleutian Arc. *Geological Society of America Bulletin*, v. 84, p. 2169–2192.

Lescinsky, D. T., and T. W. Sisson. 1998. Ridge-forming, ice-bounded flows at Mount Rainier, Washington. *Geology*, v. 26, p. 351–354.

Lingley, W. S., Jr. 1995. Preliminary observations on marine stratigraphic sequences, central and western Olympic Peninsula. *Washington Geology*, v. 23, p. 9–20.

McCarthy, J., and D. W. Scholl. 1985. Mechanisms of subduction accretion along the central Aleutian Trench. *Geological Society of America Bulletin*, v. 96, p. 691–701.

Moran, S. C., D. R. Zimbelman, and S. D. Malone. 2000. A model for the magmatic-hydrothermal system at Mount Rainier, Washington, from seismic and geochemical observations. *Bulletin of Volcanology*, v. 61, p. 425–436.

Pazzaglia, F. J., and M. T. Brandon. 2001. A fluvial record of long-term steady-state uplift and erosion across the Cascadia forearc high, western Washington State. *American Journal of Science*, v. 301, p. 385–431.

Shreve, R. L., and M. Cloos. 1986. Dynamics of sediment subduction, melange formation, and prism accretion. *Journal of Geophysical Research*, v. 91, p. 10229–10245.

Snavely, P. D., Jr., H. C. Wagner, and D. L. Lander. 1980. *Geologic Cross Section of the Central Oregon Continental Margin*. Boulder, CO: Geological Society of America, Map and Chart Series MC-28J.

Suczek, C. A., R. S. Babcock, and D. C. Engebretson. 1994. Tectonostratigraphy of the Crescent Terrane and related rocks, Olympic Peninsula, Washington, in *Geologic Field Trips in the Pacific Northwest*, v. 1, edited by D. A. Swanson and R. A. Haugerud. Boulder, CO: Geological Society of America, Annual Meeting, p. 1H1–1H11.

Tabor, R. W., and W. M. Cady. 1978. Geologic map of the Olympic Peninsula. Washington: *U. S. Geological Survey*, Map I-994.

Vallance, J. W., and K. M. Scott. 1997. The Osceola Mudflow from Mount Rainier: Sedimentology and hazard implications of a huge clay-rich debris flow. *Geological Society of America Bulletin*, v. 109, p. 143–163.

Walder J. S., and C. L. Driedger. 1993. *Geomorphic Change Caused by Outburst Floods and Debris Flows at Mount Rainier, Washington, with Emphasis on Tahoma Creek Valley*. U.S. Geological Survey, Water-Resources Investigations Report 93–4093. 93 pp.

Walder J. S., and C. L. Driedger. 1994. Rapid geomorphic change caused by glacial outburst floods and debris flows along Tahoma Creek, Mount Rainier, Washington, USA. *Arctic and Alpine Research*, v. 26, p. 319–327.

Yelin, T. S., A. C. Tarr, J. K. Michael, and J. S. Weaver. 1994. *Washington and Oregon Earthquake History and Hazards*. Washington: U.S. Geological Survey Open-file Report 94-226B. 10 pp.

Yorath, C. J., et al. 1985. Lithoprobe, southern Vancouver Island: Seismic reflection sees through Wrangellia to the Juan de Fuca plate. *Geology*, v. 13, p. 759–762.

Zimbelman, D. R., R. O. Rye, and G. P. Landis. 2000. Fumaroles in ice caves on the summit of Mount Rainier: Preliminary stable isotope, gas, and geochemical studies. *Journal of Volcanology and Geothermal Research*, v. 97, p. 457–473.

WEBSITES

National Park Service: www.nps.gov

Park Geology Tour:
www2.nature.nps.gov/grd/tour/index.htm

Cascadia Subduction Zone
45. Olympic NP:
www2.nature.nps.gov/grd/parks/olym/index.htm
46. Oregon Caves NM:
www2.nature.nps.gov/grd/parks/orca/index.htm
47. Redwood N&SP:
www2.nature.nps.gov/grd/parks/redw/index.htm
49. Crater Lake NP:
www2.nature.nps.gov/grd/parks/crla/index.htm
50. Lassen Volcanic NP:
www2.nature.nps.gov/grd/parks/lavo/index.htm
51. Mount Rainier NP:
www2.nature.nps.gov/grd/parks/mora/index.htm

Southern Alaska
48. Kenai Fjords NP:
www2.nature.nps.gov/grd/parks/kefj/index.htm
53. Aniakchak NM:
www2.nature.nps.gov/grd/parks/ania/index.htm
54. Katmai NP:
www2.nature.nps.gov/grd/parks/katm/index.htm
55. Lake Clark NP:
www2.nature.nps.gov/grd/parks/lacl/index.htm

Sierra Nevada
56. Kings Canyon NP:
www2.nature.nps.gov/grd/parks/seki/index.htm
57. Sequoia NP:
www2.nature.nps.gov/grd/parks/seki/index.htm
58. Yosemite NP:
www2.nature.nps.gov/grd/parks/yose/index.htm

U. S. Forest Service: www.fs.fed.us

Geology:
www.fs.fed.us/geology/mgm_geology.html
    52. Mt. St. Helens NVM:
    www.fs.fed.us/gpnf/mshnvm

U. S. Geological Survey: www.usgs.gov

Geology in the Parks:
www2.nature.nps.gov/grd/usgsnps/project/
home.html

Volcanoes:
volcanoes.usgs.gov

   Historical Eruptions:
   volcanoes.usgs.gov/Volcanoes/Historical.html

Volcano Monitoring:
volcanoes.usgs.gov/About/What/Monitor/
monitor.html

Cascades Volcano Observatory:
vulcan.wr.usgs.gov

Alaska Volcano Observatory:
www.avo.alaska.edu/avo4/index.htm

Volcano Photos:
   Cascades:
   vulcan.wr.usgs.gov/Photo/framework.html
   Wrangell Mountains and Cook Inlet, Alaska:
   wrgis.wr.usgs.gov/dds/dds-39/index.html

   Alaska Peninsula and Aleutian Islands:
   wrgis.wr.usgs.gov/dds/dds-40/index.html

6

Collisional Mountain Ranges

▲▲▲▲▲ *"National park lands follow the mountain ranges that lie inland from the Atlantic and Gulf coasts of the United States. A visit to one of those parks reveals not only beautiful mountain scenery, but also rocks and topographic features that tell a story of ancient episodes of drifting plates and crashing continents."*

The Himalayas are the loftiest mountains on Earth, at the juncture of the Indian subcontinent and Asia (Fig. 6.1). Yet the highest peak, Mount Everest, has layers of limestone that were deposited in an ocean. How did those layers rise to 5½ miles (8.8 kilometers) above sea level? A key lies in recognizing that Mount Everest and the Himalayas are part of an extensive region of high topography that includes the Tibetan Plateau. The region is so high and broad, in fact, that it affects climate, forcing the great monsoons on the south, a vast desert on the north, and in its center, a region of glaciers and snowfields known as the "third pole."

The high region between India and Asia is part of the Himalayan-Alpine chain that extends all the way to Western Europe. The Himalayas, Hindu Kush, Zagros, Alps, and other impressive mountain ranges are forming as India, Saudi Arabia, and Africa move northward and collide with Asia and Europe. The Appalachian Mountains in the United States and Canada, along with the Caledonide Mountains that stretch across the British Isles into Scandinavia, are parts of a similar continental collision zone that formed 300 to 400 million years ago. Later rifting opened the Atlantic Ocean, isolating the mountains as separate ranges on different continents. National parks in the Appalachian Mountains preserve evidence of the ancient continental collision (Table 6.1). Shenandoah and Great Smoky Mountains national parks, along with the Blue Ridge Parkway, no longer have mountains as lofty as the Himalayas, but their rocks and structures reveal an earlier ocean basin and continental margin that were deformed and uplifted. The southwestern continuation of the Appalachian collision zone, the Ouachita and Marathon mountains, includes

FIGURE 6.1 The highest mountains on Earth, the Himalayas, form as India crashes into Asia.
a) Map showing Indian subcontinent about to collide with Asia as the intervening Tethys Ocean closed about 50 million years ago. The brown area north of India represents the amount of continental crust that has been shoved beneath Asia, uplifting the Himalayas and the Tibetan Plateau. **b)** Perspective from space, looking west. On the left are low-lying plains on the northern part of the Indian subcontinent. The band of snow-capped peaks is the Himalayas, the dry region to the north the Tibetan Plateau.

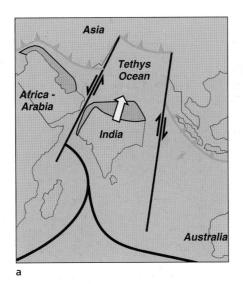

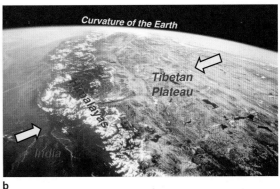

Hot Springs National Park in Arkansas and part of Big Bend National Park in Texas. Gates of the Arctic National Park and Preserve lies in a younger continental collision zone, the Brooks Range in northern Alaska.

Collisional mountain ranges are consequences of the difference in **buoyancy** between continental and oceanic crust (Fig. 6.2a). Subduction of thin oceanic crust is analogous to a swimmer putting a tennis ball beneath his or her belly. The ball offers little support to the weight of the swimmer. A soccer ball is larger; its buoyancy would tend to lift the swimmer out of the water.

Subduction occurs where a plate capped by thin oceanic crust converges with one that has thick continental crust (Fig. 6.2b). But there's a problem if the subducting plate includes a continent. Once all of the oceanic crust subducts, the thick crust of the continent collides with the continent on the overriding plate, lifting the region to high elevation (Fig. 6.2c). The colliding continents deform by compression; their rocks are metamorphosed, folded, and uplifted. Plate convergence eventually stops, so that one continent is "sutured" to the other along the collisional mountain range.

TABLE 6.1 National parks in collisional mountain ranges

APPALACHIAN-OUACHITA-MARATHON			NORTHERN ALASKA
APPALACHIAN MOUNTAINS	**OUACHITA MOUNTAINS**	**MARATHON MOUNTAINS**	**BROOKS RANGE**
59. Acadia NP, ME 60. Big South Fork NRRA, KY/TN 61. Blue Ridge Parkway, VA/NC 62. Bluestone NSR&RA, WV 63. Chattahoochee River NRA, GA 64. Delaware Water Gap NRA, PA 65. Gauley River NRA, WV 66. Great Smoky Mountains NP, NC/TN 67. Obed WSR, TN 68. Russell Cave NM, AL 69. Shenandoah NP, VA	70. Hot Springs NP, AR	71. Big Bend NP, TX 72. Rio Grande WSR, TX	73. Cape Krusenstern NM, AK 74. Gates of the Arctic NP, AK 75. Kobuk Valley NP, AK 76. Noatak NPres, AK

(a) Crustal Buoyancy

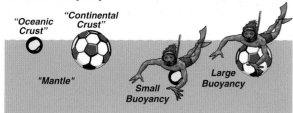

(b) Subduction

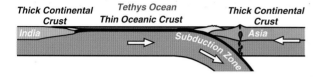

(c) Collisional Mountain Range

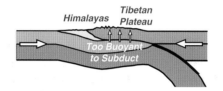

FIGURE 6.2 Crustal buoyancy and continental collision.
a) A tennis ball will not support a swimmer's weight, but a larger soccer ball might. **b)** Continents approach as thin oceanic crust subducts. **c)** When the ocean closes, continental collision results in a broad region of uplifted topography, such as the Himalayas and Tibetan Plateau (Fig. 6.1).

APPALACHIAN-OUACHITA-MARATHON MOUNTAIN CHAIN

National park lands follow the mountain ranges that lie inland from the Atlantic and Gulf coasts of the United States (Fig. 6.3). A visit to one of those parks reveals not only beautiful mountain scenery, but also rocks and topographic features that tell a story of ancient episodes of drifting plates and crashing continents.

Soft versus Hard Collisions

Differences in the height, breadth, and form of portions of the Appalachian-Ouachita-Marathon mountain chain are due partly to the amount of continental

FIGURE 6.3 The Appalachian Mountains are part of a collisional mountain range that includes the Ouachita Mountains of Arkansas and Oklahoma, and the Marathon Mountains of West Texas.
After 300 million years, mountains have eroded deeply and are covered in places by young sediments of the Atlantic and Gulf coastal plains. Numbers refer to some of the parks in Table 6.1. The Blue Ridge Parkway follows the crest of the mountains in Virginia and North Carolina, connecting Shenandoah and Great Smoky Mountains national parks. Hot Springs National Park is in the Ouachita Mountains, while the Marathon Mountains extend into Big Bend National Park. Acadia National Park is in the northern Appalachians in Maine. Cross section G–G' is shown in Fig. 6.8.

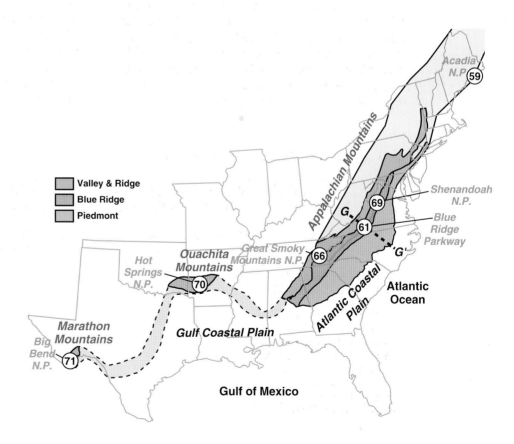

FIGURE 6.4 Soft versus hard collisions. a) Soft collision (Ouachita and Marathon mountains). Plate convergence stops shortly after the ocean closes, with minimal amounts of deformation and crustal thickening. The mountains are low and consist primarily of deformed sedimentary layers of the ocean and continental margin. (Inset: Deformed sandstone and shale layers in Ouachita Mountains, Arkansas, near Hot Springs National Park.) **b)** Hard collision (Himalayas, Alps, Appalachians, and Brooks Range). Plate convergence continues for some time after the continents collide, causing substantial deformation, crustal thickening, and high mountains. (Inset: Uplifted strata from the ocean are exposed in the Alps in Switzerland.)

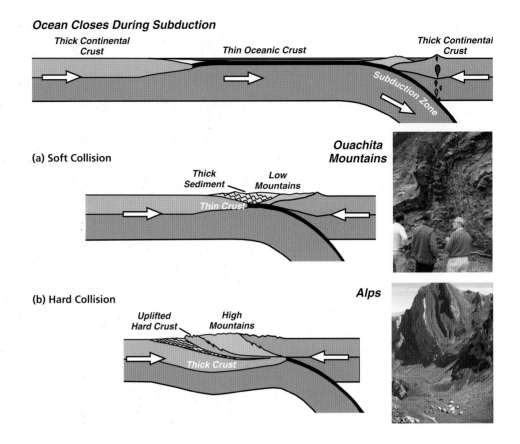

Ocean Closes During Subduction

Thick Continental Crust — Thin Oceanic Crust — Thick Continental Crust — Subduction Zone

(a) Soft Collision

Thick Sediment — Low Mountains — Thin Crust — *Ouachita Mountains*

(b) Hard Collision

Uplifted Hard Crust — High Mountains — Thick Crust — *Alps*

convergence that occurred after the ocean closed—that is, whether the collision was soft or hard. In a **soft collision**, convergence ceases soon after the ocean basin closes (Fig. 6.4a). The topography remains low because the crust is still thin, allowing the remnants of the ocean basin to fill with thick sedimentary layers. Such a collision is considered soft because deformation of continental crustal rocks was minor and uplift was

slight. The Ouachita and Marathon ranges consist of small mountains developed in deformed sedimentary layers, without exposure of deeper metamorphic rocks such as those found in the Appalachians.

During a **hard collision** a continent continues to converge for perhaps hundreds of miles after the initial collision (Fig. 6.4b). The crust may thicken to twice the normal continental thickness, resulting in high topography as well as uplift and exposure of the hard (igneous and metamorphic) rocks of the deeper crust. High mountains such as the Himalayas and Alps are a consequence not only of compressional deformation due to convergence, but also of uplift resulting from the buoyancy of the thickened crust (Fig. 6.5a). As high topography erodes, weight is gradually taken off the thick crust, so that it bobs upward—an effect known as **isostatic rebound**. Somewhat lower mountains, the Brooks Range in northern Alaska, were formed by hard collision, but they have eroded downward over time (Fig. 6.5b). The southern Appalachian Mountains are also the product of hard collision; in their prime, those mountains may have been as high as Mount Everest. Prolonged erosion with accompanying

---

MOLTEN GLURP?

"Earth scientists had just discovered something fascinating about the continent Patty Keene was standing on, incidentally. It was riding on a slab about forty miles thick, and the slab was drifting around on molten glurp. And all the other continents had slabs of their own. When one slab crashed into another one, mountains were made."

Kurt Vonnegut, Jr.,
Breakfast of Champions.

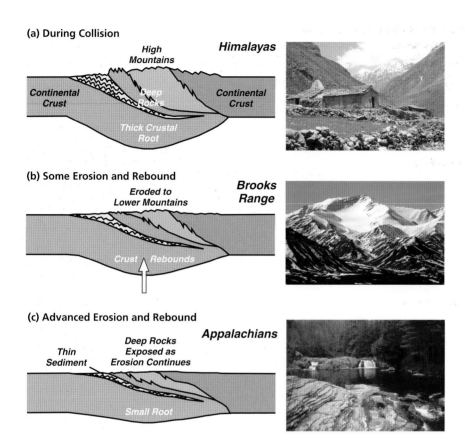

(a) During Collision

High Mountains

Himalayas

Continental Crust

Deep Rocks

Continental Crust

Thick Crustal Root

(b) Some Erosion and Rebound

Eroded to Lower Mountains

Brooks Range

Crust Rebounds

(c) Advanced Erosion and Rebound

Appalachians

Thin Sediment

Deep Rocks Exposed as Erosion Continues

Small Root

FIGURE 6.5 Erosion and isostatic rebound. a) Young mountain range. Soon after continents collide, the crust may be so thick and buoyant that it elevates topography as high as the Himalayas and the Tibetan Plateau (Inset: Himalayas, looking northward into Tibet from Badrinath, India.) **b)** Moderate erosion. As erosion removes some of the weight of the mountains, the buoyant crust rebounds upward. (Inset: Brooks Range in Alaska, where rocks that were buried a few miles below the surface are now exposed in Gates of the Arctic National Park and Preserve.) **c)** Old mountain range. The crust rebounds to near normal continental thickness as mountains erode to smaller hills. (Inset: Rocks that were buried and metamorphosed deep within the crust are now at the surface in the southern Appalachian Mountains, at Linville Falls along the Blue Ridge Parkway in North Carolina.)

rebound of the crust explains how rocks that were metamorphosed because of high temperature and pressure at 10 to 15 miles (15 to 25 kilometers) depth are now at the surface (Fig. 6.5c).

The Appalachian Story

Appalachian geologists have envisioned the evolution of eastern North America as a metaphor based on Greek mythology. **Iapetus** was the son of **Gaia** and the father of **Atlas**. An ancient sea bordering eastern North America was called the Iapetus Ocean (Fig. 6.6a). As the ocean closed, Europe, Africa, and South America collided with North America, forming a mountain range that was, in places, as lofty as the Himalayas (Fig. 6.6b). The sutured continents later ripped apart, opening the Atlantic Ocean (Fig. 6.6c). The metaphor is a representation of the events that shaped the Appalachian Mountains and adjacent coastal plain: Gaia (Mother Earth) gave birth to Iapetus (an ocean), who in turn fathered Atlas (the Atlantic Ocean).

The current Himalayan-Alpine collision zone is a more-or-less continuous chain of mountains, 6,000 miles (10,000 kilometers) long and as much as 1,000 miles (1,500 kilometers) wide. The closing of the Iapetus and other oceans, and the accompanying collision of continents some 300 to 400 million years ago, formed a chain

---

OUR NATION'S HIGHEST OFFICE—AN ACCIDENT OF PLATE TECTONICS!

The presidential election of 2000 was the closest in the history of the United States. In fact, it all hinged on the outcome of an extremely close race in one state—Florida. Al Gore had a total of 267 electoral votes sewn up, compared to George W. Bush's 246. Had Florida remained in its original position, as part of Africa, then that would have been the result. But continental collision, and later rifting, has changed not only the position of our coastline, but the political landscape as well. Final outcome: Al Gore 267 electoral votes, George W. Bush 271.

(a) 500 Million Years Ago

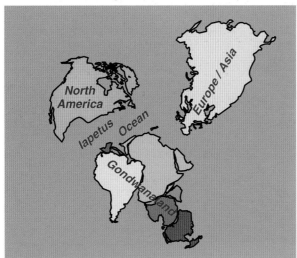

(b) 250 Million Years Ago

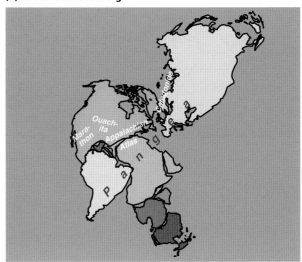

(c) Today

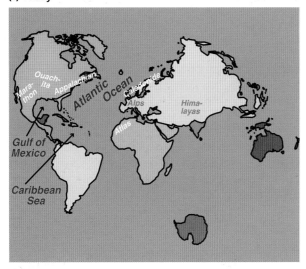

FIGURE 6.6 Continental bumper cars and development of Appalachian-Ouachita-Marathon collisional mountain range. a) The Iapetus Ocean opens. Land that will later become Florida is part of Africa. The Yucatán Peninsula and Cuba may have been on the north side of South America. **b)** The Iapetus Ocean closes. Pangea forms as continents collide. The Appalachians are part of a larger zone of continental collision that includes the Marathon and Ouachita mountains, the Atlas Mountains in Africa, and the Caledonide Mountains in Greenland, the British Isles, and Scandinavia. **c)** The Atlantic Ocean and Gulf of Mexico open. The modern oceans originated about 200 million years ago when Europe, Africa, and South America ripped away from North America. Fragments of the collision zone mountains were left on three continents: the Appalachians in North America, the Atlas in Africa, and the Caledonides in Europe. The Atlantic continues to widen along the Mid-Atlantic Ridge (Fig. 3.27). The Gulf of Mexico ceased opening about 100 million years ago, when tectonic activity shifted to the Caribbean region. Continental collision is occurring today where Africa and India ram into Europe and Asia, forming the Alpine-Himalayan chain.

of mountains as long and broad as the Himalayan-Alpine chain. In its prime, the ancient chain of mountains paralleling the Atlantic and Gulf of Mexico coasts of the United States probably included ranges every bit as high as the Himalayas and as majestic as the Alps. The difference is time. Time has opened the Atlantic, segregating fragments of the mountain chain on three continents. It has worn down the lofty peaks into mountains that are smaller but no less inspiring. And the story of an ocean opening, closing, and reopening is evident in the rocks and landscapes of national parks in the eastern United States (Fig. 6.7).

Appalachian Mountains

National Park Service lands in the Appalachian Mountains offer an amazing opportunity for visitors to experience both the overall landscape and the local geology of an ancient continental collision zone. The Blue Ridge Parkway stretches 469 miles (755 kilometers), from Shenandoah National Park on the northeast to Great Smoky Mountains National Park on the southwest (Fig. 6.3). Although those park lands are mostly confined to one physiographic province, the Blue Ridge, their heights provide vistas of other parts of the collision zone: the Valley and Ridge Province to the west, and the Piedmont to the east (Fig. 6.8). Outcroppings along roadcuts and valley walls reveal deformed and metamorphosed rocks that had been along the edge of ancient North America. Some of those rocks were buried more than 10 miles (15 kilometers) during collision, then uplifted and exposed by a combination of thrust faulting, erosion, and isostatic rebound.

The Blue Ridge is an old piece of the North American continent that broke off and was pushed westward over younger rocks when Africa came crashing in about 300 million years ago. At viewpoints along the Blue Ridge Parkway, as well as its northward continuation as Skyline Drive in Shenandoah National Park, you can see evidence of how the Appalachians formed, how the Blue Ridge evolved during uplift and erosion of the mountains, and how the Appalachians affect the people and plants of the region.

DEVELOPMENT OF GEOLOGIC PROVINCES. The tectonic history of the Appalachian Mountains involves the opening of an ancient ocean along a divergent plate boundary, the closing of the ocean during plate convergence, and then more divergence that opened the Atlantic Ocean. Tracing this large-scale development helps us understand the place of origin of the various

FIGURE 6.7 Acadia National Park, Maine.
The rugged coastline lies where the Appalachian Mountains extend into the Atlantic Ocean.

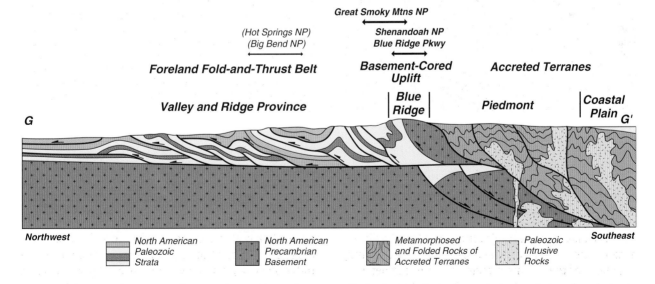

FIGURE 6.8 Simplified cross section of the tectonic provinces of the southern Appalachian Mountains. (Section G–G' in Fig. 6.3). The Blue Ridge Parkway and Shenandoah National Park lie entirely within the Blue Ridge Province. Great Smoky Mountains National Park is mostly within the Blue Ridge Province, with a portion extending into the Valley and Ridge Province. The Ouachita and Marathon mountains are foreland fold-and-thrust belts similar to the Valley and Ridge Province, and include Hot Springs National Park and a small part of Big Bend National Park.

geologic provinces of the southern Appalachian Mountains, as well as the rocks and structures observed along the Blue Ridge Parkway and in the two national parks it connects.

Grenville Orogeny. Long before the Appalachians were around, and before Pangea formed, an old continent existed that was nothing like the North America we know today. About 1.1 billion (1,100 million) years ago, a very ancient ocean closed, leading to collision of the old continent with another. The resulting period of mountain building is known as the **Grenville Orogeny**. It deformed and metamorphosed the rocks known today as the North American Precambrian basement. Those ancient rocks now lie buried beneath younger layers in the Valley and Ridge Province, and are exposed at the surface in the Blue Ridge Province. The granite of Old Rag Mountain in Shenandoah National Park is some of the hard crust of ancient North America (Fig. 6.9a).

Continental Rifting and the Opening of the Iapetus Ocean. About 750 million years ago, the continents that had coalesced during the Grenville Orogeny began to rip apart (Fig. 6.10a). The resulting

continental rift zone was much like today's Basin and Range Province, with long mountain ranges separated by down-dropped valleys (Chap. 3). Sedimentary and volcanic layers that were deposited in continental rift valleys are exposed in Shenandoah National Park and along the Blue Ridge Parkway (Figs. 6.9b and 6.11). The old continent completely rifted apart, opening the Iapetus Ocean (Fig. 6.10b). As the ocean widened, the edge of ancient North America subsided, and a blanket of sedimentary strata buried the rift valleys and eroded mountain ranges (Fig. 6.12). The edge of ancient North America was a passive continental margin, much like the East Coast today (Chap. 4). Fig. 6.13 shows examples of the passive margin strata along the Blue Ridge Parkway and in Great Smoky Mountains National Park.

Closing of the Iapetus Ocean and Continental Collision. From about 400 to 300 million years ago, the Iapetus Ocean gradually closed during subduction (Fig. 6.10c). In the process, volcanic islands, continental fragments, and finally the African continent collided with the ancient continental margin of North America (Fig. 6.10d). The western part of the Appalachians contains

---

CLOSE YOUR EYES AND IMAGINE

A stop at a viewpoint along the Blue Ridge Parkway or Skyline Drive in Shenandoah National Park can provide an opportunity to observe a landscape shaped by the collision of continents, and contemplate events that occurred long ago. Look down from the hard rock of the Blue Ridge, the ancient edge of North America. Gaze eastward upon the rolling hills of the Piedmont, remnants of the rocks and islands of a former ocean. Now close your eyes and imagine the ancient Iapetus Ocean. Watch as, like a ship on the horizon, Africa moves closer and closer. The mighty Iapetus narrows. Materials from its sea bottom are gradually pushed over the edge of North America, along with volcanic islands and pieces of smaller continents. The ocean completely disappears as Africa comes crashing in. The edge of your continent rumbles and is shoved westward. You feel cold and breathe heavily as you rise toward the stratosphere.

Around you are massive, snow-capped peaks as high as Mount Everest and K-2. The commotion stops as the African bumper car starts to pull away. Glaciers and rivers erode away the rugged landscape, and you find yourself at lower elevation as the warmth returns to your body and you once again can breathe without laboring. The Atlantic Ocean grows wider and wider as Africa fades into the distance. You've experienced the cycle of closing and opening an ocean. Open your eyes and see the results. You're standing on the Blue Ridge, the part of North America's edge that was shoved upward and westward. The Piedmont in front of you is a remnant of the ancient ocean. Turn around and look west, where you'll notice the gently folded carpet of the Valley and Ridge Province, sedimentary layers of North America that were rumpled during the collision.

a   b

FIGURE 6.9 Shenandoah National Park, Virginia. a) View of the Piedmont, looking east from Old Rag Mountain. The mountain is composed of 1.1-billion-year-old granite, some of the Precambrian basement rock of ancient North America that was uplifted as part of the Blue Ridge Province. **b)** Sedimentary strata along the southern portion of Skyline Drive. The layers were deposited by rivers and lakes in a rift valley as the ancient continent rifted apart from 750 to 600 million years ago.

(a) 750 Million Years Ago

Rift Valley   Mountain Range
Continental Crust
Lithosphere

(b) 500 Million Years Ago

Passive Continental Margin of Ancient North America   Ancient Ocean   Gondwanaland

(c) 400 Million Years Ago

North America   Gondwanaland

(d) 300 Million Years Ago

Pangea

(e) Today

Southern Appalachian Mountains
Valley and Ridge   Blue Ridge   Piedmont   Coastal Plain   Atlantic Ocean   Africa

FIGURE 6.10 Simplified plate-tectonic evolution of the southern Appalachian Mountains. a) Old continent rips apart. The long mountain ranges and rift valleys were similar to those forming today in East Africa and the Basin and Range Province (Chap. 3). **b)** Ancient ocean opens (Fig. 6.6a). Continental blocks destined to become North America and Gondwanaland drift apart. The eastern edge of ancient North America developed into a passive continental margin, similar to the modern East Coast (Chap. 4). As the margin cooled and subsided, thick sedimentary strata covered the earlier mountains and rift valleys. **c)** Ocean narrows during subduction. Oceanic sediments and volcanic islands were at times added to the edge of North America. **d)** Ocean completely closes. The southern Appalachians form as the African portion of Gondwanaland collides with North America (Fig. 6.6b). **e)** Atlantic Ocean opens. Ancient ocean rocks are left behind as the Piedmont Province, along with a sliver of Africa that now lies beneath the Coastal Plain of Florida and offshore regions of Georgia and the Carolinas (Fig. 6.6c). The Blue Ridge Province is part of North America's ancient continental margin, while the Valley and Ridge Province contains sedimentary layers of North America that were folded and faulted during the collision.

a

b

c

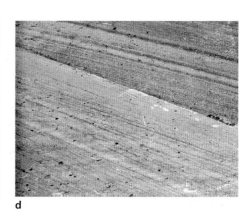

d

FIGURE 6.11 Rift valley strata in the Grandfather Mountain area of the Blue Ridge Parkway. a) View of the parkway near Linn Cove Viaduct in North Carolina. Rock in foreground is conglomerate of the Grandfather Mountain Formation. Its resistance to erosion is shown by prominence of MacRae Peak (middle left), along the high ridge of Grandfather Mountain. **b)** Close-up of conglomerate in Fig 6.11a reveals pebbles that were deposited by mountain streams flowing into the rift valley. **c)** Tilted sandstone layers south of Grandfather Mountain. **d)** Close-up of Fig. 6.11c shows that layering and crossbedding within the sandstone is still evident after slight metamorphism.

FIGURE 6.12 Passive continental margin of ancient North America. During breakup of a larger continent, the older igneous and metamorphic rocks (ancient North America) were cut by normal faults and overlain by younger strata. The Grandfather Mountain and similar formations now exposed in the Blue Ridge Province consist of sedimentary and volcanic layers filling continental rift valleys (Figs. 6.9b and 6.11). Sediment deposited on the continental shelf includes sandstone, siltstone, and shale of the Chilhowee Formation (Fig. 6.13). Shales of the Ashe Formation formed in much deeper water of the Iapetus Ocean.

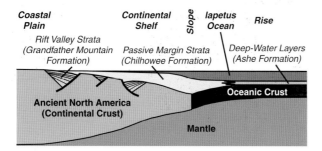

OUT OF AFRICA

Imagine the Grandfather Mountain area 700 million years ago. The region was in the zone where a very ancient continent was ripping apart. As in East Africa (see Part II), rift valleys subsided and were filled with sand, mud, and gravel deposited in rivers and lakes. At times volcanoes erupted, so that the sedimentary deposits are interstratified with lava flows. Some of the erupting material solidified before reaching the surface as intrusive dikes. The sequence of rocks, known as the Grandfather Mountain Formation (Fig. 6.11), consists of the remnants of the rift valley deposits, after they were deformed and metamorphosed during the later continental collision. The bulk of the Grandfather Mountain Formation is coarse sandstone, but in places it contains pebbles, hence it is a conglomerate. Interlayered volcanic flows include light-colored, high-silica rhyolite, as well as dark, low-silica basalt.

a　　b

FIGURE 6.13 **Sedimentary strata deposited along the ancient passive continental margin of North America as the Iapetus Ocean opened during the Cambrian Period (about 500 million years ago). a)** Sandstone of the Chilhowee Formation on the spur road to Linville Falls along the Blue Ridge Parkway in North Carolina. **b)** Quartzite (metamorphosed sandstone) of the Cochran Formation along the Little River in Great Smoky Mountains National Park.

the rocks that were originally part of North America. The Valley and Ridge Province is folded and faulted sedimentary strata of the former continental margin (Fig. 6.14), and the Blue Ridge Province is a piece of the deeper, hard crust that broke off and was shoved westward. Farther east, the rocks formed elsewhere and were attached (accreted) to the edge of North America as the ocean closed. The Piedmont Province is an array of volcanic islands, sedimentary rocks, and crust of the Iapetus Ocean that were caught up in the collision. In places the Atlantic Ocean opened some distance east of the zone of "suturing" between the continents. Rocks beneath young sediments on the Coastal Plain of Florida and areas offshore from Georgia and the Carolinas are stranded pieces of Africa, left behind when the Atlantic opened.

Great Smoky Mountains National Park is the most visited unit in the entire National Park System, attracting more than 7 million people per year (Fig. 6.15). After Mount Mitchell, the park contains the highest mountains in the Appalachians, including Clingman's Dome (6,642 feet; 2,024 meters) and Mount Guyot (6,621 feet; 2,018 meters). The spectacular scenery of the park lies within the Blue Ridge Province, with a small portion on the northwest in the Valley and Ridge Province. But the scenery at the surface is only part of the story, because the old rocks of the Blue Ridge were pushed a considerable distance over younger rock of the Valley and Ridge Province (Fig. 6.16). A roadcut just outside the park reveals the Great Smoky Fault, along which metamorphic rocks were thrust over folded and faulted sedimentary strata (Fig. 6.17).

Continental Rifting and the Opening of the Atlantic Ocean. The final events in the Appalachian story have been occurring over the past 250 million years, as Pangea ripped apart, the Atlantic Ocean opened, and the Appalachian Mountains wore down

FIGURE 6.14 **The Valley and Ridge Province of the Appalachian Mountains formed from compression during continental collision. a)** Rock layers fold, developing a pattern similar to that formed by pushing on a carpet. **b)** Layers break along thrust faults as they're squeezed. **c)** Deformation of Valley and Ridge sedimentary strata occurs through a combination of folding and thrust faulting. **d)** Layers are tilted as well as folded. **e)** The compression has deformed the sedimentary layers into a complex pattern of plunging folds and thrust faults.

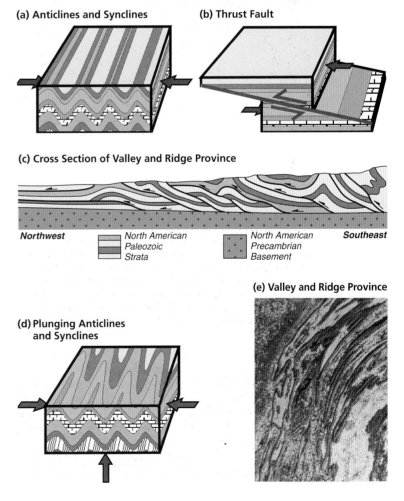

(a) Anticlines and Synclines

(b) Thrust Fault

(c) Cross Section of Valley and Ridge Province

Northwest　　*Southeast*

North American Paleozoic Strata

North American Precambrian Basement

(d) Plunging Anticlines and Synclines

(e) Valley and Ridge Province

RUMPLING THE CARPET

During the late stages of continental collision, the sedimentary rocks of the Valley and Ridge Province were detached from the hard rocks underneath and pushed westward. As the strata compressed, they deformed in two different ways. First, similar to pushing a rug across the floor (or squeezing an accordion), the compression resulted in a series of anticlines and synclines (Fig. 6.14a). Second, the rocks were cut by thrust faults, stacking up like roof shingles (Fig. 6.14b). The overall structure is a series of folds separated by thrust faults (Fig. 6.14c). The term *Valley and Ridge* originates from the pattern of erosion that occurs in the folded and faulted strata. Erosion has cut downward into the maze of folds and fault blocks, but not uniformly. Sandstone layers resist erosion, so they remain high as ridges. Soft shales and limestones dissolve in the wet climate and form valleys. Viewed from airplanes or satellites, the plunging anticlines and synclines form an amazing pattern of elongated ridges that sometimes end in *V*s (Fig. 6.14d,e). The torn, rumpled-carpet topography of the Valley and Ridge Province lies in front of the Appalachian Mountain Range. Geologists call such features **foreland fold-and-thrust belts (Chap. 10)**.

a

b

FIGURE 6.15 Great Smoky Mountains National Park, Tennessee and North Carolina. a) The thick foliage, highlighted by spectacular fall colors, obscures the spectacular geology beneath the park. **b)** Precambrian Thunderhead Sandstone at Laurel Falls.

FIGURE 6.16 Models of deep Appalachian structure. a) High-angle faults. The broken crust is pushed more-or-less straight upward. Major breaks in the crust (faults) are at high angles, implying that the deformation is due to vertical forces. **b)** Low-angle detachment. The broken crust is thrust laterally along a low-angle "detachment." Major forces are horizontal, as in a plate-tectonic scenario.

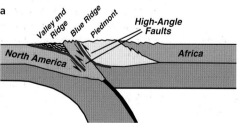

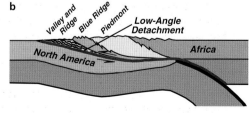

FIGURE 6.17 Continental collision resulted in folding and faulting of Blue Ridge and Valley and Ridge rocks in Great Smoky Mountains National Park. a) Roadcut on U.S. Highway 321 northwest of Townsend, Tennessee. Older metamorphic rocks (quartzite) were thrust northwestward over younger sedimentary layers along the Great Smoky Fault. **b)** Deformed sandstone and shale layers, 300 feet (100 meters) northwest of the photo in part (*a*). The hard layers of quartzite folded the layers of sandstone and shale below, as if they were soft butter.

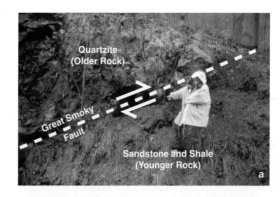

by erosion (Fig. 6.10e). Sedimentary and volcanic strata are found in rift valleys in the eastern part of the Piedmont Province, relics of the rifting of Pangea about 200 million years ago. Coastal Plain sediments are the landward part of the passive margin sequence that extends offshore across the continental shelf, slope, and rise (Chap. 4). New oceanic crust is being created as the Atlantic Ocean widens at the current plate boundary, the Mid-Atlantic Ridge. The southern Appalachian Mountains appear to have evolved from a very hard continental collision (Fig. 6.4b). Individual mountains may have attained heights comparable to Mount Blanc in the Alps (15,781 feet; 4,810 meters) or

WHAT'S BENEATH THE APPALACHIANS?

The deep structure of the Appalachian Mountains has sparked a long history of debate among geologists. In the 1940s and 1950s, the controversy revolved around the nature of faulting beneath the Valley and Ridge Province. Where faults cut the surface, they commonly dip eastward, at fairly steep angles. One school of thought, known as **thick-skinned**, suggested that faults continue at high angles, cutting the old igneous and metamorphic rocks beneath. A **thin-skinned** group suggested that the faults flatten and merge into a nearly horizontal "detachment" zone above the old, hard rocks. Geologists exploring for oil and gas used a technique of sending sound waves into the ground and observing the echoes that returned from different rock layers. Such **seismic reflection profiles** revealed that thrust faults merged with a nearly horizontal zone above hard rocks beneath the Valley and Ridge Province, demonstrating the thin-skinned nature of the deformation. The softer sedimentary layers had been detached from the rigid rocks beneath and thrust westward (Fig. 6.8). A later argument centered around the Blue Ridge and Piedmont. How far had those provinces moved westward over the North American continent?

One group felt that the rocks moved mostly vertically along steep faults, and had not moved very far westward (Fig. 6.16a). Another group thought that the low-angle detachment zone known to underlie the Valley and Ridge Province continued eastward beneath the Blue Ridge, and perhaps even farther. An implication of the latter interpretation is that rocks at the surface were pushed westward tens of miles, and maybe even hundreds of miles, over North America. The debate was part of the controversy of plate tectonics. Did mountain ranges result from down-warping and vertical uplift (**geosynclinal theory**), or from large horizontal movements (**plate-tectonic theory**)? In the late 1970s, the southern Appalachian Mountains served as a natural laboratory to test the two hypotheses. New seismic reflection profiles revealed that strong reflections from the detachment zone could be traced eastward, entirely beneath the Blue Ridge and the western portion of the Piedmont. It was thereby revealed that the Blue Ridge and part of the Piedmont were thrust westward for a considerable distance over the North American continent (Fig. 6.16b).

The Blue Ridge Parkway crosses unusual terrain around Grandfather Mountain, between Gillespie Gap and Boone, North Carolina. Most of the parkway lies on old rocks that were metamorphosed to such a degree that the original rocks are difficult to identify. Between parkway mileposts 286 and 316, relatively young rocks are clearly sedimentary and volcanic in origin. Cades Cove, in the northwest portion of Great Smoky Mountains National Park, is another such "window," where limestones are completely surrounded by harder, older rocks. The sedimentary rocks around Grandfather Mountain formed when an ancient continent ripped apart about 700 million years ago. They are primarily conglomerate, sandstone, and shale layers deposited in lakes and streams within continental rift valleys (Fig. 6.11). The Iapetus Ocean then opened, forming a broad continental shelf off the eastern edge of ancient North America. Sandstone around Linville Falls, as well as dolomite (a rock similar to limestone) in nearby Linville Caverns, was deposited on the subsiding shelf about 550 million years ago (Fig. 6.13a). The older rocks surrounding the Grandfather Mountain Window are gneiss, metamorphosed deep within the Earth 1.1 billion years ago during the Grenville Orogeny. As Africa came crashing into North America 300 million years ago, these rocks broke off along a thrust fault and were pushed westward (Fig. 6.18a). They came to rest over younger rocks of ancient North America's continental margin, including rift valley sediments and lava flows of the Grandfather Mountain Formation, and shelf deposits of the Chilhowee Formation and Shady Dolomite (Fig. 6.18b). After thrusting, the region was bowed upward as a large anticline. Erosion wore through the old rocks on top, allowing a peep through the "window" at the younger rocks beneath (Fig. 6.18c).

(a) Africa Collides with North America

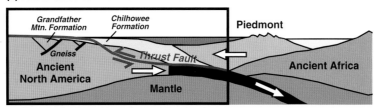

(b) Older Rocks (Gneiss) Thrust over Younger

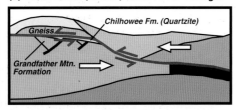

(c) "Window" Erodes through Older Gneiss, Exposing Younger Quartzite beneath

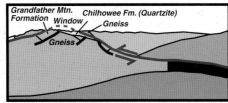

FIGURE 6.18 Simplified diagram showing development of the Grandfather Mountain Window. a) As Africa came crashing in (Fig. 6.10d), a piece of North America's continental margin broke off along a thrust fault. b) Older metamorphic rocks (gneiss) were pushed westward over younger rift valley deposits (Grandfather Mountain Formation) and continental shelf strata (Chilhowee Formation). c) After the rock layers were bent upward into a giant anticline, the top portion eroded away, exposing a "window" through the thrust fault into the underlying, younger rocks. d) Simplified geologic map of the Grandfather Mountain region. Numbers are mileage markers along the Blue Ridge Parkway. e) In the background, Linville Falls flows over high-grade metamorphic rock (gneiss) that is over 1.1 billion (1,100 million) years old. The rock in the foreground is 550-million-year-old sandstone of the Chilhowee Formation, metamorphosed to quartzite. The older rocks were pushed over the younger ones along a thrust fault that lies approximately in the position of the pool of water.

(d) Geologic Map

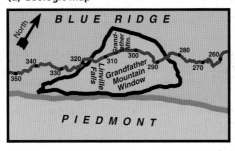

(e) Linville Falls

even Mount Everest in the Himalayas (29,035 feet; 8,850 meters). Similar to the Alps and Himalayas, the southern Appalachians owed their initial height to the great thickness of crust that resulted from the collision (Fig. 6.5a). That great thickness cannot last for long, however, because the topography erodes, taking the weight off the buoyant crustal root. Rebound of the root rebuilds just enough topography to maintain isostatic equilibrium[1] (Fig. 6.5b). With each step in the process, the crustal root is a little shallower and the overall topography a bit lower. After 300 million years, the overall height of the crest of the southern

[1]*Isostatic equilibrium* results when the weight of the mountains, pushing downward, balances the upward, buoyant force provided by a thick crust "floating" in the more dense mantle.

Appalachians is about 3,000 feet (1,000 meters), and the highest mountain, Mount Mitchell in North Carolina, stands 6,684 feet (2,037 meters) above sea level (Fig. 6.20). The erosion and accompanying isostatic rebound removed 10 to 15 miles (15 to 25 kilometers) of rock strata from portions of the southern Appalachian Mountains (Fig. 6.5c). The metamorphic rocks schist and gneiss found along the Blue Ridge Parkway formed because of high temperatures and pressures at such depths.

OUACHITA MOUNTAINS

Ancient North America did not extend nearly as far east as it does today. The Appalachian Mountains are roughly in the position of the earlier continental

---

OH, MY ACHING JOINTS!

Throughout much of the Appalachian Mountains, rocks weather and break apart along very straight surfaces. These cracking surfaces, called **joints**, commonly have vertical orientations. As rock of the Appalachian Mountains uplifted, erosion removed the weight of the overlying layers. The release in pressure caused the rock to break along vertical joints that intersect at 90° angles. This dramatic pattern is evident in cliffs and valley walls along the Blue Ridge Parkway (Fig. 6.19). In places, such as Linville Falls, creeks follow the distinctive sets of cracks.

a b c

FIGURE 6.19 Erosion along the Blue Ridge Parkway in North Carolina is influenced by cracks in the rocks, known as joints. a) Vertical joints in conglomerate layers on Grandfather Mountain. Two sets of joints are nearly perpendicular, so that rock slabs break off as rectangular blocks. **b)** Linville Falls flows through cracks carved along vertical joints. **c)** The walls of Linville Gorge are vertical because slabs of rock break off along the perpendicular joint sets.

FIGURE 6.20 Hoarfrost along the Blue Ridge Parkway, where it crosses the Black Mountains of North Carolina. At 6,684 feet (2,037 meters), nearby Mount Mitchell is the highest point in the eastern United States.

In its mission and administration, it might arguably be considered the first site functioning as a "national park." The first formal national park, Yellowstone, was established by act of Congress 40 years later.

The thermal waters for which Hot Springs National Park is named are not due to the presence of shallow magma beneath the area, as at Yellowstone (Chap. 9). Rather, the hot springs are a result of cracks in Paleozoic chert layers outside the park that allow rainwater to circulate slowly downward, where it is heated because of the normal increase in temperature with depth in the Earth. More prominent systems of faults and cracks directly beneath the park allow the water to return back to the surface much quicker, while it is still about 140 °F (60 °C)—more than enough for a refreshing, therapeutic soak!

margin (Fig. 6.3). The southern part of the continent had a large embayment. Like a safe harbor, that embayment prevented blocks of continental crust from crashing very hard into the edge of North America.

The Ouachita Mountains are a classic example of soft continental collision (Fig. 6.4a). Plate convergence ceased soon after the ocean closed, so that the crust remained relatively thin. The mountains never attained heights anywhere near those of the Himalayas and Alps, nor of the ancient Appalachians. In their prime, the Ouachitas were perhaps similar to the modest height (~10,000 feet; 3,000 meters) of the Carpathian Mountains in Central Europe. The low buoyancy of thin crust allowed the accumulation of thick sedimentary layers as the ancient ocean basin closed near the end of the Paleozoic Era (about 280 million years ago). Without thick crust and high mountains, there has been no extensive erosion and isostatic rebound, so that a pile of sedimentary layers as much as 12 miles (20 kilometers) thick still lies beneath the Ouachita Mountains.

Hot Springs National Park in Arkansas preserves evidence of the soft collision. Rocks are not the hard igneous and metamorphic variety found in the Appalachian Mountains, but rather softer sedimentary strata scraped off the seafloor and thrust over the edge of the continent as the ocean between North and South America closed (Fig. 6.21). The originally horizontal rocks are folded and faulted, much like the strata in the Valley and Ridge Province of the Appalachians (Fig. 6.14).

In 1832, Congress established Hot Springs National Reservation to preserve and protect the region of thermal waters around Hot Springs Mountain in Arkansas.

FIGURE 6.21 Hot Springs National Park, Arkansas. The rock in the foreground is Arkansas Novaculite, a form of high-silica chert deposited on the seafloor. The rock was thrust over the edge of North America during the continental collision that formed the Ouachita Mountains.

FIGURE 6.22 The Marathon Mountains extend into Big Bend National Park, Texas. The collisional mountain range rocks and structures have been obscured by later compression that formed the Laramide Uplifts (Chap. 10), and then by continental rifting of the Basin and Range Province (Chap. 3).

MARATHON MOUNTAINS

The area of Southwest Texas where the Rio Grande River takes a sharp turn to the north displays the effects of three superimposed tectonic episodes: 1) Paleozoic continental collision that formed the Marathon Mountains; 2) Mesozoic reverse faulting during the Laramide Orogeny (Chap. 10); and 3) continental rifting that is currently forming the Basin and Range Province (Chap. 3). The northernmost part of Big Bend National Park, around Persimmon Gap, includes Paleozoic-age rocks and structures similar to those found in the Ouachita Mountains (Fig. 6.22). The park thus preserves the westernmost vestiges of the oceanic closure and continental collision found not only in the Appalachians of the United States and Canada, but also in widely separated regions of Africa and Northern Europe (Fig. 6.6b).

BROOKS RANGE

Northern Alaska was the site of continental collision that progressed to a stage somewhere between the soft collision seen in the Ouachita and Marathon mountains, and the hard collision observed in the Appalachians and Himalayas (Fig. 6.23). In its prime, about 100 million years ago, the Brooks Range probably had mountains as high as the Alps of Europe (Fig. 6.24). Gates of the Arctic National Park and Preserve reveals mountains much higher than those seen in the Appalachians, because the Brooks Range is still young—erosion has not worn the landscape down quite as much (Fig. 6.25).

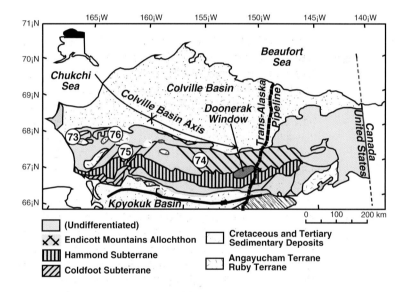

FIGURE 6.23 Map showing tectonic elements of the Brooks Range in northern Alaska. The range consists of various terranes that were added to North America when an ancient ocean closed about 100 million years ago. Gates of the Arctic National Park and Preserve (number 74 on map) spans the central part of the Brooks Range west of the Trans-Alaska Pipeline. Other parks in the Brooks Range include Cape Krusenstern National Monument (73), Kobuk Valley National Park (75), and Noatak National Preserve (76).

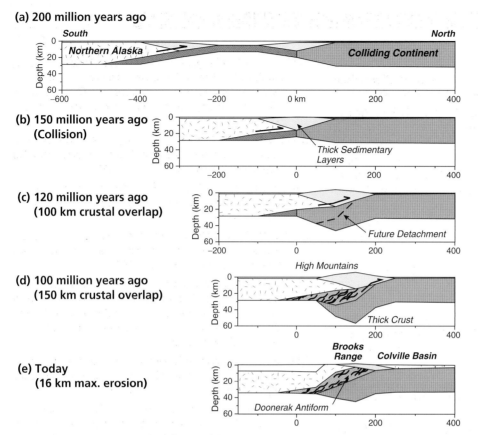

(a) 200 million years ago

(b) 150 million years ago
(Collision)

Thick Sedimentary
Layers

(c) 120 million years ago
(100 km crustal overlap)

Future Detachment

(d) 100 million years ago
(150 km crustal overlap)

High Mountains

Thick Crust

(e) Today
(16 km max. erosion)

Brooks
Range   Colville Basin

Doonerak Antiform

FIGURE 6.24 Tectonic evolution of the Brooks Range.
a) The ocean separating northern Alaska and a continental fragment begins to close. **b)** Continents collide but continue to converge. **c)** The northern continent extends for more than 60 miles (100 kilometers) beneath northern Alaska, thickening the crust. **d)** The mountains reach heights similar to those of the modern-day Alps, as the hard crust of northern Alaska breaks and uplifts. **e)** After eroding away up to 10 miles (16 kilometers) of rock, there is still a significant crustal root that maintains mountain heights much greater than those seen today in the Appalachians.

FIGURE 6.25 Gates of the Arctic National Park and Preserve, Alaska. The Brooks Range formed during a relatively hard collision about 100 million years ago. In their prime, peaks were probably as high as Mount Blanc, the loftiest point in the Alps at 15,771 feet (4,807 meters). Since then, the topography has eroded downward, but Mount Michelson, outside the park, is still 9,239 feet (2,816 meters) high, and Mount Doonerak, the highest point within the park, rises to 8,806 feet (2,684 meters).

FURTHER READING

GENERAL

Badger, R. L. 1999. *Geology along Skyline Drive, Shenandoah National Park, Virginia*. Helena, MT: Falcon. 100 pp.

Beyer, F. 1991. *North Carolina: The Years before Man: A Geologic History*. Durham, NC: Carolina Academic Press. 244 pp.

Carter, M. W., C. E. Merschat, and W. F. Wilson. 1999. *A Geologic Adventure along the Blue Ridge Parkway in North Carolina*. Raleigh: North Carolina Geological Survey. 52 pp.

Clark, S. H. B. 2001. *Birth of the Mountains: The Geologic Story of the Southern Appalachian Mountains*. Washington: U.S. Geological Survey. 23 pp.

Cook, F. A., L. D. Brown, and J. E. Oliver. 1980. The southern Appalachians and the growth of continents. *Scientific American*, v. 243, p. 156–168.

Lillie, R. J. 1996. Mountains and mountain building, in *Macmillan Encyclopedia of Earth Sciences*, edited by E. J. Dasch. New York: Simon and Schuster Macmillan. p. 693–698.

Margolin, P. 1997. The Parkway traverses many mountain ranges—but only one province. *Parkway Milepost*, Summer, p. 6.

McPhee, J. 1998. In suspect terrain, in *Annals of the Former World*, Book 2. New York: Farrar, Straus and Girous. p. 147–275.

Schultz, A., and S. Southworth. 2000. *Geology, Great Smoky Mountains National Park*. Gatlinburg, TN: Great Smoky Mountains Natural History Association. Geologic map and text.

TECHNICAL

Ando, C. J., F. A. Cook, J. E. Oliver, L. D. Brown, and S. Kaufman. 1983. Crustal geometry of the Appalachian orogen from seismic reflection studies, in *Contributions to the Tectonics and Geophysics of Mountain Chains*, edited by R. D. Hatcher, H. Williams, and I. Zietz. Boulder, CO: Geological Society of America, Memoir 158, p. 113–124.

Bryant, B. *Geology of the Linville Quadrangle, North Carolina–Tennessee: A Preliminary Report*. Washington: U.S. Geological Survey, Contributions to Geology 1121-D. 30 pp.

Bryant, B., and J. C. Reed, Jr. 1960. *Road Log of the Grandfather Mountain Area* (Field Trip Guidebook). Raleigh, NC: Carolina Geological Society. 21 pp.

Bryant, B., and J. C. Reed, Jr. 1970. *Geology of the Grandfather Mountain Window and Vicinity, North Carolina and Tennessee*. Washington: U.S. Geological Survey, Professional Paper 615. 190 pp.

Cook, F. A., D. S. Albaugh, L. D. Brown, S. Kaufman, J. E. Oliver, and R. D. Hatcher, Jr. 1979. Thin-skinned tectonics in the crystalline southern Appalachians: COCORP seismic-reflection profiling of the Blue Ridge and Piedmont. *Geology*, v. 7, p. 563–567.

Davis, D. M., F. A. Dahlen, and J. Suppe 1983. Mechanics of fold-and-thrust belts and accretionary wedges. *Journal of Geophysical Research*, v. 88, p. 1153–1172.

Dewey, J. F., and J. M. Bird. 1970. Mountain belts and new global tectonics. *Journal of Geophysical Research*, v. 75, p. 2625–2647.

Hatcher, R. D., Jr. 1987. Tectonics of the Southern and Central Appalachian Internides. *Annual Review of Earth and Planetary Sciences*, v. 15, p. 337–362.

Hatcher, R. D., Jr., and S. A. Goldberg. 1991. The Blue Ridge geologic province, in *The Geology of the Carolinas*, edited by J. W. Horton, Jr., and V. A. Zullo, Carolina Geologic Society, 15th anniversary volume. Knoxville: University of Tennessee Press, p. 11–35.

Lillie, R. J. 1984. Tectonic implications of subthrust structures revealed by seismic profiling of Appalachian/Ouachita orogenic belt. *Tectonics*, v. 3, p. 619–646.

Lillie, R. J., G. D. Johnson, M. Yousuf, A. S. H. Zamin, and R. S. Yeats. 1987. Structural development within the Himalayan foreland fold-and-thrust belt of Pakistan, in *Sedimentary Basins and Basin Forming Mechanisms*, edited by C. Beaumont and A. Tankard. Calgary, Alberta: Canadian Society of Petroleum Geologists, Memoir 12, p. 379–392.

Lillie, R. J., K. D. Nelson, B. deVoogd, J. A. Brewer, J. E. Oliver, L. D. Brown, S. Kaufman, and G. W. Viele. 1983. Crustal structure of Ouachita Mountains, Arkansas: A model based on integration of COCORP reflection profiles and regional geophysical data. *American Association of Petroleum Geologists Bulletin*, v. 67, p. 907–931.

Merschat, C. E., and L. S. Wiener. 1990. *Geology of Grenville-Age Basement and Younger Cover Rocks in the West Central Blue Ridge* (Field Trip Guidebook, North Carolina Geological Survey). Raleigh, NC: Carolina Geological Society. 42 pp.

Stewart, K. G., M. G. Adams, and C. H. Trupe (editors). 1997. *Paleozoic Structure, Metamorphism, and Tectonics of the Blue Ridge of Western North Carolina* (Field Trip Guide-book). Raleigh, NC: Carolina Geological Society. 101 pp.

Stockmal, G. S., C. Beaumont, and R. Boutilier. 1986, Geodynamic models of convergent margin tectonics: Transition from rifted margin to overthrust belt and consequences for foreland-basin development. *American Association of Petroleum Geologists Bulletin*, v. 70, p. 181–190.

Weiner, L. S., and C. E. Merschat. 1990. *Field Guidebook to the Geology of the Central Blue Ridge of North Carolina and the Spruce Pine Mining District* (revised edition). Raleigh: North Carolina Geological Survey. 24 pp.

Wilson, J. T. 1966. Did the Atlantic close and then re-open? *Nature*, v. 211, p. 676–681.

WEBSITES

National Park Service: www.nps.gov

Park Geology Tour:
www2.nature.nps.gov/grd/tour/index.htm
   Appalachian-Ouachita-Marathon:
      59. Acadia NP:
      www2.nature.nps.gov/grd/parks/acad/index.htm
      60. Big South Fork NRRA:
      www2.nature.nps.gov/grd/parks/biso/index.htm
      61. Blue Ridge Parkway:
      www2.nature.nps.gov/grd/parks/blri/index.htm
      62. Bluestone NSR&RA:
      www2.nature.nps.gov/grd/parks/neri/index.htm
      63. Chattahoochee River NRA:
      www.nps.gov/chat/index.htm
      64. Delaware Water Gap NRA:
      www2.nature.nps.gov/grd/parks/dewa/index.htm
      65. Gauley River NRA:
      www2.nature.nps.gov/grd/parks/neri/index.htm

      66. Great Smoky Mountains NP:
      www2.nature.nps.gov/grd/parks/grsm/index.htm
      67. Obed WSR:
      www.nps.gov/obed/index.htm
      68. Russell Cave NM:
      www2.nature.nps.gov/grd/parks/ruca/index.htm
      69. Shenandoah NP:
      www2.nature.nps.gov/grd/parks/shen/index.htm
      70. Hot Springs NP:
      www2.nature.nps.gov/grd/parks/hosp/index.htm
      71. Big Bend NP:
      www2.nature.nps.gov/grd/parks/bibe/index.htm
      72. Rio Grande WSR:
      www.nps.gov/rigr/index.htm
   Brooks Range:
      73. Cape Krusenstern NM:
      www2.nature.nps.gov/grd/parks/noaa/index.htm
      74. Gates of the Arctic NP:
      www2.nature.nps.gov/grd/parks/gaar/index.htm
      75. Kobuk Valley NP:
      www2.nature.nps.gov/grd/parks/noaa/index.htm
      76. Noatak NPres:
      www2.nature.nps.gov/grd/parks/noaa/index.htm

U.S. Geological Survey: www.usgs.gov

Geology in the Parks:
www2.nature.nps.gov/grd/usgsnps/project/home.html

PART IV

Transform Plate Boundaries

Dramatic landscapes form where one plate slides past another at a transform plate boundary. Chapter 7 presents the sheared up mountain ranges and effects of earthquakes seen in parks along the San Andreas Fault in California, and at parks along a similar boundary in the Caribbean Sea.

Long, sheared-up mountain ranges; narrow valleys that sometimes pond water; offset stream courses; jolting earthquakes—these are some of the dramatic results of one huge plate of Earth's outer shell sliding past another. National park lands provide an opportunity to observe evidence of such transform plate boundaries. In California, two national parks, two national monuments, a national seashore, and two national recreation areas are located in the broad zone of deformation between the North American and Pacific plates. U.S. park lands are also located in the Virgin Islands, where the Caribbean and North American plates slide past one another.

◀ **[Overleaf]**
Pinnacles National Monument, California. The volcanic rocks have been transported northwestward almost 200 miles (300 kilometers) along the San Andreas Fault, a major component of the transform plate boundary.

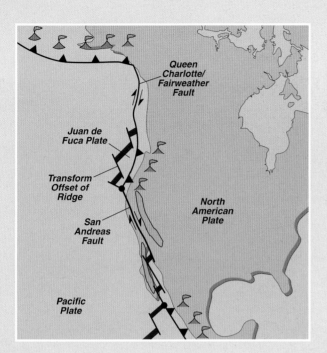

Transform plate boundaries occur where the Pacific Plate slides northward past the North American Plate. The San Andreas Fault cuts through western California, while the Queen Charlotte/Fairweather Fault affects western British Columbia and southeast Alaska. Smaller transform boundaries connect mid-ocean ridge segments between the Juan de Fuca and Pacific plates.

TEN QUESTIONS: Transform Plate Boundaries

1. What happens to the lithosphere at transform plate boundaries?

2. Why are such boundaries called "transform"?

3. Where are prominent transform plate boundaries that disrupt continental crust?

4. How destructive are earthquakes at transform plate boundaries?

5. How far can blocks of crust move at transform plate boundaries?

6. Will California fall into the ocean?

7. Is the San Andreas Fault the actual boundary between the North American and Pacific plates?

8. What is meant by the recurrence interval of earthquakes?

9. How much is the land surface offset during a big earthquake along the San Andreas Fault?

10. Why isn't there much volcanic activity at transform plate boundaries?

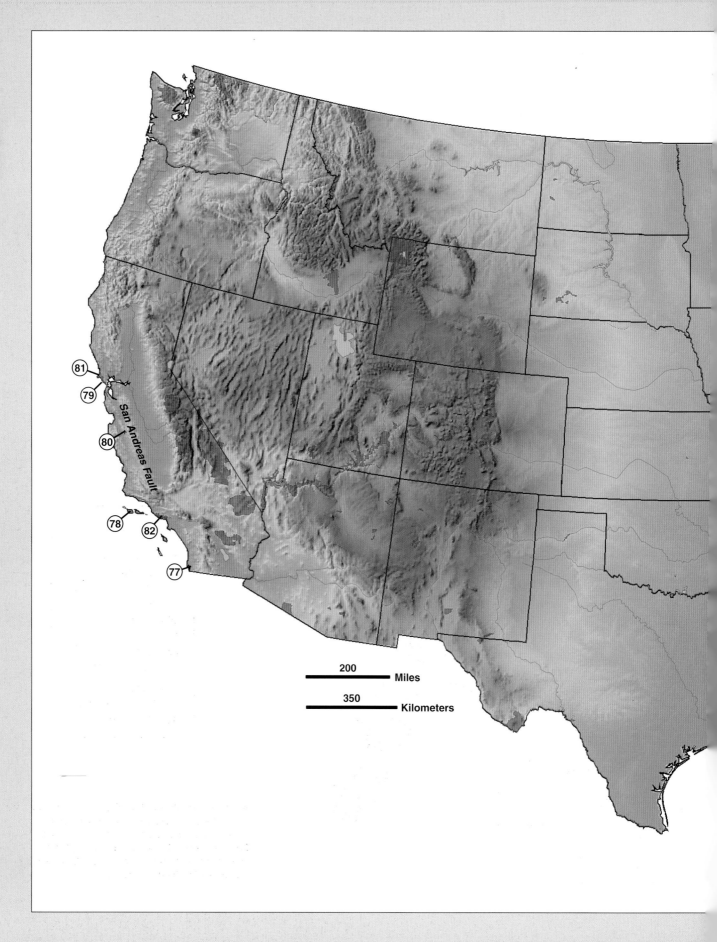

San Andreas Fault

81

79

80

78

82

77

200 Miles

350 Kilometers

Map Key

San Andreas Fault Zone:
77. Cabrillo NM
78. Channel Islands NP
79. Golden Gate NRA
80. Pinnacles NM
81. Point Reyes NS
82. Santa Monica
    Mountains NRA

Caribbean:
83. Buck Island Reef NM
84. Virgin Islands NP

US Virgin Islands

84

8
— Miles   **Caribbean**
20
— Kilometers

83

National park sites at transform plate boundaries.
Parks in western California lie along a broad zone of
deformation that includes the famous **San Andreas
Fault**. In the **U.S. Virgin Islands** region, the plate
boundary is in an oceanic realm as the Caribbean Plate
slips by the *North American Plate*.

7

San Andreas and Other Transform Boundaries

▲▲▲▲▲▲ *"The San Andreas Fault is creeping along at about 2 inches per year, accompanied by earthquakes, landslides, and the development of long slivers of mountain ranges. No, California will not fall into the ocean. But if you're patient, amazing and somewhat hilarious things will happen!"*

Where plates slide past one another, the landscape is sheared up into long slivers of mountain ranges with deep valleys in between. Perhaps nowhere on Earth is such a landscape more dramatically displayed than along the San Andreas Fault in western California (Fig. 7.1). A transform plate boundary is manifest by more than just a single strike-slip fault (Fig. 2.7c) on the surface. Rather, it is commonly a zone of many parallel faults that is tens to a hundred or so miles wide (Fig. 7.2). The landscapes of Channel

FIGURE 7.1 San Andreas Fault northeast of Santa Barbara, California. a) View looking southeast along the surface expression of the fault where it crosses the Carrizo Plain. **b)** View looking northeast. Offset stream channels reveal the right-lateral motion along the strike-slip fault.

a

b

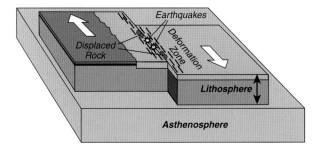

FIGURE 7.2 Transform plate boundary. The grinding action between the plates results in shallow earthquakes, large lateral displacement of rock, and a broad zone of crustal deformation.

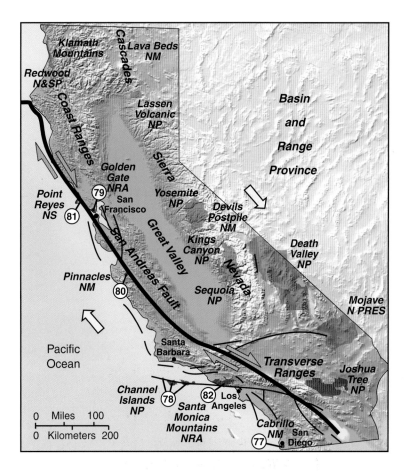

Islands National Park, Pinnacles National Monument, Point Reyes National Seashore, and other California park lands are products of such a broad zone of deformation, where the Pacific Plate moves north-northwest past North America (Table 7.1). In the Caribbean Sea, Virgin Islands National Park and Buck Island Reef National Monument are located on another transform plate boundary.

WESTERN CALIFORNIA

National park lands along the transform plate boundary in California contain rocks formed during earlier subduction that occurred in western North America (Fig. 7.3). As in modern subduction zones, the region had an accretionary wedge (Coast Range), a forearc basin (Great Valley), and a volcanic arc (Sierra Nevada). The accretionary wedge rocks are found in Channel Islands National Park, Golden Gate and Santa Monica Mountains national recreation

FIGURE 7.3 Shaded-relief map of California showing the broad zone of deformation caused by the north-northwest motion of the Pacific Plate past North America. Point Reyes National Seashore, Golden Gate National Recreation Area, and Pinnacles National Monument present landscapes affected by the main line of movement, the San Andreas Fault. Channel Islands National Park, Santa Monica Mountains National Recreation Area, and Joshua Tree National Park are within or near the Transverse Ranges, a block of crust that rotated as a result of the shearing motion. Cabrillo National Monument also lies within the broad zone of deformation between the two plates. The shearing motion along the transform boundary may be one of the factors responsible for continental rifting in the adjacent Basin and Range Province. Rocks in the Coast Ranges and Sierra Nevada formed during an earlier episode of subduction that affected all of California. The parks are numbered as in Table 7.1. Joshua Tree National Park was included with continental rift zone parks (Chap. 3).

TABLE 7.1 National parks at transform plate boundaries

CONTINENTAL	OCEANIC
SAN ANDREAS FAULT	**CARIBBEAN**
77. Cabrillo NM, CA	83. Buck Island Reef NM, US Virgin Islands
78. Channel Islands NP, CA	84. Virgin Islands NP, US Virgin Islands
79. Golden Gate NRA, CA	
80. Pinnacles NM, CA	
81. Point Reyes NS, CA	
82. Santa Monica Mountains NRA, CA	

FIGURE 7.4 Cabrillo National Monument. Tilted sedimentary layers were originally part of the accretionary wedge related to the Farallon Plate subduction.

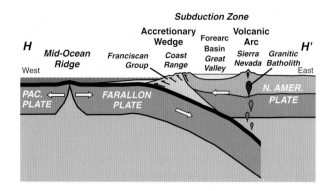

FIGURE 7.6 Plate-tectonic cross section of western California 40 million years ago. (Line H–H' in Fig. 7.5a.) National park lands contain rocks and structures formed during subduction of the Farallon Plate beneath North America. The Coast Range was an accretionary wedge of uplifted oceanic rocks known in northern California as the Franciscan Group. The Sierra Nevada was a volcanic arc, where lava flows and other volcanic materials were deposited on the surface while coarse-grained, granitic-batholith rocks developed in magma chambers within the Earth.

areas, and Cabrillo National Monument (Fig. 7.4). Point Reyes National Seashore and Joshua Tree National Park have granitic, magma-chamber rocks of the eroded arc, and Pinnacles National Monument preserves volcanic rocks. Rocks have been disrupted by shearing and other forces associated with the transform plate motion and, in some instances, transported northward a long distance from where they originally formed.

The San Andreas and other strike-slip faults accommodate the relative motion between the Pacific and North American plates in California. But the effects of the motion were first manifested only about 25 million years ago when those two plates first touched.

FIGURE 7.5 Development of the San Andreas Fault. a) Subducting plate boundary. The mid-ocean ridge system approaches North America as subduction consumes the Farallon Plate. H–H' is the line of the cross section in Fig. 7.6. **b)** Transform plate boundary. The mid-ocean ridge reaches the subduction zone. Pacific and North American plates make initial contact as the Farallon Plate separates into the Juan de Fuca and Cocos plates. The north-northwestward motion of the Pacific Plate past North America begins to form the sheared-up landscape of western California. **c)** Modern plate boundary. The transform boundary gets longer as the Juan de Fuca and Cocos plates grow smaller. The San Andreas Fault takes up most of the relative motion in the broad zone of shearing between the Pacific and North American plates in California.

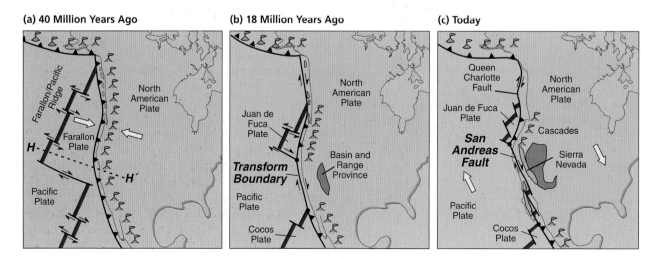

Prior to that contact, another plate was sandwiched between the Pacific and North American plates (Fig. 7.5a). The Farallon Plate was growing along a mid-ocean ridge system on its western edge and was being recycled into the deeper mantle at a subduction zone on its eastern side (Fig. 7.6). The subduction resulted in a volcanic mountain range paralleling the entire West Coast, from Canada all the way to Central America. The granitic rocks of Sequoia, Kings Canyon, and Yosemite national parks are the eroded remnants of the magma chambers that fed the volcanoes (Chap. 5).

The relative speeds of the three plates involved were such that North America began to overtake the mid-ocean ridge about 25 million years ago (Fig. 7.5b). That was the first time the North American and Pacific plates actually touched along the West Coast, creating the San Andreas transform plate boundary. The Farallon/Pacific Ridge system was not parallel to the subduction zone, so it did not subduct all at once. Rather, the Farallon Plate divided into two pieces, the Juan de Fuca Plate on the north and the Cocos Plate to the south. As more of the ridge system was consumed, the transform plate boundary grew longer; it now stretches 750 miles (1,200 kilometers) from the Gulf of California to Cape Mendocino (Fig. 7.5c).

A map of earthquake activity in California reveals that the San Andreas and other faults are related to motion along the transform plate boundary (Fig. 7.7). The patterns show that, like most of the mountain

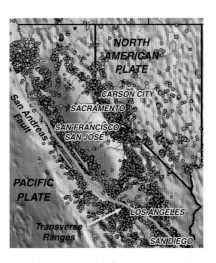

FIGURE 7.7 California earthquakes (1977–1996). Though it is a feature that involves considerable horizontal displacement, the San Andreas Fault is just one of many structures that accommodate the movement of the Pacific Plate past North America. The distribution of earthquakes highlights the broad zone of deformation between the two plates. Earthquakes related to uplift of the Transverse Ranges occur at the broad, east-west bend in the San Andreas Fault.

ranges and valleys, the fault lines trend in northwest-southeast directions, highlighting a broad zone that is deforming as the Pacific and North American plates grind past one another. Those trends are apparent in the topography of Point Reyes National Seashore, Golden Gate National Recreation Area, and Pinnacles National Monument, along the northern portion of the San Andreas Fault. In southern California, the earthquake fault lines bend to an east-west orientation, paralleling the topography of the Transverse Ranges, a block of rotated, compressed, and uplifted crust. Joshua Tree National Park, Santa Monica Mountains National Recreation Area, and Channel Islands National Park lie within or near the Transverse Ranges and their westward extension out into the Pacific Ocean.

Northern San Andreas Fault: A Sheared-Up Landscape

To visualize deformation along the transform plate boundary in California, place a deck of cards between your hands in a praying position (Fig. 7.8). Imagine that your left hand is the undeformed Pacific Plate, your right hand the intact North American Plate. Notice what happens as you move your left hand away and slide your right hand toward you. The cards slip along their faces, forming a broad zone of shearing between your unaffected hands. For western California, each slipping card face would be a fault surface—specifically, a strike-slip fault. The broad zone of transform motion between the Pacific and North American plates formed numerous slivers of mountain ranges with narrow valleys in between. The valleys are com-

---

THE GEOGRAPHIC HILARITY OF WESTERN CALIFORNIA

The San Andreas Fault is creeping along at about 2 inches (5 centimeters) per year, accompanied by earthquakes, landslides, and the development of long slivers of mountain ranges. No, California will not fall into the ocean. But if you're patient, amazing and somewhat hilarious things will happen!

- In about 10 million years, baseball purists can rejoice. The Dodgers and Giants will once again be in the same city, as Los Angeles will be a suburb of San Francisco!
- In 30 million years, residents of Oregon and Washington will no longer have to travel by interstate highway to get to Disneyland. Instead they can take a ferry boat, because Disneyland will be part of a resort island just off their coast!

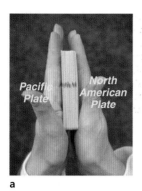

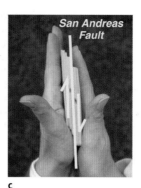

a   b   c

FIGURE 7.8 Zone of shearing at transform plate boundary. a) Hold deck of cards upright, with hands in praying position. Your left hand represents the Pacific Plate, and your right hand represents the North American Plate. **b)** Numerous strike-slip faults form as you move your hands laterally past one another. **c)** Imagine most of the motion being taken up along one of the card faces; that face would be the San Andreas Fault in California.

monly due to erosion along individual fault lines. Sometimes the valleys are partially filled with water, as at Point Reyes National Seashore, where Tomales Bay and Olema Valley follow the main trace of the San Andreas Fault (Fig. 7.9).

Strike-slip faults are classified as **right-lateral** or **left-lateral**, depending on which way one block moved relative to the other (Fig. 7.10). In western California, most of the strike-slip faults are right-lateral, mimicking the motion of the Pacific Plate relative to North America. The San Andreas Fault is a spectacular example, as you can see by taking a walk along the Earthquake Trail in Point Reyes National Seashore. Standing next to a broken fence line, you can see that the other part of the fence moved 16 feet (5 meters) to the right (Fig. 7.11).

The San Andreas Fault is just one of numerous faults in the broad transform boundary between the North American and Pacific plates (Fig. 7.12). But in recent times, movement along the San Andreas Fault has taken up the bulk of the motion between the two plates. The San Andreas Fault cuts through the San Francisco Bay area in such a way that Point Reyes

National Seashore is on the west side of the fault, while Golden Gate National Recreation Area (Fig. 7.13) is on the east. Because of right-lateral motion on the San Andreas Fault, if you could stand long enough in one of the parks, you could look across the fault and see the other park moving to the right!

ROCKS OUT OF PLACE. Both Point Reyes National Seashore and Pinnacles National Monument highlight the great movement along the San Andreas Fault because they contain rocks that seem out of place compared to nearby rocks. Both parks have rocks formed by igneous activity during the earlier subduction phase. Pinnacles has volcanic rocks, while Point Reyes has remnants of magma chambers much like those found in parks today in the High Sierras (Chap. 5).

A visit to Point Reyes National Seashore represents a step from North America onto the Pacific Plate. In fact, 20 million years ago, the land that is now Point Reyes Peninsula was actually much farther south. Tomales Bay represents the active plate boundary (Fig. 7.9b). Sedimentary and metamorphic rocks of the Franciscan Group are on the east side of the bay. Those

a   b

FIGURE 7.9 Point Reyes National Seashore. a) Point Reyes Peninsula is a mass of granite known as the Salinian block, part of the Sierra Nevada Batholith. **b)** View looking south-southeast from Tomales Point. Tomales Bay is the surface expression of the San Andreas Fault. The granite in the foreground is similar to rocks in the Tehachapi Mountains 310 miles (500 kilometers) to the southeast. Sedimentary and metamorphic rocks across the bay are part of the Franciscan Group accretionary wedge (Fig. 7.6).

(a) Right-Lateral

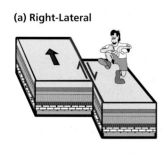

(b) Left-Lateral

FIGURE 7.10 To determine the sense of motion on a strike-slip fault, stand on one side of the fault and see which direction the other side moves. a) For a right-lateral strike-slip fault, the block on the other side moves to the right. **b)** In a left-lateral strike-slip fault, the block on the opposite side moves left. The transform plate boundary in western California has right-lateral motion, as do the San Andreas and most of the other strike-slip faults within the broad zone of deformation between the two plates (Figs. 7.3 and 7.12).

a

b

c

FIGURE 7.11 The April 18, 1906, San Francisco Earthquake—surface effects observed in and near Point Reyes National Seashore. a) Fresh fault surface about 6 miles (10 kilometers) south of Olema. **b)** Barn at Skinner Ranch, one mile west of Olema, looking northwest. The San Andreas Fault passes under the barn. The shed on the right moved about 15 feet (4½ meters) during the magnitude 7.8 earthquake. **c)** Disrupted ground along the fault line north of Skinner Ranch. **d)** A walk along the park's Earthquake Trail might be thought of as strolling back and forth between the North American and Pacific plates. The fence line was offset about 16 feet (5 meters) during the earthquake. The offset shows the right-lateral motion characteristic of most faults along the transform plate boundary.

d

THE "WORLD SERIES EARTHQUAKE": WAS THIS "THE BIG ONE"?

Just before Game Three of the 1989 World Series baseball game between the Oakland Athletics and the San Francisco Giants, an earthquake struck the Bay Area of California. The magnitude 6.9 earthquake was centered south of San Francisco, beneath a mountain called Loma Prieta, near Los Gatos. The earthquake resulted in 62 deaths and caused about $1 billion in property damage. Was this the big earthquake residents of California have been expecting since the last "big one," the magnitude 7.8 San Francisco Earthquake of 1906? If we consider the amount of energy released by a magnitude 7.8 earthquake compared to a 6.9, the answer is obvious: No way! It would take about 25 magnitude 6.9 earthquakes to release the same energy as a 7.8 (Fig. 2.24). In other words, the 1989 earthquake relieved only about 4% (or four years' worth) of the stress that had built up on the northern part of the San Andreas Fault system since the 1906 quake. The World Series Earthquake, while devastating in its own right, was only a wake-up call compared to the huge earthquake that will probably occur in the next few decades. (By the way, the A's won the series four games to none.)

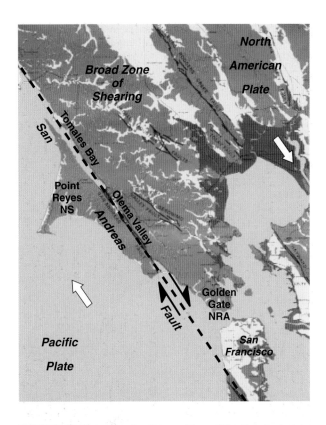

FIGURE 7.12 Map showing the position of the San Andreas Fault in the region north of San Francisco. The fault extends out to sea through the Olema Valley and Tomales Bay, the low-lying areas separating the Point Reyes Peninsula from the mainland. Most of Point Reyes National Seashore lies on the west (Pacific Plate) side of the fault, while Golden Gate National Recreation Area is mostly on the east (North American Plate) side.

rocks were originally sedimentary and volcanic layers deposited on top of the oceanic Farallon Plate. They were deformed and metamorphosed as the plate subducted tens of millions of years ago, forming a coastal mountain range (accretionary wedge) along the edge of North America (Fig. 7.6). The rocks to the west of Tomales Bay, however, are quite different—they include a lot of granite. Those rocks formed within magma chambers that lay beneath the volcanic mountain range (volcanic arc) that formed at the same time, but farther inland. When the Pacific and North American plates touched 25 million years ago, a sliver of western California began to slide northwestward along with the rest of the Pacific Plate. The granitic rocks that are now on the west side of Tomales Bay were plucked from a more southerly part of the volcanic arc and transported some 310 miles (500 kilometers) to their current position at Point Reyes National Seashore.

A similar story can be told for Pinnacles National Monument, about 125 miles (200 kilometers) southeast of San Francisco. But instead of granitic magma-chamber rocks, Pinnacles preserves remnants of ancient volcanoes. The rocks are mostly of high-silica, rhyolite composition. Age dating shows that they formed about 23 million years ago, during the transition from volcanic arc to transform motion in central and southern California. Rocks of similar age and composition near Lancaster and Gorman, called the Neenach Volcanics, suggest that the Pinnacles were transported about 190 miles (305 kilometers) along the San Andreas Fault (Fig. 7.14).

FIGURE 7.13 Golden Gate National Recreation Area. The park contains Franciscan Group rocks that were manufactured in the ocean, then deformed and uplifted as part of the accretionary wedge during the earlier subduction phase (Fig. 7.6). **a)** View across the Golden Gate Bridge, from the Marin Headlands to the city of San Francisco. **b)** The Marin Headlands consist of sandstone, chert, and pillow basalt formed on top of the crust of the oceanic Farallon Plate. **c)** The view of the Golden Gate from downtown San Francisco. Alcatraz Island is made of Early Cretaceous–age sandstone.

a
b
c

FIGURE 7.14 Parks in western California contain blocks of crust that have moved great distances northwestward along the San Andreas Fault. Volcanic rocks at Pinnacles National Monument were displaced about 190 miles (305 kilometers). The granitic rocks of Point Reyes National Seashore have moved about 310 miles (500 kilometers) northward along the San Andreas Fault.

Approaching Pinnacles National Monument, you can see impressive spires rising from the farmland of west-central California (Fig. 7.15). A hike along the park's spectacular trail system reveals that much of Pinnacles is massive rock with cobblestone-sized, angular blocks. The rock layers are similar to the sedimentary rock conglomerate (Table 2.4), but their angular form and volcanic origin reveal that they are really a rock called **volcanic breccia**. The rock fractures along vertical cracks, resulting in the inspiring Pinnacles in the High Peaks area of the park. Huge boulders have broken off and created roofs over the cracks, forming Balconies Caves.

Transverse Ranges: A Ball-Bearing Landscape

The **Transverse Ranges** are so named because they trend in an east-west direction, contrary to the northwest-southeast orientation typical of other ranges within the transform plate boundary region. The orientation illustrates how blocks of crust can rotate, like ball bearings, when they are caught in the vise between the sliding Pacific and North American plates (Fig. 7.16). Block rotation of the Transverse Ranges began with faulting about 16 million years ago. The rotated block corresponds to an abrupt east-west bend in the San Andreas Fault (Fig.

FIGURE 7.15 Pinnacles National Monument. a) The Pinnacles are volcanic rocks that have been cracked along vertical joints and eroded into steep spires. **b)** Balconies Caves resulted from erosion of the vertical joints in the volcanic breccia; high boulders have fallen across the cracks, forming the cave roof. **c)** Spires of volcanic breccia along the High Peaks Trail. **d)** Close-up of volcanic breccia, showing angular, cobblestone-sized particles.

(a) 20 Million Years Ago

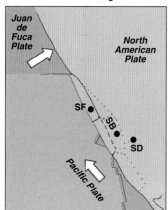

(b) 16 Million Years Ago

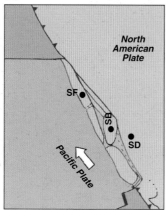

(c) 8 Million Years Ago

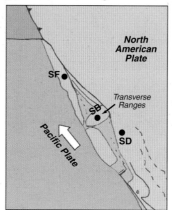

(d) Today

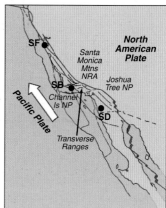

FIGURE 7.16 Development of Transverse Ranges. The crustal block marked "SB" (for the city of Santa Barbara) rolls like a ball bearing, trapped within the zone of deformation between the Pacific and North American plates. Clockwise rotation of the block explains why the Transverse Ranges run east-west, unlike the northwest-southeast orientation of most mountain ranges in western California. Santa Monica Mountains National Recreation Area lies within the Transverse Ranges, while Channel Islands National Park is part of the range that has been covered with rising waters of the Pacific Ocean since the last ice age. Part of Joshua Tree National Park lies on the eastern end of the range. "SF" = San Francisco; "SD" = San Diego.

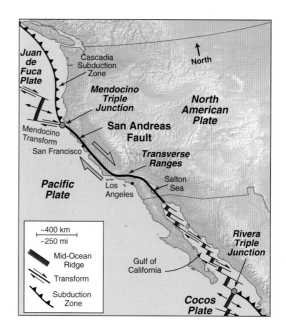

FIGURE 7.17 The Transverse Ranges form because of north-south compression where the San Andreas Fault bends to an east-west orientation. The fault currently takes up most of the transform plate motion as the Pacific Plate slides north-northwestward past the North American Plate. The fault extends from the Mendocino Triple Junction, where the Juan de Fuca, Pacific, and North American plates meet, through southwestern California to the Salton Sea. Beyond that, the Gulf of California is a new ocean basin opening along small mid-ocean ridge and transform segments. The Rivera Triple Junction is the common point among the Pacific, North American, and Cocos plates.

EARTHQUAKES: A FRAME-BY-FRAME ANALYSIS

Geologists draw diagrams of the western sliver of California gradually sliding by the rest of the state along the San Andreas fault zone. But motion is not so gradual—it's more like a movie. In a movie, one frame every ⅛ second is enough to create the perception that motion is smooth. But in reality, we are sensing the discrete progression of eight of those frames every second. The San Andreas Fault acts in much the same way, only each "frame" appears about every 100 to 150 years. During that time, called the **recurrence interval**, there is

little or no motion along the fault. The jump to the next frame is abrupt, with one side of the fault moving 10 to 20 feet (3 to 6 meters). That abrupt jump from one frame to another is a colossal earthquake, like the one that broke the northern segment of the fault during the 1906 San Francisco Earthquake. The offset fence line along the Earthquake Trail in Point Reyes National Seashore (Fig. 7.11d) reveals only the motion from one frame to the next in a grand "geologic movie."

a

b

c

FIGURE 7.18 Channel Islands National Park. a) The sedimentary layers were deposited on the ocean floor, on top of the ancient Farallon Plate. **b)** They were later scraped off the subducting plate as part of the accretionary wedge (Coast Range), and then **c)** rotated, compressed, and uplifted as part of the Transverse Ranges (Fig. 7.16).

7.17). The east-west orientation of the block means that the region has recently been subjected to north-south compression, uplifting the Transverse Ranges along reverse faults. Uplift of the offshore region of the block began about 2 million years ago, causing mountains to rise out of the sea as the Channel Islands.

Channel Islands National Park includes five islands off the coast of southern California: Anacapa, Santa Cruz, Santa Rosa, San Miguel, and Santa Barbara (Fig. 7.18). The islands are tops of a submerged portion of the Transverse Ranges, which include Santa Monica National Recreation Area and, on their eastern end, part of Joshua Tree National Park. The Channel Islands contain sedimentary layers and pillow lavas that formed on the ocean floor. Like many of the rocks that are caught up in the zone of transform motion between the Pacific and North American plates, the rocks at Channel Islands were deformed as part of the accretionary wedge during earlier subduction of the Farallon Plate. Onshore, Santa Monica Mountains National Recreation Area spans the Transverse Ranges north of Los Angeles (Fig. 7.19). Farther east, Joshua Tree National Park contains rocks and structures related to three tectonic provinces (Fig. 7.20). Granitic rocks are part of the Sierra Nevada Batholith, a product of ancient subduction (Chap. 5). Those rocks have been cut by normal faults related to Basin and Range continental rifting (Chap. 3), as well as strike-slip faulting and block rotation caused by the transform plate motion. The region has been further deformed by the compression that formed the Transverse Ranges.

FIGURE 7.19 Santa Monica Mountains National Recreation Area, Circle X Ranch. The east-west orientation of the Transverse Ranges was caused by rotation of a crustal block at the bend in the San Andreas Fault.

FIGURE 7.20 Joshua Tree National Park. a) The park landscape combines effects of subduction (Chap. 5), followed by ongoing continental rifting of the Basin and Range Province (Chap. 3) and transform motion along the San Andreas Fault. **b)** The prominent granite boulders are remnants of magma chambers that formed during subduction of the Farallon Plate (Figs. 7.5a and 7.6).

a

b

CARIBBEAN REGION

The San Andreas Fault Zone is not the only active transform plate boundary within U.S. national park lands. Southeast of Florida, the Caribbean Plate is sliding east-northeast at about 0.8 inch (2 centimeters) per year relative to the North American Plate (Fig. 7.21). Both plates are capped by oceanic crust. Farther east, the North American Plate is subducting, forming the Lesser Antilles Island Arc. Transform boundaries occur on the north and south sides of the Caribbean Plate. The motion on the north is not pure transform; there is some convergence that contributes to uplift of the topography. The long mountain ridges and narrow bays in the region surrounding U.S. Virgin Islands National Park (Fig. 7.22) are a product of compression (due to the convergence) in addition to left-lateral, strike-slip faulting (due to the transform motion).

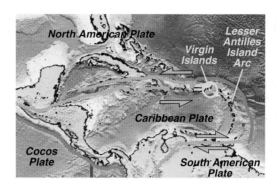

FIGURE 7.21 Tectonics of the Caribbean Sea region. The Virgin Islands lie along a zone of left-lateral, strike-slip faults defining the transform plate boundary between the Caribbean and North American plates.

FIGURE 7.22 National park lands in the Virgin Islands.
a) Cinnamon Bay in U.S. Virgin Islands National Park. The mountainous landscape is a product of both strike-slip faulting and compression, as the Caribbean Plate slides east-northeast past the North American Plate. **b)** Buck Island Reef National Monument, where coral and fish life thrive in the warm tropical waters.

a

b

FURTHER READING

GENERAL

Alt, D. D., and D. W. Hyndman. 1975. *Roadside Geology of Northern California*. Missoula, MT: Mountain Press. 244 pp.

Anderson, D. L. 1971. The San Andreas Fault. *Scientific American*, v. 225, no. 5, p. 52–68.

Bergen, F. W., H. J. Clifford, and S. G. Spear. 1997. *Geology of San Diego County: Legacy of the Land*. San Diego: Sunbelt Publications. 175 pp.

Bolton, P. A. (editor). 1993. *The Loma Prieta, California, Earthquake of October 17, 1989—Public Response*. Washington: U.S. Geological Survey. Professional Paper 1553-B, 69 pp.

Chronic, H. 1986. *Sierra Nevada, Cascades, and Pacific Coast. Pages of Stone: Geology of Western National Parks and Monuments*, Book 2. Seattle: Mountaineers. 170 pp.

Collier, M. 1999. *A Land in Motion: California's San Andreas Fault*. San Francisco: Golden Gate National Parks Association. 118 pp.

Decker, B., and R. Decker. 1994. *Road Guide to Joshua Tree National Park*. Mariposa, CA: Double Decker Press. 48 pp.

Johnson, E. R., and R. P. Cordone. 1994. *Pinnacles Guide: Pinnacles National Monument, San Benito County, California*. Stanwood, WA: Tillicum Press. 64 pp.

Konigsmark, T. 1998. *Geologic Trips: San Francisco and the Bay Area*. Guallala, CA: GeoPress. 174 pp.

Lacopi, R. L. 1995. *Earthquake Country*. Boulder, CO: Perseus. 152 pp.

McPhee, J. 1993. *Assembling California*. New York: Farrar, Straus, and Giroux. 304 pp.

Thomas, G., and M. M. Witts. 1971. *The San Francisco Earthquake*. New York: Dell. 301 pp.

Yanev, P. I. 1991. *Peace of Mind in Earthquake Country: How to Save Your Home and Life*. San Francisco: Chronicle Books. 218 pp.

Yeats, R. S. 2001. *Living with Earthquakes in California: A Survivor's Guide*. Corvallis: Oregon State University Press. 406 pp.

TECHNICAL

Atwater, T. 1970. Implications of plate tectonics for the Cenozoic tectonic evolution of western North America. *Geological Society of America Bulletin*, v. 81, p. 3513–3536.

Brune, J. N., T. L. Henyey, and R. F. Roy. 1969. Heat flow, stress, and rate of slip along the San Andreas Fault, California. *Journal of Geophysical Research*, v. 74, p. 3821–3827.

Galloway, A. J. 1977. Geology of the Point Reyes Peninsula, Marin County, California. *California Division of Mines and Geology Bulletin*, v. 202, 72 pp.

Healy, J. H., and L. G. Peake. 1975. Seismic velocity structure along a section of the San Andreas Fault near Bear Valley, California. *Bulletin Seismological Society of America*, v. 65, p. 1177–1197.

Hill, M. L. 1981. San Andreas Fault: History of concepts. *Geological Society of America Bulletin*, v. 92, p. 112–131.

Huftile, G. J. 1991. Thin-skinned tectonics of the upper Ojai Valley and Sulphur Mountain area, Ventura Basin, California. *American Association of Petroleum Geologists Bulletin*, v. 75, p. 1353–1373.

Ladd, J., T. Holcombe, G. Westbrook, and N. T. Edgar. 1990. Caribbean marine geology: Active margins of the plate boundary, in *The Geology of North America*, v. H: *The Caribbean Region*, Boulder, CO: Geological Society edited by G. Dengo and J. Case. America. p. 261–290.

Lemiszki, P. J., and L. D. Brown. 1988. Variable crustal structure of strike-slip fault zones as observed on deep seismic reflection profiles. *Geological Society of America Bulletin*, v. 100, p. 665–676.

Luyendyk, B. P., M. J. Kammerling, and R. Terres. 1980. Geometric model for Neogene crustal rotations in southern California. *Geological Society of America Bulletin*, v. 91, p. 211–217.

Mann, P., E. Calais, J. Ruegg, C. DeMets, P. E. Jansma, and G. S. Mattioli. 2002. Oblique collision in the northeastern Caribbean from GPS measurements and geological observations. *Tectonics*, v. 21, p. 7.1–7.26.

Mooney, W. D., and R. H. Colburn. 1985. A seismic-refraction profile across the San Andreas, Sargent, and Calaveras faults, west-central California. *Bulletin Seismological Society of America*, v. 75, p. 175–191.

Norris, R. M., and R. W. Webb. 1990. *Geology of California*. New York: Wiley. 541 pp.

Schiffman, P., and D. L. Wagner (editors). 1992. *Field Guide to the Geology and Metamorphism of the Franciscan Complex and Western Metamorphic Belt of Northern California*. Sacramento, CA: California Division of Mines and Geology, Special Publication 114, 78 pp.

Simpson, R. (editor). 1994. *The Loma Prieta, California Earthquake of October 17, 1989: Tectonic Processes and Models*. U.S. Geological Survey, Professional Paper 1550-F. 131 pp.

Sylvester, A. G. 1988. Strike-slip faults. *Geological Society of America Bulletin*, v. 100, p. 1666–1703.

Trent, D. D. 1984. Geology of the Joshua Tree National Monument, Riverside and San Bernardino Counties. *California Geology*, v. 37, p. 75–86.

Wallace, R. (editor). 1990. *The San Andreas Fault System, California*. U.S. Geological Survey, Professional Paper 1515. 283 pp.

Wilson, J. T. 1965. A new class of faults and their bearing on continental drift. *Nature*, v. 207, p. 343–347.

Yeats, R. S., and K. R. Berryman. 1987. South Island, New Zealand, and the Transverse Ranges, California: A seismotectonic comparison. *Tectonics*, v. 6, p. 363–376.

Yeats, R. S., K. E. Sieh, and C. R. Allen. 1997. *The Geology of Earthquakes*. New York: Oxford University Press. 568 pp.

WEBSITES

National Park Service: www.nps.gov

Park Geology Tour:
www2.nature.nps.gov/grd/tour/index.htm
  San Andreas Fault:
    77. Cabrillo NM:
    www.nps.gov/cabr/index.htm
    78. Channel Islands NP:
    www2.nature.nps.gov/grd/parks/chis/index.htm
    79. Golden Gate NRA:
    www2.nature.nps.gov/grd/parks/goga/index.htm
    80. Point Reyes NS:
    www2.nature.nps.gov/grd/parks/pore/index.htm
    81. Pinnacles NM:
    www2.nature.nps.gov/grd/parks/pinn/index.htm
    82. Santa Monica Mountains NRA:
    www2.nature.nps.gov/grd/parks/samo/index.htm
  Caribbean:
    83. Virgin Islands NP:
    www2.nature.nps.gov/grd/parks/viis/index.htm
    84. Buck Island Reef NM:
    www2.nature.nps.gov/grd/parks/buis/index.htm

U.S. Geological Survey: www.usgs.gov

Geology:
geology.usgs.gov/index.shtml

San Andreas Fault:
pubs.usgs.gov/gip/earthq3

Earthquakes:
earthquakes.usgs.gov
  National Earthquake Information Center:
  neic.usgs.gov/neis/states/states.html
  Largest U.S. Earthquakes:
  wwwneic.cr.usgs.gov/neis/eqlists/10maps_usa.html
  1906 San Francisco Earthquake:
  wwwneic.cr.usgs.gov/neis/eqlists/USA/
    1906_04_18.html
  1989 Loma Prieta Earthquake:
  geopubs.wr.usgs.gov/dds/dds-29

University of California–Santa Barbara

Plate Tectonic Animations:
transfer.lsit.ucsb.edu/geol/projects/emvc/
cgi-bin/dc/list.cgi?list

University of Texas

Institute for Geophysics:
www.ig.utexas.edu

Plates Project:
www.ig.utexas.edu/research/projects/plates/
plates.htm

PART V
Hotspots

Previous chapters illustrated geological features in national parks formed by processes occurring where plates interact along their boundaries. Spectacular scenery, as well as volcanic and earthquake activity, can also occur within the interiors of plates, far from their boundaries. Chapter 8 shows volcanic landscapes formed as a result of the oceanic Pacific Plate riding over the Hawaiian Hotspot. Parks highlighted in Chapter 9 reveal geologic features formed where magmas from the Yellowstone Hotspot melted their way through the much thicker continental crust of the North American Plate.

167

Some chains of volcanoes lie within the interiors of lithospheric plates rather than along the edges. The farther the volcanoes are from the largest and most active volcano, the older they are. Such chains are thought to form over narrow "plumes" of hot material rising from deep within the Earth. A **hotspot** is a region of hot mantle where magma forms above a plume. A line of volcanoes develops as a lithospheric plate moves over a hotspot, much as a line of melted wax forms as a sheet of waxed paper is moved slowly over a burning candle.

Hotspots move very little relative to deeper parts of Earth's mantle. Hence they provide a simple framework to track the motion of lithospheric plates. One volcano after another forms on a plate's surface as the plate moves over the hotspot. The resulting chain of volcanoes is: 1) parallel to the direction of plate motion; and 2) progressively older the farther away it is from the hotspot.

Geologists can calculate the direction and speed of the plate by mapping the orientation and changing age of volcanism. A hotspot thus provides a framework by which the **absolute motion** of a plate can be calculated (that is, motion with respect to the nearly stationary deep mantle). Such calculations are important because at plate boundaries, we can determine only the **relative motion** of one plate compared to that of the adjoining plate.

◀ [Overleaf]
Hawai`i Volcanoes National Park. Fluid lava erupts as the Pacific Plate moves over the Hawaiian Hotspot.

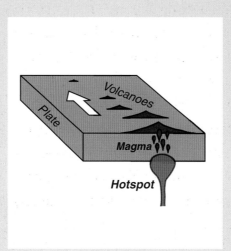

Hotspot volcanism is like moving a sheet of waxed paper over a burning candle. As a lithospheric plate moves over a hotspot, a trail of volcanoes develops on the surface of the plate.

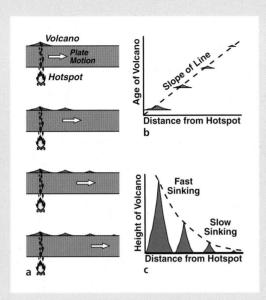

A hotspot reveals absolute plate motion. a) A volcano develops on top of a plate, directly above a hotspot. Other volcanoes form in a line as the plate moves, defining the direction of plate motion. **b)** In a graph showing the age of volcanism compared to the distance from the hotspot to each volcano, the slope of the line reveals the speed that the plate has moved over the deeper mantle. **c)** Graph showing that volcanoes initially sink at a fast rate as they move off a hotspot, then more slowly as they drift away.

Two prominent hotspot tracks appear on a map of the 50 United States, one involving a plate with thin oceanic crust, and one with thicker continental crust. The Hawaiian Islands are broad and high at the southeast, becoming smaller and lower to the northwest. This pattern continues as submerged seamounts extending far beyond the islands. The Hawaiian Ridge–Emperor Seamount chain reveals that the Pacific Plate has been moving over a hotspot, first in a northerly direction, then northwestward. Two national parks, Haleakalā on Maui and Hawai`i Volcanoes on the big island called Hawai`i, represent different stages of passage of the plate over the hotspot. On the North American continent, the Snake River Plain of southern Idaho connects the Columbia Plateau region of southeastern Washington and northeastern Oregon with Yellowstone National Park in the northwest corner of Wyoming. Volcanic rocks in the plain are thought to be the result of movement of the North American continent over a hotspot. The volcanic landscape of Craters of the Moon National Monument in Idaho illustrates the waning stages of the same hotspot activity that now causes the geysers and other geothermal activity so spectacularly displayed in Yellowstone National Park.

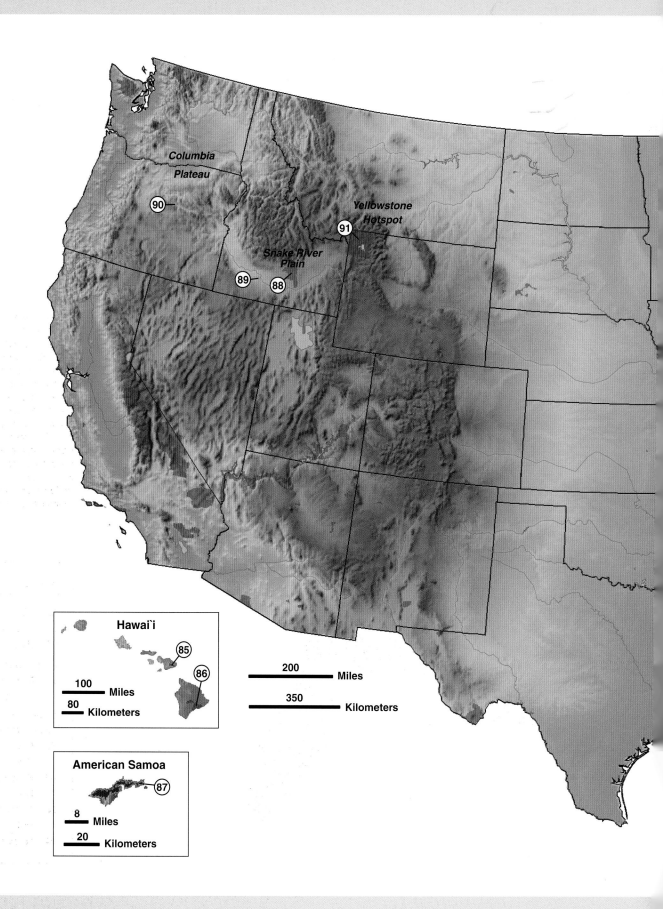

Columbia
Plateau

90

Yellowstone
Hotspot

91

Snake River
Plain

89 88

Hawai`i

85

86

100
——— Miles
80
——— Kilometers

American Samoa

87

8
——— Miles
20
——— Kilometers

200
——— Miles
350
——— Kilometers

Shaded-relief map showing national park sites that lie above hotspots, or within volcanic terranes formed as lithospheric plates moved over hotspots. Parks in **Hawai`i** and **American Samoa** are on thin oceanic crust, whereas thicker continental crust is associated with the hotspot track in the **Columbia Plateau** of Oregon and Washington, the **Snake River Plain** of Idaho, and the **Yellowstone** region where Idaho, Montana, and Wyoming meet.

Map Key
Hawai`i:
85. Haleakalā NP
86. Hawai`i Volcanoes NP
American Samoa:
87. NP of American Samoa.
Yellowstone/Snake River Plain/Columbia Plateau:
88. Craters of the Moon NM
89. Hagerman Fossil Beds NM
90. John Day Fossil Beds NM
91. Yellowstone NP

Hawai`i Volcanoes National Park.
The 1959 eruption of Kīlauea Iki. The fluid, basaltic lava is a product of hotspot activity in an oceanic area.

TEN QUESTIONS: Hotspots

1. What causes hotspots?

2. Why does material melt at hotspots?

3. Besides Hawai`i and Yellowstone, where are other hotspots in the world?

4. Why is Hawai`i a line of volcanic islands, rather than a long, continuous ridge of volcanic material?

5. Why do the Hawaiian Islands decrease in elevation from southeast to northwest?

6. How and why does the volcanism in Yellowstone National Park differ from that seen at parks in Hawai`i?

7. Why have there been recent volcanic eruptions at Haleakalā National Park and Craters of the Moon National Monument?

8. Why are the recent volcanic rocks of Craters of the Moon National Monument so different from those at Yellowstone?

9. Why are there geysers, hot springs, and other geothermal features at Yellowstone?

10. How do Yellowstone geysers work?

8

Oceanic Hotspots

"The Pacific Plate is moving west-northwestward over the Hawaiian Hotspot at about 4 inches per year. Although this movement is slow, from a human perspective, it adds up. Over a person's lifetime of 70 years, an island moves about 20 feet."

Inspiring views of island formation from Haleakalā. Red-hot lava and vast moonlike craters at Hawai`i Volcanoes. The national parks of Hawai`i provide an exceptional glimpse at landscapes developing above an oceanic hotspot (Fig. 8.1 and Table 8.1). As Mark Twain wrote of Hawai`i, "This lava is the accumulation of ages; one torrent of fire after another has rolled down

**FIGURE 8.1 Hawai`i Volcanoes National Park.** The ongoing eruption through the east rift zone of Kīlauea Volcano has produced a cinder and spatter cone, called Pu`u `Ō`ō, and fluid, basaltic lava flows. Much of the lava goes into lava tubes, where it flows to the ocean.

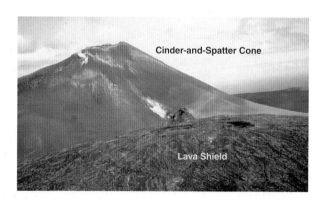

TABLE 8.1 National parks at oceanic hotspots

HAWAII	WESTERN PACIFIC
85. Haleakalā NP, HI 86. Hawai`i Volcanoes NP, HI	87. National Park of American Samoa

here in old times, and built up the island structure higher and higher. Underneath, it is honeycombed with caves . . . "

A hotspot is thermally-expanded, buoyant mantle that lifts an overlying plate, much as hot asthenosphere causes uplift of the ocean floor at a mid-ocean ridge. As hotspot material rises, it decompresses and begins to melt; the resulting magma has low-silica (basalt) composition. Broad, gently sloping shield volcanoes develop on the seafloor as the plate moves over the hotspot (Fig. 8.2). If eruptions are voluminous enough, an island eventually pokes up out of the ocean.

Three important things happen as plate motion carries a volcanic island away from a hotspot. First, volcanism wanes then stops entirely, so that the island ceases to grow. The second effect is that erosion wears away at the island. Finally, the underlying mantle cools and contracts, so that the island and surrounding ocean floor sink down. All three factors tend to make islands smaller and lower as they move away from a hotspot. A large volcanic island progresses to a smaller, lower island, then to an atoll barely above the sea, and finally to a submerged seamount.

HAWAI'I-EMPEROR SEAMOUNT CHAIN

In the nineteenth century, geologists could see that the Hawaiian Islands were higher and less eroded toward the southeast. Most of the early geologists thought that volcanic activity had begun simultaneously along the entire island chain. By that interpretation, all the volcanoes would have been active for tens of millions of years, and then they progressively shut down from Midway Island toward Hawai'i.

It was not until the middle of the twentieth century that geologists agreed that volcanic activity appeared to have initiated from the northwest and advanced to the southeast. That meant that the life span of a particular volcano was only a million years or so, not tens of millions as the earlier idea would suggest. Mapping of the seafloor also revealed that the volcanic chain is much more extensive than just the islands (Fig. 8.3). It includes ancient, submerged volcanoes, known as the Hawaiian Ridge, and then the Emperor Seamount Chain that extends all the way to the Aleutian Trench southwest of Alaska. In the 1960s, with the advent of plate-tectonic theory and its hotspot corollary, the reason for the southeast progression of both initiation and cessation of volcanism became apparent. Volcanic activity begins, and an island grows, as a region of the Pacific Plate moves over the

Hawaiian Hotspot; volcanism gradually wanes, then ceases, as the island moves away from the hotspot.

Several observations make a convincing argument for the hotspot origin of the Hawaiian Ridge–Emperor Seamount Chain (Fig. 8.4). Three trends are apparent as you look northwestward: 1) The volcanic activity becomes progressively older; 2) the degree of erosion and development of vegetation and coral reefs in-

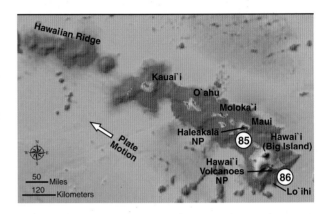

FIGURE 8.2 The Hawaiian Islands are only the tips of enormous shield volcanoes. Hawai'i Volcanoes National Park on Hawai'i (also known as "the Big Island") lies directly above the hotspot and displays high elevation and extensive volcanic activity. Haleakalā National Park displays the waning stages of volcanism that occurs on the lower topography of Maui, an island that has recently moved off the hotspot. The volcanoes get lower and older toward the northwest, sinking below sea level along the Hawaiian Ridge as they are carried farther from the hotspot by the motion of the Pacific Plate.

FIGURE 8.3 The Hawai'i–Emperor Seamount Chain. Prior to 42 million years ago, the Pacific Plate was moving northward, forming volcanic islands that are now the Emperor Seamounts. The plate motion has shifted to northwestward, forming the Hawaiian Ridge and Hawaiian Islands.

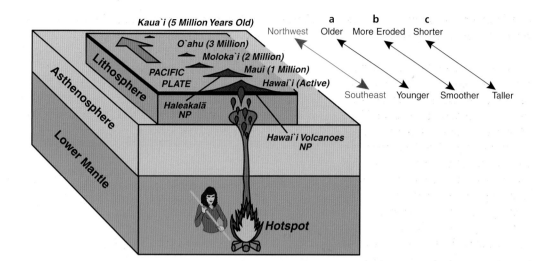

FIGURE 8.4 Evidence for the formation of the Hawaiian Islands over a hotspot. a) Volcanism becomes older toward the northwest. **b)** Erosion and vegetation increase from southeast to northwest. **c)** Elevations become lower toward the northwest. The Big Island of Hawai'i, including Hawai'i Volcanoes National Park, is currently directly above the hotspot, where it experiences periods of extensive volcanic activity. Haleakalā National Park, on the island of Maui, has moved off the hotspot but still has some eruptions.

creases; and 3) the islands, coral reefs, and seamounts become progressively lower.

The Pacific Plate is moving west-northwestward over the Hawaiian Hotspot at about 4 inches (10 centimeters) per year. Although this movement is slow from a human perspective, it adds up. Over a person's lifetime of 70 years, an island moves about 20 feet (7 meters). Geologists can estimate the distance across the top of the Hawaiian Hotspot by examining the re-

gion of the three most active volcanoes: Kīlauea and Mauna Loa on the Big Island, and the still-submerged Lo'ihi, off the island's south coast (Fig. 8.5). Those volcanoes lie within a circle of about 60-mile (100-kilometer) diameter. The time it takes for a region of the plate to move that distance over the hotspot is about a million years, the approximate period of fast, shield-building activity of a volcano. The distance from the Big Island to Kaua'i is about 300 miles (500

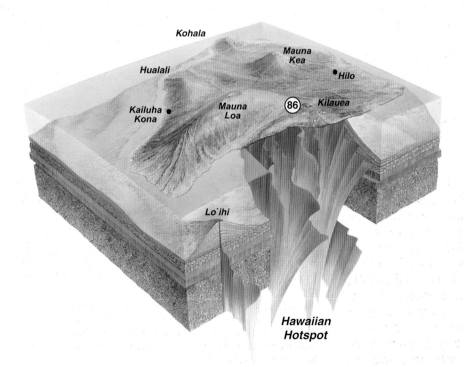

FIGURE 8.5 Hawaiian hotspot magma. Kīlauea Volcano lies directly above the hotspot and has most of the volcanic activity. Magma rising in vertical, sheetlike patterns results in fissure eruptions through rift zones out of the flanks of the volcano. Magma surfacing in other places results in eruptions of Mauna Loa, as well as development of a new volcano, Lo'ihi, off the southern coast of the island. 86 = Hawai'i Volcanoes National Park.

(a) Mauna Kea, Hawai`i

(b) Mauna Loa, Hawai`i

(c) Mt. Everest, Nepal/Tibet

(d) K-2, Pakistan/Tibet

(e) Haleakalā, Hawai`i

FIGURE 8.6 The world's five tallest mountains. a) The ~32,000-foot (9,750 meter) Mauna Kea (~18,000 feet below sea level to 13,796 feet above). **b)** The ~32,000-foot Mauna Loa (~18,000 feet below sea level to 13,679 feet above). **c)** The 29,035-foot (8,850 meters) Mount Everest. **d)** The 28,250-foot (8,611 meters) K-2. **e)** The ~28,000-foot Haleakalā (~18,000 feet below sea level to 10,023 feet above).

kilometers). At 4 inches (10 centimeters) per year, the island of Kaua`i moved over the hotspot about 5 million years ago. Over a span of 75 million years, the Pacific Plate moved about 4,500 miles (7,500 kilometers), the approximate length of the Hawai`i-Emperor Seamount Chain (Fig. 8.3).

Why Islands? Why Seamounts?

Plate-tectonic forces form large mountains in different ways. The highest elevations above sea level, the Himalayas, form where continents collide (Chap. 6). Those high elevations result from the buoyancy of Indian continental crust as it protrudes beneath Asia, as well as local relief due to compression of the continents as they collide, uplifting peaks such as Mount Everest and K-2 (Fig. 8.6). But a completely different process forms mountains within the Hawaiian Islands. Massive amounts of lava erupt as the Pacific Plate moves over the hotspot, building broad shield volcanoes on the ocean floor (Fig. 8.7). Unlike the Himalayas, formed via isostatic uplift and structural deformation, the mountains of the Hawaiian Islands formed from the stacking of lava flows during thousands of volcanic eruptions.

Discrete islands (or clusters of islands) form because of the "plumbing system" of a hotspot. As a plate moves over a hotspot, which is rooted in the deeper

TALLEST MOUNTAIN ON EARTH

From one perspective, the Hawaiian Islands contain three of the five tallest mountains on Earth (Fig. 8.6). Hawaiian volcanoes actually start on the ocean floor, some 18,000 feet (5,500 meters) below sea level. On the Big Island of Hawai`i, Mauna Kea and Mauna Loa rise 13,796 feet (4,205 meters) and 13,679 feet (4,169 meters) above sea level, respectively. Those two mountains are each about 32,000 feet (9,750 meters) from base to top. The top of the highest point on Earth, Mount Everest, is only 29,035 feet (8,850 meters) above sea level. If we consider that the base of Mount Everest is near sea level, then Mauna Kea and Mauna Loa are about 3,000 feet (1,000 meters) taller. Mount Everest is thereby the third-tallest mountain in the world, followed by K-2 at 28,250 feet (8,611 meters). Haleakalā, on the island of Maui, extends from 18,000 feet below sea level to 10,023 feet (3,055 meters) above, making it the fifth-tallest mountain in the world, about 28,000 feet (8,500 meters) high.

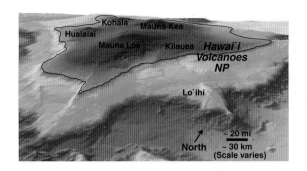

FIGURE 8.7 Oblique view of the island of Hawai`i from the south. When you look at the Hawaiian Islands, you see only a small portion of the volcanoes—the part above sea level. Compared to other types of mountain ranges, which typically have much steeper slopes from their bases to summits, the Hawaiian mountain chain contains an enormous amount of material. The Big Island is less than 10% of the surface area of a huge mass of interlocking shield volcanoes. The slopes are so gentle that the base of this intermingled volcano has a diameter of about 120 miles (200 kilometers), extending about 20 miles (30 kilometers) offshore from the island. The entire Sierra Nevada mountain range in California and Nevada contains about the same amount of material as the shield volcanoes that make up the Big Island of Hawai`i, from the submerged base to the summits of Mauna Kea and Mauna Loa. Hawai`i Volcanoes National Park, shown in red, includes parts of Kīlauea and Mauna Loa volcanoes.

mantle, the magma conduit bends, much like the flame of a candle in the wind (Fig. 8.8). As an active volcano growing on the plate moves away from the hotspot, the path up the conduit is so curved that it becomes difficult to continue erupting there. A conduit to the surface lasts 1 to 2 million years, until the hotspot magma takes a more direct, vertical path to the surface, forming a new island.

After they move away from their hotspot source, Hawaiian islands get progressively lower and smaller because of two factors, **thermal subsidence** and **erosion**. A hotspot causes the overlying mantle to expand like a hot-air balloon, elevating the nearby seafloor as an island grows (Fig. 8.9). When the island moves away from the hotspot, the mantle cools and shrinks, so that the seafloor and island subside. Islands lose height as they move away from the hotspot because the cooling plate becomes denser and sinks, similar to the ocean floor sinking as a cooling plate moves away from the heat of a mid-ocean ridge. Erosion also lowers the islands, as there are no more eruptions to supply a new coat of volcanic material.

POLYNESIANS AS INTERPRETERS

Ancient folklore about the formation of the Hawaiian Islands is not unlike what we envision today from scientific observation, at least in terms of the age progression of volcanism, island growth, and erosion:

Pele, the goddess of fire and volcanism, was originally born a mortal, the daughter of Haumea and Ku-waha-ilo. From her uncle, Kanaloa, she learned the art of fire making. But Pele was also mischievous. She was given a digging stick, called an ōō (pronounced "oh-oh"). She would dig and make fire from the Earth. One day her playing got out of hand and she burned down her family's house. Her sister, Na-maka-o-kaha`i, got very angry, exiling Pele to sea in a small boat. Using the ōō as a paddle, Pele made fire from the bottom of the ocean. The rock from the fire eventually built a pile above the sea, so that Pele had an island to live on, Kaua`i. Hearing what Pele was up to, Na-maka-o-kaha`i sent a large wave across the ocean, extinguishing the fire. Pele again rubbed her stick on the bottom of the ocean, forming another island, O`ahu. Na-maka-o-kaha`i sent another wave, extinguishing that island's fire.

This argument between sisters continued, as the fires built Moloka`i, Lana`i, and Maui. But each time Na-maka-okaha`i would send a wave to extinguish the fire. On Maui, Pele finally decided to stand up to her sister. A fierce battle ensued between these mortals, with Na-maka-o-kaha`i eventually killing Pele. But that was not enough for Na-maka-o-kaha`i. So intense was her anger and hatred for her sister that she tore Pele's body apart, depositing the remains on a hill near Hana, now called Ka`iwiō Pele ("Bones of Pele"). This was a serious mistake. Tearing Pele's body apart freed her spirit, so that she was no longer mortal, but a goddess. Pele's spirit lived on, as she built Mauna Kea, Mauna Loa, and other volcanoes on the Big Island of Hawai`i; she now resides in the fiery crater of Kīlauea.

It appears that ancient Hawaiians made observations, and provided some interpretations, that are still accurate by today's scientific account of the formation of the Hawaiian Islands: volcanic activity, extinction, and erosion have progressed in such a way that islands become older and lower from southeast to northwest.

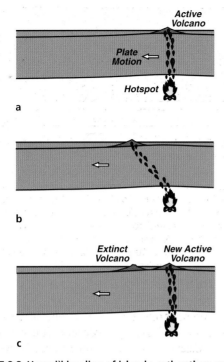

FIGURE 8.8 **Hawai`i is a line of islands, rather than one long, continuous ridge. a)** Hotspot magma rises through vertical conduit in the Pacific Plate. A volcanic island (or cluster of islands) forms above the conduit. **b)** As the Pacific Plate moves northwestward, it gradually bends the conduit away from the vertical. **c)** After a million years or so, it is easier for magma to rise vertically along a new conduit, rather than follow the indirect route up the older one.

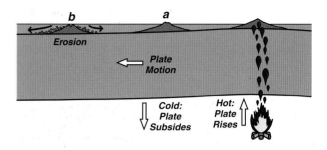

FIGURE 8.9 **Reasons why Hawaiian islands get lower and smaller through time. a)** Thermal subsidence. The surface of the plate, including the volcanic island, lowers as the hot mantle beneath the plate cools and contracts. **b)** Erosion. With no continued volcanic activity, water and wind gradually wear away the island and carry the material to the deep ocean.

Stages of Island Development

The islands, atolls, and seamounts of the Hawaiian Ridge–Emperor Seamount Chain illustrate stages of development as the Pacific Plate moves over the hotspot. To the southeast, the still-submerged Lo`ihi represents the initial stage. Hawai`i Volcanoes and Haleakalā national parks have features that reflect the second and third stages, respectively, as volcanic islands develop over the hotspot and then are transported away by the moving plate. Low-lying atolls and submerged volcanoes along the Hawaiian Ridge and Emperor

HOTSPOTS AND SPECIES EVOLUTION

An understanding of plate tectonics is important for appreciating natural selection and species population, propagation, and evolution. The Hawaiian Islands are the most isolated landmass on Earth, some 2,000 miles (3,000 kilometers) from the nearest continent. That makes it difficult for plants and animals to reach and populate the islands. When species do reach the islands, they are isolated in a unique environment, evolving along distinct lines. Before the arrival of the first people, ancient Polynesians about 1,700 years ago, many species on Hawai`i were **endemic**, meaning they existed nowhere else on Earth.

The Hawaiian Hotspot presents an intriguing twist to thoughts on species evolution. First, the islands have developed near the center of the huge Pacific Plate that is, to a large degree, covered by water. Second, the time that a given island grows and remains emergent (above the water) is limited. Studies show that an island remains emergent for perhaps 10 million years, never more than about 30 million. Species that arrived 50 million years ago, for example, cannot remain on that same island because the island is gone, submerged beneath about 13,000 feet (4,000 meters) of water as one of the Emperor Seamounts. If a species evolved to something that was strictly land-loving, it would die out unless it had a way to fly or swim southeastward to younger, emerging islands. It has been shown that there were times when no islands were emergent along the Hawai`i–Emperor chain. Many land plants and animals that had evolved would have become extinct during that time. An implication of hotspot theory is that certain land plants and animals had to have arrived on the Hawaiian Island chain in the past 30 million years, because there was a gap between 34 and 30 million years ago when no land was emergent.

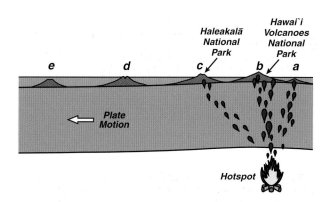

FIGURE 8.10 **Uplift, subsidence, and erosion of volcanoes as they override an oceanic hotspot. a) Seafloor volcano.** As the plate approaches the hotspot, lava pours out on the ocean floor. (Example: Lo'ihi Volcano.) **b) Broad, high island.** Land emerges above sea level as shield volcanoes develop. (Example: the island of Hawai'i.) **c) Smaller, lower island.** As the island moves away from the hotspot, volcanic activity wanes, then ceases. The island has lower topography because the cooler lithospheric plate contracts and sinks, and because erosion strips away material. (Example: the islands of Maui, Moloka'i, O'ahu, and Kaua'i.) **d) Atoll stage.** Thermal subsidence, erosion, and wave action lower the island to near sea level, where all is submerged except coral reefs. (Example: the Hawaiian Ridge.) **e) Seamount stage.** Continued plate cooling and contraction lowers the island below sea level. (Example: Emperor Seamounts.)

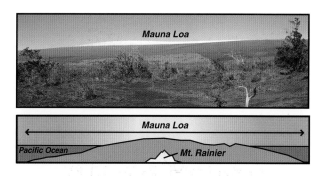

FIGURE 8.11 **Hawaiian shield volcanoes have enormous height and volume.** Mauna Loa (the "Long Mountain") starts 18,000 feet (5,500 meters) below sea level, and rises to nearly 14,000 feet (4,300 meters) above. It is about 100 miles (160 kilometers) wide. Mount Rainier, a composite volcano in the Cascade Mountains of Washington State (Chap. 5), is about 14,000 feet tall, and only about 10 miles (16 kilometers) wide.

Seamount Chain represent the last two stages caused by erosion and thermal subsidence some distance from the hotspot.

STAGE 1: SEAFLOOR VOLCANO. The surface of an old oceanic plate typically lies about 18,000 feet (5,500 meters) below sea level. The seafloor lifts up as it moves toward the hotspot, because the mantle expands as it's heated. At the same time, basaltic magma begins to pour out on the seafloor because of the drop in pressure on the rising hotspot material. Layer upon layer of lava gradually builds a volcano. About 20 miles (30 kilometers) off the south coast of the island of Hawai'i, thousands of eruptions on the seafloor have built a pile of lava to within 3,000 feet (1,000 meters) of the ocean surface (Fig. 8.10a). The subsea volcano, called Lo'ihi, will probably break the surface in about 60,000 years, becoming the youngest island in the Hawaiian Hotspot chain.

STAGE 2: BROAD, HIGH ISLAND. The Big Island of Hawai'i provides a glimpse of Hawaiian islands at the time they form (Fig. 8.10b). When directly above the Hawaiian Hotspot, a volcano can erupt continuously for years and pour out at a rate of 1 cubic mile (4 cubic kilometers) of fresh lava per century. The volume of the shield volcanoes comprising the island of Hawai'i is staggering. It is represented roughly by a cone with a base about 120 miles (200 kilometers) across, rising from about 18,000 feet (5,500 meters) below sea level to nearly 14,000 feet (4,300 meters) above (Fig. 8.7). Such a cone has a volume of about 25,000 cubic miles (100,000 cubic kilometers)—about the same as the entire Sierra Nevada mountain range, and twice that of the Appalachians! The volume of Hawai'i illustrates that hotspots are capable of generating enormous amounts of volcanic material over a relatively short period of time (Fig. 8.11).

The Big Island of Hawai'i is composed of five major volcanoes: Kohala, Hualalai, Mauna Kea, Mauna Loa, and Kīlauea (Fig. 8.12). Kohala last erupted about 60,000 years ago, and is considered extinct. Activity there ceased after the volcano moved off the hotspot. Mauna Kea, the highest peak on the Hawaiian Islands, has had several eruptions in the past 10,000 years, the last one about 3,500 years ago. Still near the hotspot, it is dormant but likely to erupt sometime in the future. Hualali last erupted in 1801. It is truly active, having erupted several times in the past 10,000 years.

The last activity on Mauna Loa was in 1984. Eruptions commonly start at the summit, then migrate to a flanking rift zone; flows often travel long distances. The 1984 eruption lasted three weeks, with lava reaching within 4 miles (6 kilometers) of the city of Hilo. Kīlauea is one of the world's most active volcanoes. An eruption from the east rift zone that was still occurring as of 2004 began in January 1983. Activity is

**FIGURE 8.12 Volcanoes on the island of Hawai`i.** The volcanic activity is most intense on the southeast part of the island, directly above the hotspot. Hawai`i Volcanoes National Park (86) includes the south flank of Mauna Loa and about half of Kīlauea. Frequent eruptions occur from rift zones on the flanks of Mauna Loa and Kīlauea.

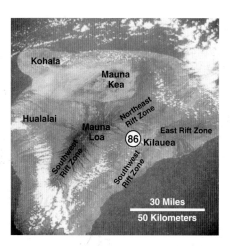

widespread, the summit having last erupted in 1982 and the southwest rift zone in 1974.

STAGE 3: SMALLER, LOWER ISLAND. While Hawai`i Volcanoes National Park illustrates a hotspot island in the prime of its life, Haleakalā National Park shows processes that develop with maturity (Fig. 8.10c). Haleakalā Volcano rose from the seafloor about a million years ago. Like other Hawaiian volcanoes, it built up a shield of layer upon layer of basalt lava flows,

through thousands of eruptions. Building up of the broad Haleakalā Volcano continued until about 360,000 years ago (Fig. 8.14a). In its prime, Haleakalā Volcano was perhaps, like Mauna Kea and Mauna Loa, over 13,000 feet (4,000 meters) above sea level. That height can be envisioned by projecting the surface slopes, as well as the dipping layers of basalt, upward to where they meet at a point (Fig. 8.14b).

Since Maui moved off the hotspot, Haleakalā has been less active, spewing forth far less volcanic material. During Earth's ice ages, freezing and thawing at high elevation broke rock apart. Streams eroded the valleys at Haleakalā's summit and reduced the mountain's height by more than 3,000 feet (1,000 meters). The so-called "crater" at the summit of the mountain is not volcanic, but rather the product of erosion of the top of Haleakalā (Fig. 8.14c). Even though Maui has moved about 80 miles (130 kilometers) from the hotspot, some volcanic activity continues at Haleakalā. Those eruptions have partially filled in the crater region with cinder cones and lava flows, some as young as a few hundred years (Fig. 8.14d).

Once a volcano has moved more than about 100 miles (160 kilometers) away from a hotspot, volcanic activity is essentially done. Over time, the island gets lower in height and smaller in area, as thermal subsidence and erosion take over (Fig. 8.9). On a clear day, views from

A HOTSPOT WITHOUT PLATE TECTONICS?

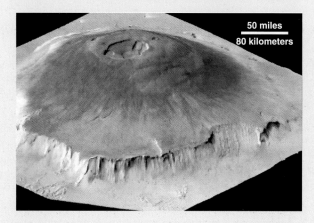

Olympus Mons is the largest volcano in the solar system. With a diameter of about 400 miles (600 kilometers), it rises 70,000 feet (21,000 meters) over the Martian landscape (Fig. 8.13). But what causes a volcano that is over twice the height of Mount Everest, and about 20 times the mass of volcanoes comprising the Big Island of Hawai`i (above and below the ocean)? One idea is that, like the Hawai`i–Emperor Seamount Chain, Olympus Mons formed as basaltic lava poured out from a hotspot. But unlike Earth, Mars does not have moving plates. So, instead of a chain of volcanoes, Olympus Mons developed one giant shield. It would be as if the Hawaiian Hotspot just kept pumping out lava without the Pacific Plate moving above it.

FIGURE 8.13 Olympus Mons, Mars. The nearly 70,000-foot (21,000-meter) volcano is the tallest in the solar system. Like the Hawaiian Islands, it is a broad shield volcano composed of numerous basaltic lava flows. Note the prominent collapse caldera at the top of the mountain, similar to those at the summits of Kīlauea and Mauna Loa in Hawai`i Volcanoes National Park. Topographic relief exaggerated two times.

FIGURE 8.14 Haleakalā National Park displays a late stage of hotspot activity. a) The 360,000-year-old layers of basalt near the top of Haleakalā Volcano represent the ending phases of the broad, high island stage. **b)** Projection of the dipping layers upward to the summit supports an interpretation that, in its prime, Haleakalā was about 3,000 feet (1,000 meters) higher. **c)** The "crater" region of Haleakalā is really an erosional feature that has developed since the island of Maui moved off the hotspot, drastically reducing the level of volcanic activity. The valleys on Haleakalā are thought to have been about 3,000 feet (1,000 meters) deep before being filled with young volcanic material. Eruptions have continued until modern times, as Haleakalā is still an active volcano. **d)** Cinder cones and lava flows, some less than 1,000 years old, cover the Haleakalā crater floor. **e)** Silverswords in the foreground illustrate some of the unique plants that have developed in such a dry, isolated, volcanic landscape.

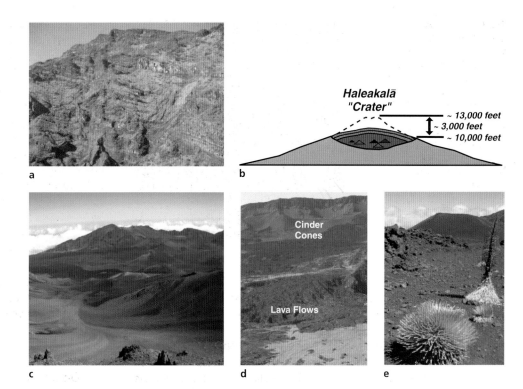

the summit area of Haleakalā National Park highlight such effects (Fig. 8.15). To the southeast, Mauna Kea and Mauna Loa volcanoes rise to nearly 14,000 feet (4,300 meters) because of the enormous volume of volcanic material that has poured onto the surface of the Pacific Plate. Looking northwestward, West Maui Volcano and the island of Molokaʻi are less than 6,000 feet (2,000 meters) above sea level because, without renewed volcanic activity, erosion and thermal subsidence gradually diminish the volcanoes. This trend continues northwestward, as Oahu and Kauaʻi show more advanced stages of erosion and vegetation cover (Fig. 8.16).

GEOLOGISTS ARE WEIRD—THEY DATE ROCKS!

Geologists date lava flows using different methods, depending on how ancient the flow might be (Table 2.2). For recent eruptions, they use the **radiocarbon method**. Organic material can be preserved as charcoal within volcanic ash beds or beneath lava flows. Radiocarbon dating determines the timing of the ash deposition or of the flow covering and burning wood or roots. Tissues of organic material contain the isotope carbon-14 (^{14}C). Through time, the material decays to nitrogen. The half-life of ^{14}C is 5,700 years, meaning that during that amount of time half of the remaining ^{14}C decays to nitrogen. After one half-life (5,700 years), the organic material contains just as much nitrogen as ^{14}C. After two half-lives (11,400 years), the ratio will be ¼ ^{14}C, ¾ nitrogen; after three half-lives (17,100 years), there will be ⅛ ^{14}C, ⅞ nitrogen. After about seven half-lives, there is not enough original ^{14}C to make accurate age determinations; the radiocarbon method is only good for dating material formed less than about 40,000 years ago.

For older lava flows, the **potassium/argon (K/Ar) method** can be used to date the time of solidification of minerals within lava flows or ash beds. The method can be used on actual rocks or ash beds, provided they have not been subjected to weathering. The half-life of the radioactive isotope of potassium (^{40}K) is 1.3 billion years, making the K/Ar method useful to date eruptions older than about 100,000 years. There is therefore a gap, making it difficult to date eruptions that occurred between 40,000 years ago (the upper limit of radiocarbon dating) and 100,000 years ago (the lower limit of K/Ar dating).

FIGURE 8.15 Views from the summit region of Haleakalā National Park reveal the lowering of Hawaiian volcanoes from southeast to northwest. a) Looking southeastward at Mauna Kea, on the Big Island directly above the hotspot, rising nearly 14,000 feet (4,300 meters) above the sea. In the foreground, Haleakalā, on Maui, lies just over 10,000 feet (3,000 meters) above sea level. **b)** Looking in the other direction, West Maui Volcano and the island of Moloka`i are much lower, displaying more advanced subsidence and erosion.

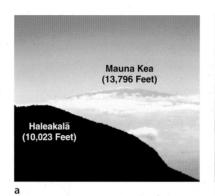

a   b

STAGE 4: ATOLL. After about 10 million years, an island gradually erodes and sinks to near sea level (Fig. 8.10d). **Coral reefs** grow in the shallow water surrounding the island. Eventually, the volcanic part of the island erodes down entirely, so that all that remains is the circular reef, called an **atoll**. Coral reefs and atolls, such as Midway Island and Kure Atoll, poke up along the Hawaiian Ridge, a line of volcanoes that is mostly submerged (Fig. 8.2).

STAGE 5: SEAMOUNT. After 20 to 30 million years, continued thermal subsidence wins out over coral reef growth, so that an island and its atoll disappear entirely (Fig. 8.10e). Once north of about 25° latitude, the ocean water is too cold to allow the growth of corals. Over time, the remnants of the volcanic islands and their capping coral reefs sink into the ocean as bumps on the seafloor, known as **seamounts**. The Hawaiian Ridge includes a series of seamounts representing a submarine continuation of the Hawaiian Island chain. Age dating along the ridge reveals a grad-

ual progression back to 45 million years ago, showing that the Pacific Plate has been moving northwestward during that time span. Beyond Daikakuji Seamount, the submerged volcanoes, known as the Emperor Seamount Chain, trend in a northward direction. Individual seamounts get older in that direction, suggesting that the Pacific Plate was moving northward prior to 45 million years ago. The oldest, Meiji Seamount, formed nearly 80 million years ago. It is the remnant of a volcano that may have been as lofty as Hawai`i's Mauna Kea, but now it lies below more than 10,000 feet (3,000 meters) of water and is about to be subducted into Alaska's Aleutian Trench.

VOLCANIC FEATURES IN HAWAIIAN NATIONAL PARKS

In 1916, the same year that the National Park Service was founded, Hawai`i National Park was established to preserve and showcase the Hawaiian Islands' volcanic

FIGURE 8.16 Nā Pali Cliffs, Kaua`i. The heavy erosion and vegetation illustrate an advanced stage of hotspot island development.

SNARING THE SUN

Haleakalā means "House of the Sun." According to Hawaiian folklore, the demigod Maui climbed to the summit of the highest point on the island that now bears his name. There he waited for dawn, when he snared the sun with a rope, forcing it to slow its journey across the sky. The days became longer, allowing the people more time to complete their tasks. In modern times, hundreds of people ascend to the summit of Haleakalā National Park each morning to observe sunrise at the place of Maui's legendary feat.

ISLAND SLICING

Erosion of the Hawaiian Islands is normally a slow process, caused by water moving particles of rock downslope to the ocean. But another process, known as **mass wasting**, can move materials downslope very fast. Huge portions of the Hawaiian Islands are thought to have slid into the ocean. On some of the older islands, large sea cliffs, such as the Nā Pali Cliffs on Kaua`i (Fig. 8.16), are remnants of such giant landslides. The landslides occur when the flanks of a volcano are oversteepened because of volcanic activity or normal erosion. The material from the island glides along the rubble of basalt that commonly lies underwater and forms the base of the volcano. Under special circumstances, such as when moving magma weakens a rift zone on the flank of a volcano, a landslide could cause a large part of an island to slide into the ocean. In Hawai`i Volcanoes National Park, there is concern that the south flank of Kīlauea may break away along the active rift zones that cut across the volcano (Fig. 8.17).

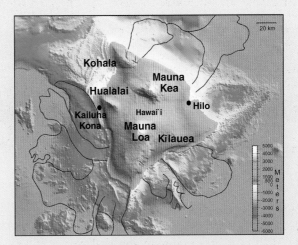

**FIGURE 8.17 Hawai`i above and below the sea.** The Island of Hawai`i is just the tip of a massive pile of volcanic material that built up on the Pacific Ocean floor. Large portions of the island sometimes collapse as gigantic landslides (outlined in black) that spread the material outward.

features. In 1961, that large park was separated into Hawai`i Volcanoes National Park on the island of Hawai`i, and Haleakalā National Park on Maui. Hawai`i Volcanoes National Park contains two massive shield volcanoes: half of Kīlauea Volcano and part of Mauna Loa (Fig. 8.7). Mauna Iki and Mauna Ulu are smaller, gently sloping volcanoes known as **lava shields**. Haleakalā National Park encompasses the summit and crater area of Haleakalā Volcano, as well as a narrow corridor extending through the rain forest to Maui's southeast coast near Kīpahulu.

The Big Island of Hawai`i is one of the most volcanically active places on Earth. Off and on for the past 700,000 years, material has been pouring out, piling flow upon flow onto the growing shield volcanoes. Unlike composite volcanoes, which may erupt sporadically for a few months, Hawaiian shield volcanoes can erupt continuously for years. The most recent eruption of Kīlauea started in 1983, and, as of 2004, has continued ever since (Fig. 8.18).

The type of volcanic eruption depends on three things: the composition of the magma, its temperature, and the amount of gas it contains. The first two factors tend to make Hawaiian eruptions fluid rather than explosive, because oceanic hotspot magmas are low in silica (~50% SiO_2) and have high temperatures (~2000 °F;

1100 °C). Lava flows are rarely more than 30 feet (10 meters) thick, and are commonly less than 10 feet (3 meters). They often travel for several miles, spreading out over older flows. From thousands of such fluid

THE GEOLOGY CLIMATE

Analogy with the "weather climate" is useful in appreciating why Haleakalā is considered an active volcano. For several centuries now, the climate of the Hawaiian Islands has been wet on the northeast (windward) sides of the islands, and dry on the southwest sides. We can expect that there will be lots of rain on the northeast sides of islands next year, and that the pattern will probably continue throughout our lifetime. There is no reason to suspect that the climate there will suddenly change to desert conditions! We can have similar expectations regarding the "geology climate." There have been periodic volcanic eruptions of Haleakalā for the past several thousand years, and it is inevitable that eruptions will continue for some time into the future. So Haleakalā is considered an active volcano; there is high probability that it will erupt sometime in the next few centuries.

a

b

c

d

FIGURE 8.18 The latest eruption of Kīlauea Volcano began in 1983. a) Lava flowed over Chain of Craters Road. As of 2004, a hike across young flows might be rewarded with spectacular views of lava pouring into the sea from lava tubes. **b)** As the lava cools and hardens, it forms a protrusion of land, called a bench, extending out into the ocean (Fig. 8.19). **c)** Glowing lava as it surfaces on top of the bench. **d)** Close-up of glowing lava.

eruptions, enormous shield volcanoes build up, from the ocean floor to high above sea level.

Volcano Plumbing System

Features at Hawai`i's national parks provide intriguing clues about the way magma flows beneath a volcano and eventually reaches the surface. Magma can accumulate in chambers beneath the summit region of a shield volcano. Although summit eruptions occur, often the magma spreads outward and erupts through

deep fractures along the flanks of the volcano, known as **rift zones** (Fig. 8.19). Some of the molten material hardens into sheetlike dikes and sills below the surface. Lava flows can crust over, allowing still hot, fluid material to travel great distances through underground **lava tubes**.

RIFT ZONES. Several prominent deep fractures cut across the landscape of Hawai`i Volcanoes National Park. The east and southwest rift zones on the flanks of Kīlauea Volcano, as well as the southwest and northeast rift zones of Mauna Loa, are quite active features (Fig. 8.12). A fissure eruption begins as steam, and other hot gasses escape through long cracks. The gasses and globs of magma can shoot out the cracks at high velocity, like material spewing out of a pressure cooker when the lid is (unwisely) opened (Fig. 8.20).

Prominent rift zones also cut the flanks of Haleakalā Volcano in Haleakalā National Park. The east rift zone coincides with one of the deep valleys (Haleakalā Crater; Fig. 8.14). Most of the cinder cones and lava flows on the crater floor erupted less than 5,000 years ago. The youngest flows in the valley have been dated as only 800 to 1,000 years old. The last eruption of Haleakalā was along the volcano's southwest rift zone, at La Perouse Bay in the mid-1600s.

LAVA TUBES. The basaltic lavas in Hawai`i are so fluid that they can travel down the flanks of volcanoes, below ground as well as above. Once a hard crust forms over a flow, the lava may be so well insulated that it

FIGURE 8.19 Current eruptions at Hawai`i Volcanoes National Park. Since 1983, lava formed a cinder and spatter cone at the east rift zone of Kīlauea Volcano. The erupting lava sometimes disappears into underground lava tubes, where it flows 7 miles (11 kilometers) to the Pacific Ocean (Fig. 8.18).

can remain hot long enough to travel more than 20 miles (30 kilometers). Lava tubes are an integral part of the distribution system of a shield volcano, allowing single eruptions to disperse lava over large areas. The most recent eruption of Kīlauea, which began in 1983, pours out of an opening at Pu`u `Ō`ō (Fig. 8.19). The lava enters a system of lava tubes, through which it flows 7 miles (11 kilometers) downslope, forming a bench where it pours out into the ocean beyond Pulama Pali (Fig. 8.18). Older lava tubes can be seen at Hawai`i Volcanoes and Haleakalā national parks (Figs. 8.21 and 8.22a).

DIKES AND SILLS. Magma can remain within fissures after eruptions from rift zones cease. That material cools and solidifies into sheetlike bodies known as **dikes** (which cut across layers) and **sills** (which are parallel to layers; Fig. 8.19). Intrusive rock is often harder and more resistant to erosion than the surrounding volcanic strata, making dikes and sills stand out. Prominent examples of dikes are seen on the eroded walls of the crater region at Haleakalā National Park (Fig. 8.22b).

Surface Features

Features on the surface of volcanic landscapes can tell us about processes occurring beneath volcanoes, as well as the interaction between erupting lava and air or water on the surface.

CINDER AND SPATTER CONES. When fluid magma has a lot of gas and the eruption is around a central vent, material may shoot high into the air like a fire hose (Fig. 8.23a). The magma cools in the air and rains down as **tephra** in sizes ranging from small (ash), to medium

---

HOT WALKS!

Two of the many fascinating hikes in Hawai`i Volcanoes National Park traverse very young lava flows. A trail across the crater of Kīlauea Iki ("Little Kīlauea") crosses hardened lava from the 1959 eruption (Fig. 8.23). In 1989 a flow from the most recent activity on the east rift zone destroyed a visitor center in the park. Later flows have covered the coastal part of the Chain of Craters Road (Fig. 8.18a). A walk starting where the hardened flow crosses the road leads to the area where the lava sometimes flows out of tubes directly into the sea.

a

b

FIGURE 8.20 Eruptions of fluid lava in Hawai`i Volcanoes National Park. a) Fountain eruption during the 1969–1974 eruption of Mauna Ulu. **b)** Line of spatter cones from the 1983 Kīlauea Volcano eruption.

a

b

**FIGURE 8.21 New and old lava tubes in Hawai`i Volcanoes National Park. a)** View through skylight at magma flowing through lava tube during 1969–1974 Mauna Ulu eruption. **b)** Entrance to Pua Po`o Lava Tube.

a   b

FIGURE 8.22 Examples of magma plumbing and lava distribution features in the crater area of Haleakalā National Park. **a)** A lava tube is a now-empty artery that once carried hot, fluid lava. **b)** A dike cutting crater walls was once magma that cooled within a fissure on the flanks of the volcano.

(cinder), to large (volcanic bomb). A large pile of such material around the vent builds up into a **cinder cone** (Fig. 8.24). Lava may also spew forth more quietly, building a **spatter cone** (Fig. 8.20b).

Like chocolate chips on top of vanilla wafers, cinder and spatter cones dot the flanks of the shield volcanoes in Hawai`i Volcanoes National Park. Most occur along rift zones. Cinder cones generally form from sin-

          b

a

FIGURE 8.23 Kīlauea Iki. a) The fountain during the 1959 eruption reached a height of 1,900 feet (580 meters), forming the Pu`u Pua`i cinder cone (Fig. 8.24b). **b)** Lava flows from the eruption coat the bottom of Kīlauea Iki Crater. **c)** Hikers can explore the crater and 1959 flows on a Hawai`i Volcanoes National Park trail.

c

(a) Pu`u `Ō`ō

(b) Pu`u Pua`i

(c) Pu`u Huluhulu

FIGURE 8.24 Cinder cones at Hawai`i Volcanoes National Park. a)** Pu`u `Ō`ō (a combination cinder and spatter cone) formed during the ongoing eruption of Kīlauea that began in 1983. It has not had time to develop much vegetation. **b) Some vegetation has been established at Pu`u Pua`i since it formed during the 1959 Kīlauea Iki eruption. **c)** Pu`u Huluhulu formed about 400 years ago, and has been covered by thick vegetation.

gle eruptions lasting weeks or months (Fig. 3.15). They can grow to as much as 1,000 feet (300 meters) high, their slopes retaining a steep angle of about 30°. Volcanic gas may continue to erupt from a cinder cone for months to years afterward. It interacts with the iron in the cinders, forming iron oxide, or rust—cinder cones in the crater region of Haleakalā National Park display distinctive red colors (Fig. 8.25).

FIGURE 8.25 Cinder cones in Haleakalā National Park. Young cinder cones within the crater area display red and other bright colors resulting from oxidation of iron minerals by steam and hot water.

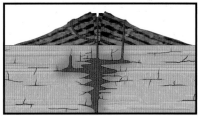

a

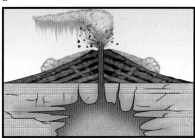

b

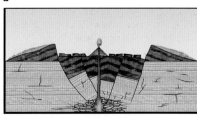

c

FIGURE 8.26 Caldera development. a) Magma accumulates in a chamber beneath the volcano. Eruptions occur not only through the summit, but also out of fissures along the flanks of the volcano. The volcano builds up over time through layer upon layer of lava flows and other volcanic materials. **b)** At times the magma inflates and erupts a large volume of material that partially empties the chamber. **c)** A large crater (caldera) can develop when the summit region of the volcano collapses as rubble into the partially-emptied magma chamber. Lava flows and cinder cones can form during later eruptions on the caldera floor.

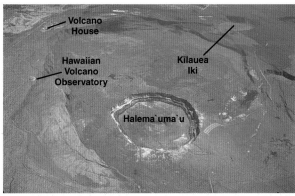

a

b

FIGURE 8.27 The summit region of Kīlauea Volcano. a) The large collapse caldera has smaller craters on its floor and along its edges, including Kīlauea Iki and Halema'uma'u. **b)** The Volcano House overlooks Kīlauea Caldera. Black outlines show the tops of large blocks of material that have broken off the steep caldera walls.

COLLAPSE CRATERS. At times, enough magma can erupt out of an underground chamber that the ground collapses into the vacated space (Fig. 8.26). Spectacular examples are preserved in national parks (Crater Lake, Chap. 5; Yellowstone, Chap. 9). Very young collapse craters on many different scales can be seen in Hawai'i Volcanoes National Park. One of the larger features, a **caldera** at the top of Mauna Loa, is 4 miles (6 kilometers) long and 1½ miles (2½ kilometers) wide. The Kīlauea summit caldera is about 2 miles (3 kilometers) in diameter; its walls are so steep that they often give way as landslides (Fig. 8.27).

Smaller versions of calderas, known as **pit craters**, can form where magma intrudes and inflates a magma chamber. As the magma drains, often without surface eruptions, the surface collapses. Prominent pit craters in Hawai'i Volcanoes National Park include Halema'uma'u and Kīlauea Iki (Fig. 8.27), which lie

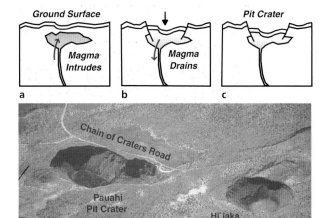

d

FIGURE 8.28 Pit crater formation. a) Intrusion of magma that may or may not erupt onto the surface. **b)** Magma drains after intrusion. **c)** Loss of support causes the surface to collapse, forming a pit crater. **d)** Pit craters along Chain of Craters Road, Hawai`i Volcanoes National Park.

within and on the fringes of Kīlauea Caldera, as well as Pauahi and Hi`iaka along Chain of Craters Road (Fig. 8.28).

SURFACE TEXTURES OF LAVA FLOWS. Basaltic lava surfaces can have different textures depending on how the lava cools (Fig. 8.29). Where the lava is hot and fluid, it forms a smooth, ropy surface, known as

pāhoehoe ("puh-hoy-hoy"). When the lava flow cools down it becomes more sluggish, forming a surface of broken, jagged blocks called `a`ā ("ah-ah"). `A`ā flows can be very difficult and dangerous to walk across, tearing boots to shreds.

Not all the gas escapes before lava solidifies, such that basalt lava flows commonly have small cavities called **vesicles** (hence the term **vesicular basalt**). If the lava erupts below water, it cools quickly, accumulating on the ocean floor in blobs, forming what look like pillows (Fig. 5.11). A smooth, glassy surface may form as a result of the quick quenching of the pillow; the glassy surface surrounds a core of granular basalt that cooled more slowly.

Hawaiian Volcanoes Wired

Besides preserving areas of natural beauty and making them accessible, national parks also serve as natural scientific laboratories. The Kīlauea summit region highlights the extensive volcanic activity within Hawai`i Volcanoes National Park (Fig. 8.30). Eyewitness accounts and geologic mapping illustrates that, in just the past 120 years, the summit has experienced at least 10 different eruption episodes (Fig. 8.31). Add to that the frequent eruptions along Kīlauea's rift zones as well as the summit and flanks of Mauna Loa, and the need to monitor signs of future volcanic activity is apparent.

The United States Geological Survey (USGS) maintains the Hawaiian Volcano Observatory (HVO), established in 1912. After years of monitoring Hawaiian volcanoes, the USGS has observed distinct warning signs that precede an eruption (Fig. 8.32). As magma

FIGURE 8.29 Surface of basalt lava flows. As lava cools, it forms a solid crust above its still molten interior. **a)** Pāhoehoe lava flows can have a ropey-textured surface, formed from hot, free-flowing lava (1996 flow on the south side of Hawai`i Volcanoes National Park). **b)** `A`ā lava flows consist of sharp, angular blocks that commonly develop when the flow is cooler and more sluggish (crater region of Haleakalā National Park).

a

b

a

b

c

FIGURE 8.30 A hike through Kīlauea Caldera reveals ongoing volcanism. a) Young lava flows coat the caldera floor.
b) Steam vents, called fumaroles, on the caldera floor demonstrate that hot material remains below the surface. **c)** Halema`uma`u is a circular pit crater, about ½ mile (1 kilometer) wide and 750 feet (230 meters) deep, within the caldera (Fig. 8.27).

FIGURE 8.31 Map of summit region of Kīlauea Volcano. The year (or month and year) for each major flow shows just how active the area has been over the past 120 years.

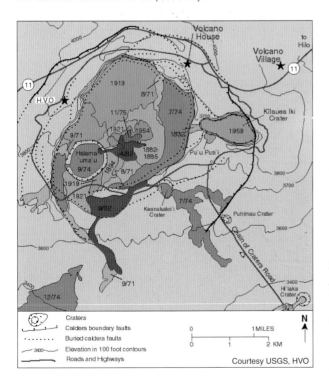

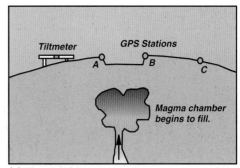

(a) Magma Rises

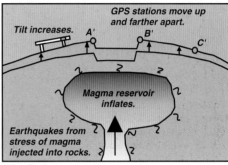

(b) Inflation

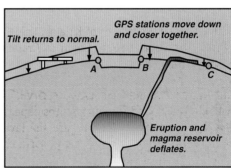

(c) Deflation

FIGURE 8.32 Measurement of magma migration. a) As magma rises, **tiltmeters** can detect changes in surface slope, while Global Positioning System (**GPS**) instruments measure the elevations and lateral positions of stations. **b)** As the magma chamber inflates, the ground tilts and stations move upward and away from one another. **c)** When magma moves out of the magma chamber, the ground surface returns to its original slope and stations move down and closer together.

a

b

c

FIGURE 8.33 National Park of American Samoa. The islands developed over a stationary hotspot beneath the South Pacific. **a)** Ta'u Island. **b)** Coral on the seafloor of the park's Ofu Unit. **c)** Ofu and Olosega islands.

works its way to a shallower position, rocks are cracked and jostled here and there so that swarms of small earthquakes occur. Tiltmeters reveal slight movements of Earth's surface above a magma chamber. The ground steepens before an eruption, as the magma chamber swells. The ground flattens out as the magma evacuates the chamber. Certain gasses, such as carbon dioxide and hydrogen sulfide, may also signal an eruption as they escape through cracks formed by the expanding magma.

ACTIVE OR NOT?

The terms *active*, *dormant*, and *extinct* are sometimes used to describe the eruptive potential of a volcano. Although subjective and often vague, these terms are nonetheless useful in comparing volcanoes. An **active** volcano is one that has erupted in recorded history and undoubtedly will erupt again in the near future. The rocks around an active volcano show a pattern of numerous eruptions in the recent geologic past, perhaps spanning thousands to tens of thousands of years, and that pattern is likely to continue. Stated statistically, an active volcano has a 95% probability of erupting in the next 200 years. A **dormant** ("sleeping") volcano has not erupted in historic times, but shows a pattern that suggests it will likely erupt in the not-too-distant future. A volcano may be considered **extinct** if it has not erupted in thousands of years, and shows no signs that it is likely to erupt in the foreseeable future.

Earthquakes suggest that Hawaiian volcanoes erupt a few weeks or months after magma begins rising from the mantle into the crust. Magma rises because it is less dense than the surrounding rock. The pressure beneath the ground lessens at shallower depths, so that the magma expands and becomes less dense. About 2 miles (3 kilometers) below the surface, the magma is about the same density as the surrounding rock, so it pools in a **magma chamber**. About 1% to 2% of the weight of Hawaiian magma is gas. As more magma tries to come up from below, the chamber swells, cracking the roof. The cracks allow trapped gas to expand and escape, starting an eruption.

SOUTHWEST PACIFIC

Hawai'i is not the only place where the Pacific Plate is currently riding over a hotspot, or has done so in the past. The floor of the ocean is scarred with numerous tracks of islands, coral reefs, and submerged seamounts. The oldest of the tracks are parallel to the northward trend of the Emperor Seamount chain, and the younger ones follow the northwestward line of the Hawaiian Islands (Fig. 8.3). Along one of the tracks is U.S. territory that includes the National Park of American Samoa (Fig. 8.33). Similar to Hawai'i, the Samoan Islands consist of interlocking shield volcanoes with basaltic lava flows, cinder cones, and collapse calderas. The heavy vegetation, erosion, coral reefs, and old volcanic rocks suggest that the Samoan Islands are between the low island and atoll positions relative to the hotspot (Fig. 8.10c,d).

FURTHER READING

GENERAL

Ashton, R. H. 2003. A dynamic landscape formed by the power of volcanoes: Geology training manual for interpreters at Hawai`i Volcanoes National Park. M.S. thesis, Oregon State University. 131 pp.

Bevens, D. 1992. *On the Rim of Kīlauea: Excerpts from the Volcano House Register 1865–1955, Hawai`i National Park*. Hawai`i National Park, HI: Hawai`i Natural History Association.

Brantley, S., and C. Heliker. 2002. *The Pu`u `Ō`ō—Kupaianaha Eruption of Kīlauea Volcano, Hawai`i, 1983–2003, Hawai`i Volcanoes National Park*. Washington: U.S. Geological Survey and the National Park Service, USGS Fact Sheet 144-02, 2 pp.

Day, A. G. 1966. *Mark Twain's Letters from Hawai`i*. Honolulu: University of Hawai`i Press. 298 pp.

Decker, R., and B. Decker. 1996. *Volcano Watching*, 5th Ed. Hawai`i National Park, HI: Hawai`i Natural History Association. 84 pp.

Decker, R., and B. Decker. 2001. *Volcanoes in America's National Parks*. New York: Norton. 256 pp.

Hazlett, R. W. 2002. *Geological Field Guide, Kīlauea Volcano, Hawai`i Volcanoes National Park*. Hawai`i Natural History Association. 127 pp.

Hazlett, R. W., and D. W. Hyndman. 1996. *Roadside Geology of Hawai`i*. Missoula, MT: Mountain Press. 304 pp.

MacDonald, G. A., and D. H. Hubbard. 1993. *Volcanoes of the National Parks of Hawai`i*. Hawai`i Natural History Association. 64 pp.

Mohlenbrock, R. H. 1996. Haleakalā, Maui, Hawai`i. *Natural History*, v. 1, p. 58–60.

Takahashi, T. J., C. Heliker, and M. F. Diggles. 2003. *Selected images of the Pu`u `Ō`ō-kupaianaha eruption*. 1983–1997, U.S. Geological Survey, Digital Data Series DDS-80, 71 pp. (http://geopubs.wr.usgs.gov/dds/dds-80).

Tilling, R. I., C. Heliker, and T. L. Wright. 1987. *Eruptions of Hawaiian Volcanoes: Past, Present, and Future*. Denver: U.S. Geological Survey, General Interest Publication. 54 pp.

TECHNICAL

Bryan, C. J., and C. E. Johnson. 1991. Block tectonics of the island of Hawai`i from a focal mechanism analysis of basal slip. *Bulletin Seismological Society of America*, v. 81, p. 491–507.

Cashman, K. V., C. Thornber, and J. P. Kauahikaua. 1999. Cooling and crystallization of lava in open channels, and the transition of pāhoehoe lava to `a`ā. *Bulletin of Volcanology*, v. 61, p. 306–323.

Clague, D. A., and G. B. Dalrymple. 1989. Tectonics, geochronology, and origin of the Hawaiian–Emperor volcanic chain, in *The Geology of North America, v. N: The Eastern Pacific and Hawai`i*. Boulder, CO: Geological Society of America, p. 188–217.

Denlinger, R. P., and P. Okubo. 1995. Structure of the mobile south flank of Kīlauea Volcano, Hawai`i. *Journal of Geophysical Research*, v. 100, p. 24499–24507.

Duncan, R. A. 1991. Hotspots in the southern oceans: An absolute frame of reference for motion of the Gondwana continents. *Tectonophysics*, v. 74, p. 29–42.

Duncan, R. A., J. Backman, and L. Peterson. 1989. Reunion hotspot activity through tertiary time: Initial results from ocean drilling program, leg 115. *Journal of Volcanology and Geothermal Research*, v. 36, p. 193–198.

Filmer, P. E., M. K. McNutt, H. F. Webb, and D. J. Dixon. 1993. Volcanism and archipelagic aprons in the Marquesas and Hawaiian islands. *Marine Geophysical Research*, v. 16, p. 385–406.

Gilluly, J. 1971. Plate tectonics and magmatic evolution. *Geological Society of America Bulletin*, v. 82, p. 2383–2396.

Heliker, C., and T. N. Mattox. 2003. *The first two decades of the Pu`u `Ō`ō-Kupaianaha eruption: chronology and selected bibliography*. U.S. Geological Survey, Prof. Paper 1676, 206 pp.

Hon, K., J. P. Kauahikaua, R. Denlinger, and K. Mackay. 1994. Emplacement and inflation of pāhoehoe sheet flows: Observations and measurements of active lava flows on Kīlauea, Hawai`i. *Geological Society of America Bulletin*, v. 106, p. 351–370.

Jackson, E. D., et al. 1980. Introduction and summary of results from DSDP leg 55, the Hawaiian–Emperor hotspot experiment. *Initial Reports Deep Sea Drilling Project*, v. 55, p. 3–31.

Kauahikaua, J. P., K. V. Cashman, T. N. Mattox, K. Hon, C. C. Heliker, M. T. Mangan, and C. R. Thornber. 1998. Observations on the basaltic lava streams in tubes from Kīlauea Volcano, Hawai`i. *Journal of Geophysical Research*, v. 103, p. 27303–27324.

Keast, A., and S. E. Miller (editors). 1996. *The Origin and Evolution of Pacific Island Biotas, New Guinea to Eastern Polynesia: Pattern and Processes*. SPB Academic Publishing, Amsterdam.

McDougall, I., and D. H. Tarling. 1963. Dating of polarity zones in the Hawaiian Islands. *Nature*, v. 200, p. 54–56.

Molnar, P., and J. Stock. 1987. Relative motions of hotspots in the Pacific, Atlantic and Indian Oceans since late Cretaceous time. *Nature*, v. 327, p. 587–591.

Morgan, W. J. 1971. Convection plumes in the lower mantle. *Nature*, v. 230, p. 42–43.

Morgan, W. J. 1983. Hotspot tracks and the early rifting of the Atlantic. *Tectonophysics*, v. 94, p. 123–139.

Natland, J. H. 1980. The progression of volcanism in the Samoan linear volcanic chain. *American Journal of Science*, v. 280A, p. 709–735.

Okal, E. A., and R. Batiza. 1987. Hotspots: The first 25 years. *American Geophysical Union*, Special Paper p. 1–11.

Sherrod, D. R., and J. P. McGeehin. 1999. *New radiocarbon ages from Haleakalā crater, Island of Maui, Hawai`i.* U.S. Geological Survey, Open-file Report 99–143, 14 pp.

Stearns, H. T. 1944. Geology of the Samoan Islands. *Bulletin of the Geological Society of America*, v. 55, p. 1279–1332.

Swanson, D. A., W. A. Duffield, and R. S. Fiske. 1976. *Displacement of the South Flank of Kīlauea Volcano: The Result of Forceful Intrusion of Magma into Rift Zones.* U.S. Geological Survey, Professional Paper 963. 36 pp.

Tilling, R. I., and J. J. Dvorak. 1993. Anatomy of a basaltic volcano. *Nature*, v. 363, p. 125–133.

Walcott, R. I. 1970. Flexure of the lithosphere at Hawai`i. *Tectonophysics*, v. 9, p. 435–446.

Watts, A. B., U. S. ten Brink, P. Buhl, and T. M. Brocher. 1985. A multichannel seismic study of lithospheric flexure across the Hawaiian–Emperor seamount chain. *Nature*, v. 315, p. 105–111.

WEBSITES

National Park Service: www.nps.gov

Park Geology Tour:
www2.nature.nps.gov/grd/tour/index.htm
  Hawai`i:
    85. Haleakalā NP:
    www2.nature.nps.gov/grd/parks/hale/index.htm
    86. Hawai`i Volcanoes NP:
    www2.nature.nps.gov/grd/parks/havo/index.htm
  Western Pacific:
    87. National Park of American Samoa:
    www.nps.gov/npsa/index.htm

U.S. Geological Survey: www.usgs.gov

Geology in the Parks:
www2.nature.nps.gov/grd/usgsnps/project/home.html

Volcanoes:
volcanoes.usgs.gov
  Hawaiian Volcano Observatory: hvo.wr.usgs.gov
    Haleakalā: hvo.wr.usgs.gov/volcanoes/haleakala
    Kīlauea: hvo.wr.usgs.gov/kilauea
    Mauna Loa: hvo.wr.usgs.gov/maunaloa
  Eruptions of Hawaiian Volcanoes:
  pubs.usgs.gov/gip/hawaii
  Historical Eruptions:
  volcanoes.usgs.gov/Volcanoes/Historical.html
  Volcano Monitoring:
  volcanoes.usgs.gov/About/What/Monitor/
    monitor.html
  Monitoring Active Volcanoes:
  pubs.usgs.gov/gip/monitor

9

Hotspots Beneath Continental Crust

▲▲▲▲▲▲ *"Yellowstone National Park is important not only because of its geysers, hot springs, and exceptional scenic beauty, but also because it is the premier place in the world to observe a landscape developing as a continent overrides a hotspot."*

John Day Fossil Beds National Monument in Oregon; Craters of the Moon National Monument in Idaho; Yellowstone National Park in the northwestern corner of Wyoming—these three parks have vastly different landscapes and rocks, but they are linked by a common thread: a hotspot rising through a moving plate capped by continental crust (Fig. 9.1). Such a feature is a bit

FIGURE 9.1 **National park lands along the Yellowstone Hotspot Track.** John Day Fossil Beds National Monument contains lava flows of the Columbia Plateau, which may represent the initial surfacing of the hotspot. Craters of the Moon National Monument has young, basaltic material characteristic of volcanism of the Snake River Plain occurring after that region of the North American Plate passed over the hotspot. The rhyolite and geothermal features of Yellowstone National Park reflect continental volcanism occurring directly above the hotspot. Hagerman Fossil Beds National Monument has sedimentary layers deposited as the Snake River Plain subsided after the region moved over the hotspot.

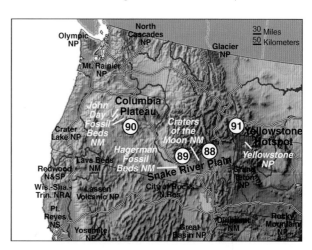

like a lava lamp (Fig. 9.2). A glass chamber filled with oil contains wax that, when cold, is dense and sinks to the bottom. A light bulb at the base raises the temperature, so that the wax expands, becomes less dense, and rises through the oil. The rising wax at times assumes the shape of a mushroom, with a broad, rounded head and slender stem. The oil in the lamp can be thought of as Earth's mantle, the wax as part of the deeper mantle, and the light bulb might be a bump on Earth's core that heats the deeper mantle.

A hotspot that encounters a moving plate capped by continental crust results in a changing pattern of volcanic activity. A rising hotspot is actually solid mantle that "flows" on a geologic time scale; that is, the heated rock moves upward a few inches per year. When it reaches the level of the shallow mantle, the material is under so little pressure that it partially melts, producing basaltic magma (Fig. 2.15c).

A rising mantle plume has a massive head as much as 500 miles (800 kilometers) in diameter (Fig. 9.3a). The volume of basaltic magma that rises initially through the overriding plate can be so enormous that it matters little what kind of crust caps the plate—huge volumes of basaltic lava pour out on the surface (Fig. 9.3b).

After a few million years, the mushroom head of the hotspot plume dissipates, leaving only a thin stem. Plate

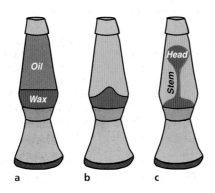

FIGURE 9.2 Lava lamp. a) Chamber filled with oil and dense wax that sinks to the bottom. **b)** The light bulb causes the wax to heat up and expand, becoming less dense than the oil. **c)** The hot wax rises like a mushroom cloud, with a large head and narrow stem.

TABLE 9.1 National parks at continental hotspots

YELLOWSTONE/SNAKE RIVER PLAIN/COLUMBIA PLATEAU
88. Craters of the Moon NM, ID
89. Hagerman Fossil Beds NM, ID
90. John Day Fossil Beds NM, OR
91. Yellowstone NP, WY/ID/MT

FIGURE 9.3 A rising hotspot plume is analogous to the "mushroom cloud" rising from a lava lamp. a) The rising plume has a large head and a thin stem. **b)** As the head penetrates the overriding plate, massive amounts of basaltic lava pour out on the surface of the plate. **c)** After the head dissipates, much smaller amounts of volcanic material from the stem penetrate the plate. To get through thick, granitic continental crust, the magma must melt high-silica minerals, so that it produces pasty and explosive rhyolite. If the plate is capped by thin, basaltic oceanic crust, the resulting magma remains fluid basalt (Chap. 8).

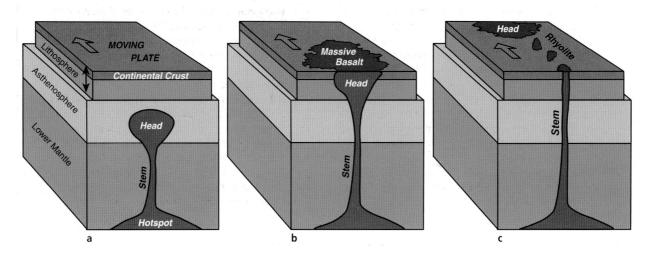

motion carries the region of extensive basalt lava away, but the magma rising from the stem must somehow work its way to the surface. That's when the thickness and composition of continental crust come into play. The basaltic melt from the hotspot stem may, in fact, form two levels of magma chambers within the overriding plate. The lowest is at the base of the crust, retaining a basalt and gabbro composition. Magma rising from that level melts its way through thick, silica-rich continental crust, forming magma chambers composed of rhyolite and granite in the shallower part of the crust. A narrow chain of explosive, rhyolite volcanoes forms on the surface of the moving plate (Fig. 9.3c).

After a region of the plate has passed over the hotspot, the rhyolite eruptions stop. But there may still be some low-silica melt within the magma chambers at the base of the crust. Fields of dark, fluid, basalt lava flows and cinder cones then cover the rhyolite volcanoes.

COLUMBIA PLATEAU, SNAKE RIVER PLAIN, YELLOWSTONE HOTSPOT TRACK

Yellowstone National Park is important not only because of its geysers, hot springs, and exceptional scenic beauty, but also because it is the premier place in the world to observe a landscape developing as a continent overrides a hotspot. About 17 million years ago, enormous volumes of lava began to pour out over what is now northeast Oregon and southeast Washington, cool-

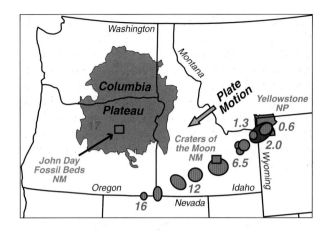

FIGURE 9.4 Ages of the massive basalt (gray) and initial rhyolite (pink) volcanic activity from the Columbia Plateau to Yellowstone. Numbers show the ages of the first volcanism, in millions of years. The North American Plate appears to be moving southwestward over a stationary hotspot, currently located beneath Yellowstone National Park. Earlier rhyolite volcanism, which occurred as the region of Craters of the Moon National Monument moved over the hotspot about 6 million years ago, is now covered by very young basalt lava flows. Much older basalt flows at John Day Fossil Beds National Monument may be part of the massive head that poured out as the hotspot plume surfaced 17 million years ago.

ing as layer upon layer of basalt (Fig. 9.4). The Columbia Plateau may be the region where the rising hotspot initially surfaced (Fig. 9.5a). By that interpretation, the

A HOTLY DEBATED TOPIC

Although the existence of hotspots is accepted widely by the scientific community, their origin is controversial. Mapping of the speed of earthquake waves through Earth's deep interior, known as **seismic tomography**, reveals that there are bumps on the top of the core up to 6 miles (10 kilometers) high. Those bumps might heat the lowermost mantle, causing it to expand and rise. Other studies suggest that hotspots may originate much shallower, within the upper mantle. The latter may be the case for Yellowstone, where recent seismic tomography studies suggest that the hotspot source could be as shallow as 125 miles (200 kilometers) below Earth's surface.

HOW MUCH LONGER TILL WE GET THERE?

For the past 16 million years, the Yellowstone Hotspot has moved very little compared to Earth's lower mantle, while the North American Plate has moved to the southwest (Fig. 9.4). The line of rhyolite volcanoes sweeping across southern Idaho is progressively younger toward Yellowstone: 16 million years old at the southwest corner of Idaho; 10 million years old near Twin Falls; 6 million years old near Idaho Falls and Craters of the Moon; and 2 million years old and younger at Island Park and Yellowstone. Creeping an inch (2½ centimeters) per year, the North American Plate has moved about 300 miles (500 kilometers) in the past 16 million years. In a few million years, people living in North Dakota won't have to come to Yellowstone, because Yellowstone will come to them!

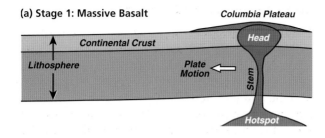

(a) Stage 1: Massive Basalt

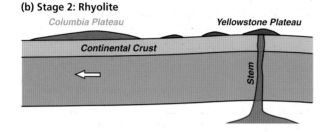

(b) Stage 2: Rhyolite

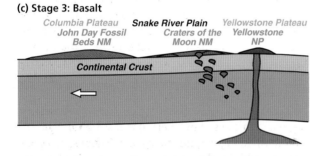

(c) Stage 3: Basalt

FIGURE 9.5 Stages of volcanic activity at a continental hotspot. a) Stage 1. As the hot mantle rises, the pressure drops. The resulting magma has a low-silica basalt and gabbro composition. The "head" of the mushroom-shaped hotspot is so massive that a huge volume of dark, fluid basalt pours out onto the surface with little change in chemistry. The 17- to- 12-million-year-old Columbia Plateau basalts in John Day Fossil Beds National Monument may be part of this initial hotspot stage. **b) Stage 2.** Once the head is gone, far smaller volumes of magma become rich in silica as they melt their way through the continental crust, resulting in explosive, rhyolite volcanoes. Huge collapse calderas, rhyolite flows, lava domes, and pumice deposits at Yellowstone National Park are products of this stage. **c) Stage 3.** After a region moves off the hotspot, the crust has been depleted of silica and has better-developed conduits to the surface. Residual magma that surfaces retains its basalt composition. Thin, fluid lava flows, lava tubes, fissures, spatter cones, and cinder cones at Craters of the Moon National Monument represent this final stage of continental hotspot volcanism.

basalt is from magma that melted from the hot "mushroom head" of mantle. The sheer volume of magma so overwhelmed the crust it penetrated that the lava flows retain their basalt composition. Basalt lava flows at John Day Fossil Beds National Monument in northeastern Oregon may be part of the initial hotspot stage. Though compelling, this interpretation is controversial because the Columbia Plateau is offset northward from the younger part of the Yellowstone Hotspot track; the hotspot head would have to be asymmetric, skewed northward relative to the stem.

South of the Columbia Plateau, near the juncture of the states of Oregon, Nevada, and Idaho, volcanism began about 16 million years ago. Since then, the North American Plate has moved southwestward over the stationary hotspot at a rate of about 1 inch (2½ centimeters) per year. The volcanism resulted from the relatively small volume of basaltic magma rising from the stem of the hotspot mushroom (Fig. 9.5b). As that magma melted its way through the continental crust, its silica content increased from the initial 50% to about 70%. A line of explosive, rhyolite volcanoes formed across the Snake River Plain of southern Idaho, all the way to Yellowstone National Park. More fluid, basalt lavas that cover the rhyolite flows at Craters of the Moon National Monument represent the final stage of volcanism (Fig. 9.5c).

The Yellowstone region is a bit like Hawai`i—it's part of a chain of volcanoes that get progressively older in one direction (Chap. 8). Only the "islands" southwest of Yellowstone are part of the North American continent. In fact, standing in the middle of Yellowstone National Park is much like standing on top of the Big Island of Hawai`i. The Yellowstone region, at 8,000 feet (2,500 meters) elevation, is about 2,000 to 3,000 feet (600 to 1,000 meters) higher than the surrounding territory. Like Hawai`i, it lies directly above a hotspot; as the hotspot rises, it expands like a hot-air balloon, elevating the Yellowstone Plateau. The landscapes and rocks of the national park lands in the region illustrate the three stages of development as the North American Plate rides over the Yellowstone Hotspot.

Stage 1: Surfacing of "Mushroom Head" (Massive Basalt)

The volume of basalt that poured out of the Columbia Plateau region between 17 and 12 million years ago was enormous. It covered about 100,000 square miles (250,000 square kilometers) to an average thickness of 10,000 feet (3,000 meters). Compare the 200,000 cubic miles (800,000 cubic kilometers) of lava to the paltry

12,000 cubic miles (50,000 cubic kilometers) that has erupted from the entire Cascade Range in the past 10 million years.

John Day Fossil Beds National Monument in northeastern Oregon contains incredible examples of mammal, plant, and other fossils. Those fossils are preserved in sedimentary layers that were deposited from 54 to 6 million years ago. Within the fossil beds are lava flows—part of the enormous volume of basalt that formed the Columbia Plateau of northeastern Oregon and southeastern Washington (Fig. 9.6). The fluid lavas poured out of long fissures, much like those seen erupting from rift zones on the flanks of shield volcanoes in Hawai`i and Iceland. More than 20 such flows, totaling about 1,600 feet (500 meters) thick, can be seen in Picture Gorge within the monument.

FIGURE 9.6 Examples of Columbia Plateau basalt layers in John Day Fossil Beds National Monument, Oregon. a) Fossil-bearing sedimentary layers of the John Day Formation are capped by hard Columbia Plateau basalt flows at Sheep Rock. **b)** The John Day River cuts through basalt layers at Picture Gorge. **c)** Columnar jointing results from shrinking as flows cool (Fig. 3.22).

a

b

c

Stage 2: Early-Phase "Stem" (Rhyolite)

The areas of rhyolite underlying the Snake River Plain form discrete areas of volcanic activity, rather than one long ridge (Fig. 9.4). This situation is analogous to that seen in the Pacific Ocean, where discrete islands form over the Hawaiian Hotspot (Fig. 8.8). "Islands" of rhyolite, progressively older to the southwest, extend across the Snake River Plain of southern Idaho. Similar to the Big Island of Hawai`i, a very high region, the Yellowstone Plateau, lies directly above the hotspot.

Yellowstone National Park was established because of its geologic wonders, especially geysers, hot springs, and mud pots. The underlying reason for these features is the Yellowstone Hotspot, which many people learn about during a visit to the park. But few of the visitors realize how so many of the park's other outstanding features result from the hotspot. For example, the park owes its high elevation to the expansion and rise of the hot mantle. As the land slowly rises, the Yellowstone River fights to stay at low elevation by carving the Grand Canyon of the Yellowstone River. Expansion of the hot mantle as it rises causes decompression melting, ultimately forming the lava flows, pumice, and obsidian found throughout the park.

Three striking features suggest that Yellowstone National Park lies above very hot mantle: 1) high elevation; 2) young volcanic rocks; and 3) hot springs and geysers. Additional scientific observations further sup-

CONTINENTAL DRIFT AND THE EVOLUTION OF HORSES

Reconstructions of the positions of continents at various times in the past, such as those shown in Fig. 6.6, help us understand how certain species of animals and plants developed. Such maps can explain the various lines of horses preserved at John Day Fossil Beds National Monument. At times, Europe and North America were joined, allowing horses to spread across both continents. But at other times, the continents were separated by an ocean, so that contact between the two groups was impossible and separate species evolved. Various times of mixing or isolation of horse species between North America and Asia also occurred. Whether or not horses could cross the land bridge between Alaska and Siberia depended not only on continental drift, but also on the effects of global sea level changes (Chap. 10).

THE WORLD'S FIRST NATIONAL PARK

On March 1, 1872, the United States Congress established Yellowstone as the world's first national park. Native Americans had known of the area's thermal features for centuries, as they hunted in the high country. The first Europeans to visit the upper reaches of the Missouri River drainage were fur trappers, beginning in the late 1700s. By the 1820s, they were reporting boiling springs and burning vapors in the Yellowstone country. The fur business was a lonely occupation—it was not uncommon for trappers to return with tall tales. They spoke not only of petrified forests, but also of petrified birds singing petrified songs! So when reports of hot springs of brilliant colors and spouts of hot water shooting a hundred feet into the air came back from the wilds, people were skeptical.

In the 1860s, prospectors were drawn to Yellowstone in the quest for gold, greatly increasing exploration of the territory. Formal expeditions were also

organized, including one led by Henry Washburn in 1870. One evening, while camping on the banks of the Madison River, the party sat in serious discussion of the need to protect the fabulous geologic features it saw in Yellowstone, and at the same time make them available for the public to see. So in the shadows of what is now National Park Mountain, the idea of a national park was born (Fig. 9.7).

In 1871, Ferdinand V. Hayden of the United States Geological Survey (USGS) led a scientific expedition to Yellowstone. Hayden was determined to bring back accurate and believable information, so he brought with him two prominent artists of the time, the painter Thomas Moran and photographer William Henry Jackson. The report and illustrations brought back by the Hayden Expedition were instrumental in influencing Congress to establish Yellowstone as a national park the following year.

FIGURE 9.7 National Park Mountain, near where the Gibbon and Firehole rivers converge to form the Madison River in Yellowstone National Park. Legend has it that the "national park idea" was born here.

port the idea of a hotspot beneath Yellowstone: 4) the **geothermal gradient** (rate the temperature increases with depth) is about 30 times the normal value; and 5) seismic waves travel slower than normal, suggesting that partially molten rock (a magma chamber) lies about 3 to 5 miles (5 to 8 kilometers) below the surface.

YELLOWSTONE VOLCANISM. Most of the rocks at Yellowstone National Park are light-colored rhyolite, pumice, and other high-silica materials. Two levels of magma chambers lie beneath Yellowstone (Fig. 9.8). After the huge mushroom head of the hotspot dissipated, the remaining stem continued to rise (Fig. 9.5b). The drop in pressure causes partial melting of the peridotite of Earth's mantle. The resulting basaltic melt flows upward, where it ponds in magma chambers at the base of the crust. Some of the magma melts its way into the crust, forming silica-rich magma just a few

FIGURE 9.8 Levels of magma chambers beneath Yellowstone National Park. The lower chamber, ponded at the base of the crust at about 20 miles (35 kilometers), contains basalt magma that melts off the mantle as the hotspot rises and decompresses. The upper chamber, only 3 to 5 miles (5 to 8 kilometers) below the surface, has rhyolite magma formed as some of the basalt magma melts its way through the crust, becoming enriched in silica.

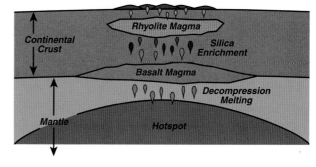

miles below the surface. The pasty, gas-rich magma from the upper chamber causes explosive eruptions and deposits of rhyolite and pumice that characterize the volcanic activity at Yellowstone.

Very large eruptions occurred in the Yellowstone region three different times: about 2 million years ago, 1.3 million years ago, and then 600,000 years ago (Fig. 9.4). Those eruptions were similar to the explosion and collapse that formed the Crater Lake Caldera in Oregon 7,700 years ago, only much bigger (Fig. 9.9). About 600,000 years ago, if you were in Texas, Missouri, or Minnesota, your world would have been covered with about half an inch (1 centimeter) of volcanic ash. The ash was increasingly thicker toward Yellowstone, where it reached hundreds of feet. That eruption produced about 1,000 to 2,000 times the volume of material as the 1980 eruption of Mount St. Helens. In fact, it may have been the biggest geologic event that occurred anywhere on Earth in the past million years. A hike up Mount Washburn, just north of Canyon Junction, provides a panoramic view of Yellowstone National Park. From the top, the Yellowstone Caldera can be seen as a depression in the center of the park, rimmed by higher mountains (Fig. 9.10).

As happened at Crater Lake, volcanic material poured out onto the floor of the Yellowstone Caldera. Rhyolite lava erupted intermittently from 600,000 to 70,000 years ago. The Yellowstone River has carved through the younger flows, exposing them along the walls of the Grand Canyon of the Yellowstone River (Fig. 9.11).

FIGURE 9.10 View of remains of the Yellowstone Caldera from the top of Mount Washburn.

FIGURE 9.9 A colossal eruption occurred at Yellowstone about 600,000 years ago. a) Map showing the extent of the huge collapse caldera relative to the park boundaries. **b)** Sketches highlighting differences in scale between the crater formed during the 1980 eruption of Mount St. Helens (Fig. 5.19), the caldera developed during the eruption of Mount Mazama (Crater Lake) 7,700 years ago (Fig. 5.28), and the 600,000-year-old caldera at Yellowstone.

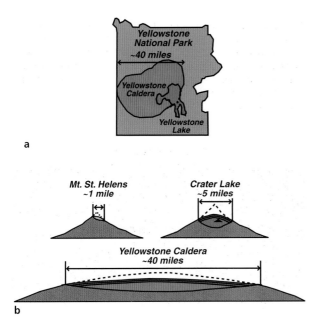

GRANITE, RHYOLITE—NEAPOLITAN ICE CREAM!

The rock on the walls of the Grand Canyon of the Yellowstone River is pink-to-gray-colored rhyolite (Fig. 9.11). Most park visitors are unfamiliar with the term *rhyolite*. But many know what granite looks like. Granite is a rock with big crystals of white quartz and pink feldspar as well as specks of black minerals that are rich in iron. You can see the crystals in granite with your eyes because, as the magma cooled slowly within the Earth, big crystals had time to grow. But suppose the same magma erupted on Earth's surface as lava flows? It would cool so fast that there would not be enough time for big crystals to form. Think of granite as sort of like Neapolitan ice cream—the kind with stripes of white (vanilla), pink (strawberry), and brown (chocolate). Now suppose you took that ice cream out of the refrigerator and let it melt. When you refroze it, it wouldn't have stripes; instead it would be all melded together as a pinkish or grayish color. That would be rhyolite; it has the same chemistry as granite, but most of the crystals are too small for you to see (Table 2.3).

FIGURE 9.11 Rhyolite exposed in the walls of the Grand Canyon of the Yellowstone River. The pink-to-gray-colored rocks are young lava flows erupted on the floor of the collapse caldera that formed 600,000 years ago.

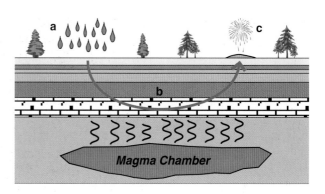

FIGURE 9.12 Hydrothermal system in Yellowstone National Park. a) Rainfall and snowmelt seep into the ground. **b)** Groundwater encounters rock heated from below by shallow magma chambers. **c)** Heated water expands and rises to the surface as geysers, hot springs, and mud pots.

THERMAL FEATURES OF YELLOWSTONE. Yellowstone National Park is a dazzling display of geysers, hot springs, mud pots, and other thermal features. But what is the source of the hot water? How does it make its way to the surface? And what causes geysers to erupt, often at such regular intervals? The answers to all of those questions relate to the presence of the hotspot directly beneath Yellowstone. The hotspot results in pockets of partially molten rock, or magma chambers, at various levels within the crust. But the hot water that surfaces as Yellowstone's geothermal features does not originate from those magma chambers. Rather, the water is rainfall and snowmelt that enters the ground from above. The magma chambers are essential, though, because they heat the surrounding rock.

As the rain and snow fall, some of the water moves down through the cracks and pore spaces of the shallow rock layers (Fig. 9.12). The groundwater descends to a depth of 1 to 2 miles (1½ to 3 kilometers), where it encounters hot rocks heated from below by a shallow, upper-crustal magma chamber. The hot water expands, rising to the surface as **hot springs** (Figs. 9.13 and 9.14). The water sometimes incorporates a lot of solid material on its way up, forming **mud pots** (Fig. 9.15). Deep within the Earth, the water is under great pressure. If it becomes superheated, the water may suddenly expand to gas, forcing the overlying water column out of a **geyser** (Fig. 9.16).

FIGURE 9.13 Some of the many hot springs at Yellowstone National Park. The white materials around the springs are silica-rich deposits called sinter, also known as geyserite. Bright orange and yellow colors are organisms capable of living under the extremely high-temperature conditions.

a

b

c

a

b

FIGURE 9.14 Mammoth Hot Springs. On the northern portion of Yellowstone National Park, the thermal waters must work their way through layers of limestone, dissolving large amounts of calcium carbonate. Layers of **travertine** precipitate out of Mammoth Hot Springs, forming beautiful terraces.

Geysers differ from hot springs in that hot springs flow continuously, whereas geyser water is forced out periodically. The reason is that the flow of geysers becomes constricted in narrow passageways. Constrictions in Yellowstone geysers are commonly due to the accumulation of **sinter** (or **geyserite**), silica-rich material that dissolves out of the rhyolite volcanic rocks and is deposited around subsurface passageways.

Like a person clearing his or her throat, the Earth uses a geyser to relieve pressure. Water within the Earth is under pressure from the weight of the water above it. As in a pressure cooker, the water can remain liquid, even when very hot. But the temperature may get so high that the water is superheated and flashes to steam. As a gas, water occupies about 1,500 times the volume it does as a liquid. That rapid expansion propels the overlying column of water upward very quickly through the constriction, causing the geyser to erupt. Geysers such as Old Faithful require some time for the water to become superheated (Fig. 9.17). Initially, puffs

LIFE UNDER EXTREME CONDITIONS

The hot springs of Yellowstone National Park contain forms of life adapted to temperatures at, and even above, the boiling point of water. Such **thermophilic bacteria** are the main source of the dazzling yellow and orange colors that are sometimes mistakenly attributed to mineral deposits. Preservation of the unique environment of Yellowstone's thermal features is critical not only for aesthetic reasons, but for some practical reasons as well. Researchers collect samples of thermophilic bacteria from Yellowstone hot springs and place them in labs for safekeeping. One form, *Thermus aquaticus,* provides the critical enzyme necessary for the duplication of molecules in DNA analysis. Such an opportunity might have been lost had Yellowstone's hot springs not been protected. It's important to preserve all of Earth's ecosystems, including the exotic ones found in Yellowstone.

a

b

c

FIGURE 9.15 Mud pots at Yellowstone National Park. The mud accumulates where thermal waters incorporate a lot of solid material on their way to the surface.

(a) Old Faithful

(b) Riverside

(c) Clepsydra

FIGURE 9.16 A few of the numerous geysers at Yellowstone National Park. a) Old Faithful Geyser in the Upper Geyser Basin. **b)** Riverside Geyser in the Upper Geyser Basin. **c)** Clepsydra Geyser in the Lower Geyser Basin.

of water vapor and small amounts of liquid water come out the geyser "throat." With the rapid expansion of water to gas below, the geyser then spouts water for a few minutes, followed by a time interval when the propelling water vapor escapes. It then takes some time for the water to recharge and heat up again, as the cycle repeats.

GRAND CANYON OF THE YELLOWSTONE RIVER. Yellowstone Caldera has partially filled with the waters of Yellowstone Lake (Fig. 9.18). The lake drains via the Yellowstone River, which flows northward across Hayden Valley. A little farther, the river pours over the Upper and Lower falls (Fig. 9.19). Below the falls lies the Grand Canyon of the Yellowstone River, in places more than 1,000 feet (300 meters) deep and ¼ mile (½ kilometer) wide. The Yellowstone River cuts into

BE CAREFUL WHAT YOU WISH FOR!

Yellowstone National Park contains about half of the world's approximately 1,000 geysers. Other areas with geysers include Iceland and New Zealand, where people have tapped into the groundwater system for geothermal energy, altering the "plumbing system," and in some cases shutting off geysers. National park status is crucial to protecting and preserving Yellowstone's geysers and other unique geothermal features from such a fate.

FIGURE 9.17 Eruption sequence of Old Faithful Geyser. a) A small amount of water vapor escapes as water heats up beneath the surface. The hot water under pressure eventually becomes superheated and flashes to water vapor. As it becomes gas, the water expands to about 1,500 times its volume. **b,c,d)** The geyser erupts as the expanding gas pushes the column of water upward. **e)** Remaining water vapor gradually escapes as the eruption ends.

a　　　　　　b　　　　　　c　　　　　　d　　　　　　e

FIGURE 9.18 Yellowstone Lake. View from West Thumb Geyser Basin.

FIGURE 9.19 Falls of the Yellowstone River. a) Upper Falls, measuring 109 feet (33 meters) in height. **b)** The Lower Falls drop 308 feet (94 meters).

(a) Upper Falls   **(b) Lower Falls**

layers of rhyolite that are about 70,000 years old. Those lava flows are some of the material that erupted onto the floor of the Yellowstone Caldera after it formed 600,000 years ago.

Yellowstone National Park has many beautiful places. Artist Point is one of them, overlooking the vivid colors of the Grand Canyon of the Yellowstone River and the 308-foot-high Lower Falls (Fig. 9.19b). From similar viewpoints, artist Thomas Moran painted some of his wonderful images of Yellowstone in 1871. But what caused the coloring seen along the walls of the canyon? The canyon area is the site of a fair-sized thermal area, similar to the Upper Geyser Basin around Old Faithful. Evidence of the hot waters can be seen as steam vents coming out of the walls of the canyon, and in the active mud pots near Clear Lake, about ½ mile (1 kilometer) south of the canyon. For thousands of years, hot water has been seeping through the layers of rhyolite. Rhyolite has a small amount of iron. When the

rhyolite is subjected to the hot water, iron oxide forms. The vivid colors are beautiful examples of such rust on the canyon walls.

When glacial ice carves a valley, it has a characteristic U shape, like many of the broad valleys of Yellowstone, Yosemite, Grand Teton, and Glacier national parks (Fig. 9.20a). But water makes a V shape, with the angle about the same all the way from the top to the bottom of the valley (Fig. 9.20b). The V-shaped profile of the Grand Canyon of the Yellowstone reveals that a stream, the Yellowstone River, is the primary carving factor.

But how did the Yellowstone River carve the canyon? Think about the Grand Canyon in Arizona. The canyon rim there is about the same elevation as Yellowstone's canyon rim, 7,000 to 8,000 feet (2,000 to 2,500 meters) above sea level. Think of holding a sharp knife over a soft wedding cake. Raise the cake up, but hold the knife in the same place. As the Colorado Plateau moved upward, the Colorado River eroded quickly, staying at an elevation of about 2,000 to 3,000 feet (600 to 1,000 meters). You can think of the Grand Canyon of the Yellowstone forming in a similar way.

HOW *DO* YOU GET A WATERFALL?

The Lower Falls of the Yellowstone River are 308 feet (94 meters) high, nearly twice as high as Niagara Falls on the New York–Ontario border. What causes such incredible falls? Consider two potatoes, one raw and one that was boiled for half an hour. Imagine what would happen if you stabbed each of the potatoes with a fork. The raw potato is hard, so it would probably bend the fork. But the boiled potato is soft—when you stab it with the fork, it flakes into pieces. Now consider the rhyolite in the Grand Canyon of the Yellowstone River. The normal rhyolite is hard; it's the darker, gray-colored rock above the Lower Falls (Fig. 9.19b). The brightly colored rhyolite downstream has been softened up by underground thermal waters, much like the boiled potato. So the Yellowstone River carved the canyon as the Yellowstone Plateau rose. But the rock at the Lower Falls is outside the thermal basin, where it stays hard like the raw potato. The Yellowstone River is having trouble cutting through that rock, so it remains higher, forming the Lower Falls.

(a) U-Shaped Glacial Valley   (b) V-Shaped River Valley

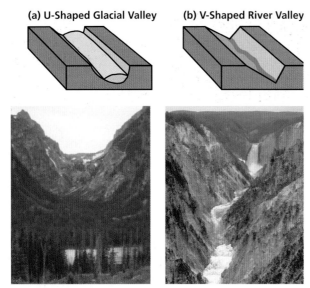

FIGURE 9.20 The shapes of valleys in Yellowstone National Park depend on whether they were carved by ice or water. a) Ice results in a flat valley floor surrounded by gradually steepening walls, producing a U shape. **b)** Water tends to carve valleys that retain the same slope from top to bottom, forming a V-shaped profile. The Yellowstone River carved the Grand Canyon of the Yellowstone River.

The Yellowstone Hotspot has been expanding and rising like a hot-air balloon, elevating the Yellowstone Plateau. So, like the knife cutting through the rising cake, the Yellowstone River carved the Grand Canyon of the Yellowstone.

GEOLOGY AND ECOLOGY: LIFE ABOVE A HOTSPOT. The overall plant and animal community, or **ecosystem**, is influenced by the local geology and climate conditions. The plants and animals found in Yellowstone National Park are there because the landscape developed over a hotspot. Rising, hot mantle has elevated the Yellowstone Plateau to 8,000 feet (2,500 meters) above sea level, resulting in long winters with deep snow cover. The rhyolite lava flows found throughout the park weather to very acidic soil. The high elevation and poor soil mean high stress for plants in Yellowstone. The dominant tree found in the park is the lodgepole pine, a tree that thrives in stressful environments (Fig. 9.21a).

Hayden Valley, in the heart of Yellowstone National Park, is a wonderful place to observe **ecology**, to see how the geologic setting affects life. Because of the high elevation of the Yellowstone Plateau, snow accumulates to depths of 5 to 10 feet (1½ to 3 meters). By late July, virtually all the snow in Yellowstone is gone. During past ice ages, Earth's average temperature was about 20 °F (12 °C) colder than it is today, so that not all the snow melted over the course of the summer. The snow accumulated year after year until an ice cap developed, similar to those seen today over Antarctica, Greenland, and part of Iceland. The Yellowstone ice cap was about 3,000 feet (1,000 meters) thick, enough to bury the top of Mount Washburn, at 10,243 feet (3,122 meters) elevation.

As the Earth started warming up about 18,000 years ago, Yellowstone Lake was filled with much more water. The remaining ice dammed the Yellowstone River at the mouth of Hayden Valley, filling it with about 250 feet (75 meters) of water. Periodically the

FIGURE 9.21 The plant and animal communities of Yellowstone National Park are part of an ecosystem that developed over a hotspot. a) Lodgepole pine forests. High elevation and acidic soil from rhyolite volcanic deposits result in stressful conditions ideal for lodgepole pines. **b)** Hayden Valley. The high elevation led to continental glacier development during ice ages. In Hayden Valley, those glaciers left impermeable deposits, where standing water now remains throughout summer. The tree line in the background roughly follows the shoreline of an ancient glacial lake that filled Hayden Valley to a depth of about 250 feet (75 meters).

a

b

FIGURE 9.22 Craters of the Moon National Monument.
The lava flows, spatter cones, and cinder cones represent late-stage hotspot volcanism—the region has subsided and basalt covers earlier rhyolite.

ice dam would break, so that huge torrents of water poured out. Such torrents greatly aided the carving of Yellowstone Canyon just downstream from Hayden Valley. The valley contains open meadows of grass, flowers, and shrubs (Fig. 9.21b). Up the slopes is a prominent tree line, marking the ancient shoreline of the glacial lake. The edge of the grassy meadows marks the extent of sediment deposited in the lake.

Glacial lake sediments come in a wide variety of sizes, from pebbles, to sand, to finer silt and clay. The pore spaces between the coarse grains get plugged up with the finer particles. Water cannot seep deeply into the ground; instead it flows on the surface. That's why Hayden Valley has so many areas of standing water. Because rain and snowmelt do not soak deeply into the ground, tree roots don't get enough water to survive. Instead, grasses, flowers, and bushes thrive. The standing water, combined with water from the Yellowstone River and smaller streams, makes Hayden Valley a lush environment for mammals and birds.

Stage 3: Late-Phase "Stem" (Basalt)

The volcanic features seen at the surface in Craters of the Moon National Monument in Idaho are 15,000 to 2,000 years old. The landscape is indeed like the moon (Fig. 9.22). Its lava flows, cinder cones, spatter cones, and lava tubes are all products of fluid, basaltic magma. The eruptions and volcanic products are quite unlike the explosive activity and light-colored rhyolite, pumice, and ash found 150 miles (250 kilometers) away in Yellowstone National Park.

Similar to Haleakalā National Park in Hawai`i, Craters of the Moon National Monument reveals late-stage volcanic activity occurring after a region of a plate passes over a hotspot. And, as at Haleakalā, more

eruptions are likely in the near future. In such a dry environment with little vegetation cover, Craters of the Moon is an ideal place to witness fresh volcanic features and to think about the plate-tectonic and magmatic processes responsible for their formation. The fact that there are plants and animals thriving within such a moonlike environment illustrates just how quickly life can establish itself on a newly-formed landscape.

Just a few million years ago, south-central Idaho had a landscape similar to that seen in Yellowstone today. The land was high above sea level, comparable to

HAYDEN VALLEY—A WEE BIT OF SCOTLAND

The heart of Yellowstone National Park may remind some visitors of Scotland. This is no coincidence. The Yellowstone Plateau is about 8,000 feet (2,500 meters) above sea level. Every 1,000 feet (300 meters) of elevation is equivalent to traveling about 300 miles (500 kilometers) northward. The climate of Yellowstone is in many respects similar to that of Scotland, which lies about 1,200 miles (2,000 kilometers) farther north. Golfers may also get a twinge to swing away as they gaze upon Hayden Valley. This, too, is no coincidence. Golf was developed in Scotland, a terrain scraped by continental glaciers and covered by their deposits. It has areas of trees, with open fairways, small lakes, and deposits of sand. Everywhere in the world, golf course designers try to replicate the glacial outwash plains of Scotland—a terrain naturally developed at Hayden Valley and other open areas of Yellowstone.

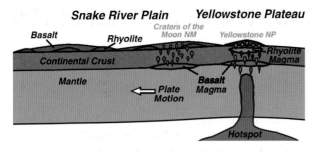

FIGURE 9.23 Late-stage hotspot volcanism covers the Snake River Plain. Partially melted mantle at the hotspot produces magma with basalt composition. Much of the magma at Yellowstone National Park was enriched in silica as it melted its way through continental crust, producing rhyolite. The region of the Snake River Plain has moved away from the hotspot; its crust has been depleted in silica, and conduits to the surface are well established. Residual basalt lava erupted at Craters of the Moon National Monument, covering earlier rhyolite deposits. Some basalt lava flows are also found at Yellowstone. The volcanism and long cracks (rift zones) seen at Craters of the Moon are also related to continental rifting of the nearby Basin and Range Province (Chap. 3).

FIGURE 9.25 Hagerman Fossil Beds National Monument. The fossil-bearing layers were deposited as the region of the Snake River Plain moved off the hotspot, cooled, and subsided.

the Yellowstone Plateau (Fig. 9.23). Pasty, silica-rich magma caused explosive volcanic eruptions and the deposition of layer upon layer of light-colored rhyolite. Since then, the Snake River Plain subsided and low-silica, fluid lava erupted, covering the earlier deposits with dark-colored basalt (Fig. 9.24). The heavy basalt layers, along with the coarse-grained gabbro that forms

at the base of the crust as the magma chambers cool, contribute to subsidence of the Snake River Plain. The subsidence has led to the deposition of sedimentary layers, including those containing the 3.7- to 3.15-million-year-old horse, mastodon, sabre-tooth cat, and other remains found at Hagerman Fossil Beds National Monument (Fig. 9.25).

FIGURE 9.24 Surface features of fluid lava flows in Craters of the Moon National Monument.
a) Pāhoehoe lava flows. **b)** `A`ā lava flows. **c)** Pressure ridges. **d)** Entrance to Great Owl Cavern, a lava tube.

a b c d

FURTHER READING

GENERAL

Brock, T. D. 1994. *Life at High Temperatures*. Yellowstone National Park, WY: Yellowstone Association for Natural Science, History and Education. 31 pp.

Bryan, T. S. 1990. *Geysers: What They Are and How They Work*. Niwot, CO: Roberts Rinehart. 24 pp.

Christopherson, E. 1962. *The Night the Mountain Fell: The Story of the Montana-Yellowstone Earthquake*. West Yellowstone, MT: Yellowstone Publications. 88 pp.

Decker, R., and B. Decker. 2001. *Volcanoes in America's National Parks*. New York: Norton. 256 pp.

Fritz, W. J. 1985. *Roadside Geology of the Yellowstone Country*. Missoula, MT: Mountain Press. 144 pp.

Good, J. M., and K. L. Pierce. 1996. *Interpreting the Landscape: Recent and Ongoing Geology of Grand Teton and Yellowstone National Parks*. Moose, WY: Grand Teton Natural History Association. 58 pp.

MacFadden, Bruce J. 1992. *Fossil Horses: Systematics, Paleobiology, and Evolution of the Family Equidae*. New York: Cambridge University Press. 369 pp.

National Park Service. 1991. *Craters of the Moon: Official National Park Handbook*. Washington: U.S. Government Printing Office, Handbook Number 139. 64 pp.

Smith, R. B., and R. L. Christiansen. 1980. Yellowstone Park as a window on the Earth's interior. *Scientific American*, v. 242, p. 104–117.

Smith, R. B., and L. J. Siegel. 2002. *Windows into the Earth: A Geologic Story of Yellowstone and Grand Teton National Parks*. New York: Oxford University Press. 242 pp.

Thayer, T. P. 1994. *The Geologic Setting of the John Day Country*. Castle Rock, WA: Northwest Interpretative Association. 23 pp.

TECHNICAL

Anderson, D. L. 1994. The sublithospheric mantle as the source of continental flood basalts: The case against the continental lithosphere and plume head reservoirs. *Earth and Planetary Science Letters*, v. 123, p. 269–280.

Anderson, D. L., T. Tanimoto, and Y. Zhang. 1992. Plate tectonics and hotspots: The third dimension. *Science*, v. 256, p. 1645–1651.

Blackwell, D. D. 1989. Regional implications of heat flow of the Snake River Plain, northwestern United States. *Tectonophysics*, v. 164, p. 323–343.

Christiansen, R. L. 1984. Yellowstone magmatic evolution: Its bearing on understanding large-volume explosive volcanism, in *Explosive Volcanism, Its Inception, Evolution and Hazards*. National Research Council Studies in Geophysics. Washington: National Academy Press, p. 84–95.

Dueker, K. G., and A. F. Sheehan. 1997. Mantle discontinuity structure from midpoint stacks of converted P to S waves across the Yellowstone hotspot track. *Journal of Geophysical Research*, v. 102, p. 8313–8327.

Duncan, R. A., and D. G. Pyle. 1988. Rapid eruption of the Deccan flood basalts at the Cretaceous/Tertiary boundary. *Nature*, v. 333, p. 841–843.

Evans, J. R. 1982. Compressional wave velocity structure of the upper 350 km under the eastern Snake River Plain near Rexburg, Idaho. *Journal of Geophysical Research*, v. 87, p. 2654–2670.

Hildreth, W., A. Z. Halliday, and R. L. Christiansen. 1991. Isotopic and chemical evidence concerning the genesis and contamination of basaltic and rhyolitic magma beneath the Yellowstone Plateau volcanic field. *Journal Petrology*, v. 52, p. 63–138.

Humphreys, E. D., and K. G. Dueker. 1994. Physical state of the western U.S. upper mantle. *Journal of Geophysical Research*, v. 99, p. 9635–9650.

Keefer, W. R. 1971. *The Geologic Story of Yellowstone National Park*. Washington: U.S. Geological Survey, Bulletin 1347. 92 pp.

Morgan, W. J. 1971. Convection plumes in the lower mantle. *Nature*, v. 230, p. 42–43.

Morgan, W. J. 1972. Deep mantle convection plumes and plate motions. *American Association of Petroleum Geologists Bulletin*, v. 56, p. 203–213.

Okal, E. A., and R. Batiza. 1987. Hotspots: The first 25 years, in *Seamounts, Islands, and Atolls,* edited by B. H. Keating, P. Fryer, R. Batiza, and G. W. Boehlert. Washington: American Geophysical Union, Monograph Series v. 43, p. 1–11.

Peng, X., and E. D. Humphreys. 1998. Crustal velocity structure across the eastern Snake River Plain and the Yellowstone swell. *Journal of Geophysical Research*, v. 103, p. 7171–7186.

Richmond, G. M. 1987. Geology and evolution of the Grand Canyon of the Yellowstone, Yellowstone National Park, Wyoming, in *Geological Society of America Centennial Field Guide, Rocky Mountain Section*. Boulder, CO: Geological Society of America. p. 155–160.

Sleep, N. H. 1990. Hotspot volcanism and mantle plumes. *Annual Review of Earth and Planetary Science*. v. 20, p. 19–43.

Smith, R. B., and L. W. Braile. 1994. The Yellowstone Hotspot. *Journal of Volcanology and Geothermal Research*, v. 61, p. 121–187.

Smith, R. B., R. T. Shuey, R. O. Freidline, R. M. Ottis, and L. B. Alley. 1974. Yellowstone hot spot: New magmatic and seismic evidence. *Geology*, v. 2, p. 451–455.

Thompson, R. N. 1977. Columbia–Snake River–Yellowstone magmatism in the context of western U.S.A. Cenozoic geodynamics. *Tectonophysics*, v. 39, p. 621–636.

WEBSITES

National Park Service: www.nps.gov

Park Geology Tour:
www2.nature.nps.gov/grd/tour/index.htm
  Columbia Plateau/Snake River Plain/Yellowstone:
    88. Craters of the Moon NM:
    www2.nature.nps.gov/grd/parks/crmo/index.htm
    89. Hagerman Fossil Beds NM:
    www2.nature.nps.gov/grd/parks/hafo/index.htm
    90. John Day Fossil Beds NM:
    www2.nature.nps.gov/grd/parks/joda/index.htm

    91. Yellowstone NP:
    www2.nature.nps.gov/grd/parks/yell/index.htm
    Old Faithful Geyser Camera: www.nps.gov/yell/oldfaithfulcam.htm

U.S. Geological Survey: www.usgs.gov

Geology in the Parks:
www2.nature.nps.gov/grd/usgsnps/project/home.html

Volcanoes: volcanoes.usgs.gov
  Volcano Monitoring:
  volcanoes.usgs.gov/About/What/Monitor/monitor.html

University of Utah, Dept. of Geology and Geophysics:

Yellowstone Hotspot Monitoring:
www.mines.utah.edu/~rbsmith/RESEARCH/YellowstoneHotspot.html
Yellowstone Caldera GPS: www.mines.utah.edu/~rbsmith/RESEARCH/Yell.Hotspot.Deformation.html

Other:

World Geyser Fields:
www.uweb.ucsb.edu/~glennon/geysers/world.htm

PART VI

Building the North American Continent

Continents grow as material is added to their edges over time. Chapter 10 presents park lands that lay over the older, and today stable, North American Craton. It also includes areas fringing the craton that have undergone deformation and uplift in recent geological times. Chapter 11 discusses parks that contain pieces of the continent that were made elsewhere, transported great distances via convergent or transform plate motion, before being accreted to North America.

◀ [Overleaf]
Denali National Park, Arizona. The highest mountains in North America result from terranes crashing into the continent.

The preceding chapters revealed the landscapes of national parks formed by processes that occur at plate boundaries and hotspots. National park lands in the Midwest rest upon a complex array of rocks and structures that hint at plate-tectonic processes responsible for the formation of ancient North America. Even where the old rocks are covered, younger sedimentary layers present compelling stories of invasion of shallow seas over the continent and the advancement and retreat of colossal ice sheets.

The ancient nucleus of North America is known as the **continental craton**. The **continental shield** is the part of the craton where old rocks are exposed at the surface, whereas the **continental platform** comprises regions of the craton covered by thin sedimentary layers. Voyageurs National Park in northern Minnesota lies on the very old igneous and metamorphic rocks on the continental shield. Mammoth Cave National Park in Kentucky and Theodore Roosevelt National Park in North Dakota represent some of the park lands that lie on continental platform strata.

Badlands National Park, South Dakota. Shallow seas that periodically covered the continental platform deposited the sedimentary layers in a foreland basin adjacent to the rising Rocky Mountains.

The craton has experienced Mesozoic and Cenozoic deformation along its western edge. Arches National Park in Utah is one of the parks that lie on the **Colorado Plateau**, a large piece of the craton that lifted up without much internal deformation. Other sites, including Rocky Mountain National Park in Colorado, reveal smaller uplifted and deformed blocks of the craton known as **Laramide uplifts**. Glacier National Park in Montana reveals deformation of sedimentary strata that were detached from the underlying hard crust and deformed as a **foreland fold-and-thrust belt**.

Many national parks in the western United States showcase rocks and structures that developed elsewhere and only recently became part of North America. Most of Alaska is like a giant jigsaw puzzle, assembled as fragments of crust formed during plate divergence, subduction, or hotspot processes, and then moved northward along transform and convergent plate boundaries. The pieces at times slammed into one another and ultimately into the rest of Alaska, where they are referred to as **accreted terranes**. Parks in other regions of the country also

North Cascades National Park, Washington. The landscape is composed of igneous, sedimentary, and metamorphic rocks that were manufactured elsewhere and later accreted to the edge of the continent.

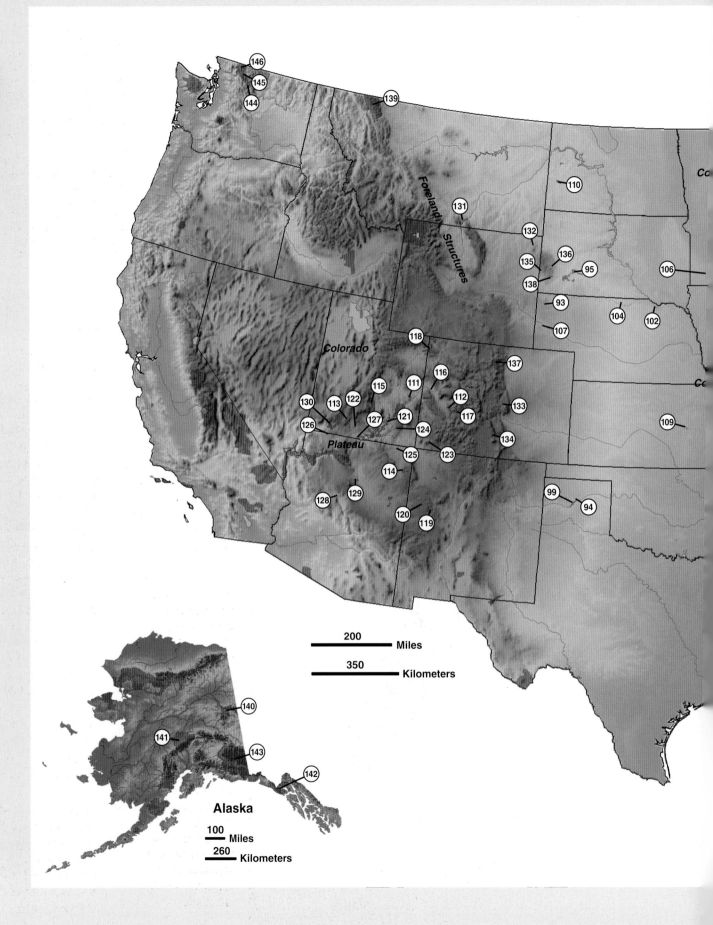

Foreland Structures

Colorado

Plateau

Alaska

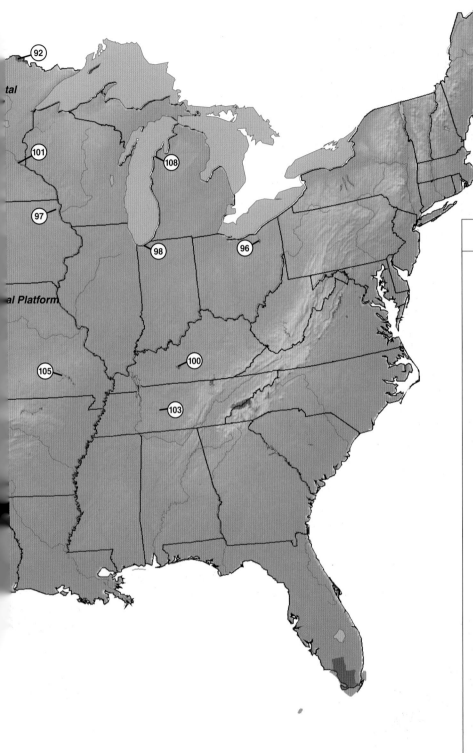

Raised-relief map of the United States highlighting national park lands that lie on the North American Craton and accreted terranes. The **continental craton** consists of a **continental shield** of exposed ancient rocks, flanked by a **continental platform** where a thin veneer of sedimentary strata covers the older rocks. The **Colorado Plateau** and **foreland structures**, including **Laramide uplifts** and **foreland fold-and-thrust belts**, are regions where the edge of the craton has recently been deformed and uplifted. Blocks of crust that were recently attached to the craton, known as **accreted terranes**, are found in parts of the West, especially Alaska.

Map Key

Continental Shield:
92. Voyageurs NP.

Continental Platform:
93. Agate Fossil Beds NM
94. Alibates Flint Quarries NM
95. Badlands NP
96. Cuyahoga Valley NP
97. Effigy Mounds NM
98. Indiana Dunes NL
99. Lake Meredith NRA
100. Mammoth Cave NP
101. Mississippi NR&RA
102. Missouri NRR
103. Natches Trace Pkwy
104. Niobrara NSR
105. Ozark NSR
106. Pipestone NM
107. Scotts Bluff NM
108. Sleeping Bear Dunes NL
109. Tallgrass Prairie NPres
110. Theodore Roosevelt NP

Colorado Plateau:
111. Arches NP
112. Black Canyon of the Gunnison NP
113. Bryce Canyon NP
114. Canyon de Chelly NM
115. Capitol Reef NP
116. Colorado NM
117. Curecanti NRA
118. Dinosaur NM
119. El Malpais NM
120. El Morro NM

121. Glen Canyon NRA
122. Grand Staircase-Escalante NM
123. Mesa Verde NP
124. Natural Bridges NM
125. Navajo NM
126. Pipe Spring NM
127. Rainbow Bridge NM
128. Tuzigoot NM
129. Wupatki NM
130. Zion NP

Laramide Uplifts:
131. Bighorn Canyon NRA
132. Devils Tower NM
133. Florissant Fossil Beds NM
134. Great Sand Dunes NM
135. Jewel Cave NM
136. Mount Rushmore N Mem
137. Rocky Mountain NP
138. Wind Cave NP

Foreland Fold-and-Thrust-Belts:
139. Glacier NP

Deformed Craton (Alaska):
140. Yukon-Charley Rivers NPres

Accreted Terranes:
141. Denali NP
142. Glacier Bay NP
143. Wrangell-St. Elias NP
144. Lake Chelan NRA
145. North Cascades NP
146. Ross Lake NRA

have terranes that were gradually accreted onto the craton, revealing that the North American continent has been growing outward over time. The **Cordillera** is the region of western North America that has experienced extensive mountain building in modern geologic times, including growth of the continent by terrane accretion. Denali, Wrangell–St. Elias, and Glacier Bay national parks in Alaska, as well as North Cascades National Park in Washington State, are made of accreted terranes that have been added to North America in the past 200 million years.

TEN QUESTIONS: Building the North American Continent

1. How do continents form?
2. How old are portions of the North American continent?
3. What do national park lands in the Midwest have to do with plate tectonics?
4. Why do shallow seas advance and retreat over continents?
5. Why does sea level rise and fall?
6. What determines the extent of continental ice sheets?
7. What is an accreted terrane?
8. How do we know where an accreted terrane came from?
9. How did Mount McKinley (Denali) develop as the highest peak in North America?
10. Why are metamorphic rocks so important in understanding the history of development of the North American continent?

10

North American Craton

▲▲▲▲▲▲ *"The Continental shield of North America contains some of the oldest rocks found anywhere on Earth. Those rocks illustrate how the continent originally formed and then grew outward through time."*

National park lands in the central part of the country rest upon the very backbone of North America, where much of the early history of the continent is preserved (Table 10.1). Earlier chapters showed park features formed by plate boundary and hotspot processes. Rocks and structures beneath parks in the continental interior represent ancient examples of those processes. In some parks, young sedimentary layers cover the old rocks, and in others, recent tectonic activity has deformed the rocks.

The old nucleus of the continent is known as the **North American Craton** and includes most of the interior of Canada and the United States (Fig. 10.1). Plate-tectonic processes are evident in the topography of areas surrounding the craton, but the interior of the craton has not experienced significant tectonic activity in hundreds of millions of years. The part of the craton where igneous and metamorphic rocks (the "crystalline basement") are exposed is the **continental shield**, whereas the **continental platform** includes regions of the craton covered by a thin veneer of sedimentary layers deposited in shallow seas or by continental glaciers. Flanking the craton are younger mountain ranges and coastal plain deposits representing the ongoing, outward growth of the continent (Chaps. 4, 5, 6, and 11).

The landscapes of many national parks relate to the ancient margins and craton of North America (Figs. 10.2 and 10.3). National park lands in northern Minnesota, the Upper Peninsula of Michigan, and parts of Wisconsin lie on very old rocks of the shield that record aspects of the early history of the continent. Chapter 3 presented features in some of those parks as part of the discussion of the Keweenawan Rift, an ancient split in the continent. Most of the platform has

UNDEFORMED CRATON			
SHIELD			
92. Voyageurs NP, MN			
PLATFORM			
93. Agate Fossil Beds NM, NE	103. Natchez Trace Pkwy MS/AL/TN		
94. Alibates Flint Quarries NM, TX	104. Niobrara NSR, NE		
95. Badlands NP, SD	105. Ozark NSR, MO		
96. Cuyahoga Valley NP, OH	106. Pipestone NM, MN		
97. Effigy Mounds NM, IA	107. Scotts Bluff NM, NE		
98. Indiana Dunes NL, IN	108. Sleeping Bear Dunes NL, MI		
99. Lake Meredith NRA, TX	109. Tallgrass Prairie N Pres, KS		
100. Mammoth Cave NP, KY	110. Theodore Roosevelt NP, ND		
101. Mississippi NR&RA, MN			
102. Missouri NRR, SD/NB			

TABLE 10.1 National Parks on the North American Craton.

DEFORMED CRATON	
COLORADO PLATEAU	**FORELAND BASEMENT UPLIFTS**
111. Arches NP, UT	131. Bighorn Canyon NRA, MT
112. Black Canyon of the Gunnison NP, CO	132. Devils Tower NM, WY
113. Bryce Canyon NP, UT	133. Florissant Fossil Beds NM, CO
114. Canyon de Chelly NM, AZ	134. Great Sand Dunes NM, CO
115. Capitol Reef NP, UT	135. Jewel Cave NM, SD
116. Colorado NM, CO	136. Mount Rushmore N Mem, SD
117. Curecanti NRA, CO	137. Rocky Mountain NP, CO
118. Dinosaur NM, UT/CO	138. Wind Cave NP, SD
119. El Malpais NM, NM	
120. El Morro NM, NM	**FORELAND FOLD-AND-THRUST BELT**
121. Glen Canyon NRA, UT	
122. Grand Staircase–Escalante NM, UT	139. Glacier NP, MT
123. Mesa Verde NP, CO	
124. Natural Bridges NM, UT	**ALASKA**
125. Navajo NM, AZ	
126. Pipe Spring NM, AZ	140. Yukon-Charley Rivers N Pres, AK
127. Rainbow Bridge NM, UT	
128. Tuzigoot NM, AZ	
129. Wupatki NM, AZ	
130. Zion NP, UT	

FIGURE 10.1 North American Craton. The continental shield is composed of igneous and metamorphic rocks forming the old nucleus of the continent. A thin veneer of younger sedimentary layers deposited where shallow seas covered the craton is known as the continental platform. The Colorado Plateau is a portion of the platform that uplifted as a coherent block. The craton is flanked by younger mountain ranges (Phanerozoic orogenic belts), formed as new crustal material accreted onto the edge of the continent (Chaps. 5, 6, and 11). The ranges include deformed edges of the craton, or foreland structures, including basement (Laramide) uplifts and fold-and-thrust belts. Young sedimentary layers cover the coastal plain as the modern passive continental margin subsides (Chap. 4).

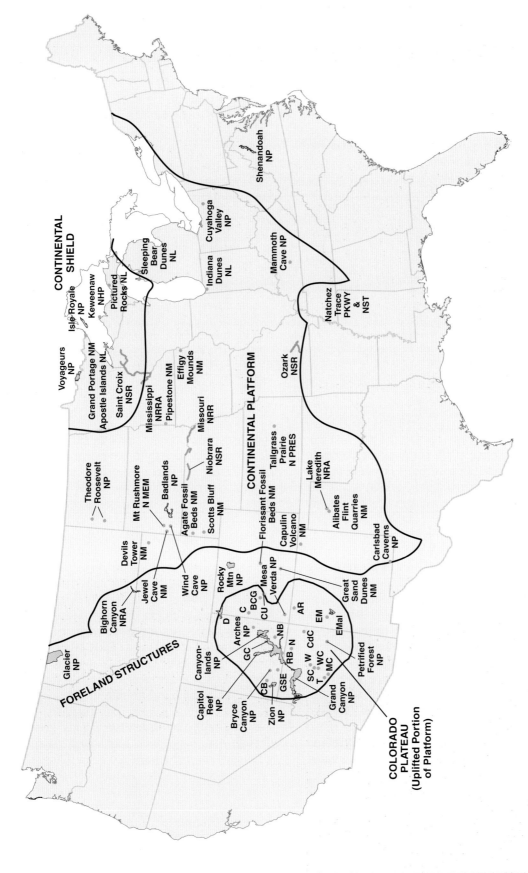

FIGURE 10.2 National park lands on the North American Craton. Voyageurs National Park contains very old igneous and metamorphic rocks of the continental shield. Some parks in the Keweenawan Rift portion of the shield, such as Isle Royale National Park, are discussed in Chap. 3. Other parks in the Midwest lie on thin strata of the stable continental platform. The Colorado Plateau is an uplifted but relatively intact portion of the platform, whereas foreland structures are more deformed. The schematic cross section in Fig. 10.3 shows positions of these park lands relative to the North American Craton and its margins. Grand Canyon National Park extends onto the ancient passive continental margin and is discussed in Chap. 4. Shenandoah National Park is along the passive continental margin that was deformed and uplifted as the Appalachian Mountains (Chap. 6). Abbreviations for national monuments on the Colorado Plateau: AR = Aztec Ruins; BCG = Black Canyon of the Gunnison; C = Colorado; CB = Cedar Breaks; CdC = Canyon de Chelly; D = Dinosaur; EM = El Morro; EMal = El Malpais; GSE = Grand Staircase–Escalante; MC = Montezuma Castle; N = Navajo; NB = Natural Bridges; RB = Rainbow Bridge; SC = Sunset Crater Volcano; T = Tuzigoot; W = Wupatki; WC = Walnut Canyon. National recreation areas: GC = Glen Canyon; Cu = Curecanti.

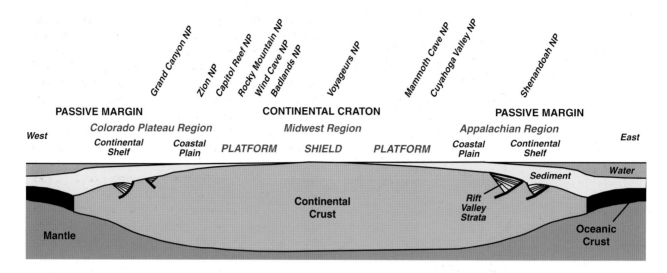

FIGURE 10.3 Diagrammatic cross section showing positions of selected national park lands relative to passive continental margin, platform, and shield areas of North America during the early part of the Paleozoic Era, about 500 million years ago. The eastern passive margin was later deformed as it collided with Gondwanaland, forming the Appalachian Mountains (Chap. 6). Part of the western passive margin and platform recently uplifted as a relatively undeformed block, the Colorado Plateau (Chap. 4), and as the more deformed Laramide uplifts and foreland fold-and-thrust belts. The continent has also grown westward through terrane accretion (Chap. 11). The remainder of the platform and shield has undergone very little uplift or internal deformation during the past 500 million years.

undergone very little deformation in the past 500 million years. A large piece of the platform, the **Colorado Plateau**, has recently lifted up, but its cover strata remain flat with little deformation. Rock layers on the western edge of the Colorado Plateau, most notably those exposed in Grand Canyon National Park, represent an ancient passive continental margin (Chap. 4). Relatively thin rock layers, revealed by erosion and in caves in parks in the Midwest as well as in canyons in some parks on the Colorado Plateau, reveal the sedimentary processes that occurred as shallow seas periodically flooded North America's interior. In other places, the craton was compressed and its hard basement rocks deformed and shoved upward as **foreland basement uplifts**, most notably in the frontal portion of the Rocky Mountains known as the **Laramide uplifts**. In places, the platform strata were detached from the underlying basement rocks and deformed as **foreland fold-and-thrust belts**. Many western parks reveal sedimentary layers that were deposited in **foreland basins** that developed adjacent to the rising mountains.

CONTINENTAL SHIELD

The continental shield of North America contains some of the oldest rocks found anywhere on Earth. Those rocks illustrate how the continent originally formed

and then grew outward through time. Early tectonic plates developed a thin veneer resembling oceanic crust. Where they converged and one plate slid beneath another at a subduction zone, a volcanic island chain formed. Similar to today's island arcs, the crust thickened by addition of volcanic material at the surface and igneous intrusives from below. As land above the sea eroded, a layer of sediment covered the surrounding oceanic crust. Some of the sedimentary layers, as well as basalt of the underlying crust, were scraped off subducting plates and formed accretionary wedges adjacent to the volcanic arcs (Chap. 5). At times, volcanic arcs collided, widening and further thickening the crustal mass. Such processes gradually formed the nuclei of ancient continents, but the new continents would not subduct because their crust was thicker and more buoyant than the crust of surrounding ocean basins (Fig. 6.2). The net result was that the ancient continental nuclei, the shields, grew outward through time as material was added via subduction, sedimentation, and terrane accretion (Chap. 11).

The North American continental shield is divided into provinces based on the types of rocks, their ages, and patterns of structural deformation. Park lands in the upper Midwest lie within the **Superior Province**, which covers a large part of Canada and extends southward into the northern parts of Michigan, Wisconsin, and Minnesota (Fig. 10.4). The rocks in the province

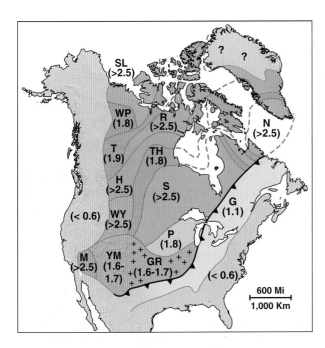

FIGURE 10.4 Outward growth of the North American continent. The map shows the ages (in billions of years) of crystalline basement rocks that are exposed at the surface in the shield and lie beneath younger sedimentary layers on the platform. Province abbreviations: G = Grenville; GR (with + pattern) = granite-rhyolite; H = Hearn; M = Mojave; N = Nain; P = Penokean; R = Rae; S = Superior; SL = Slave; T = Thelon; TH = Trans-Hudson; WP = Wopmay; WY = Wyoming; YM = Yavapai and Mazatzal.

processes occurring today beneath the Himalayan, Andean, and other mountain ranges, and are comparable to rocks exposed in Shenandoah, North Cascades, Wrangell–St. Elias, and Denali national parks. The old igneous and metamorphic rocks in the park are a couple of billion years old, covered in places by glacial deposits just a few thousand years old.

All of the rocks in Voyageurs National Park are Precambrian, meaning that they formed before the Cambrian Period, which began 543 million years ago (Table 2.1). The oldest are Archean metamorphic rocks formed about 2.7 billion years ago. They include highly metamorphosed gneiss formed at depths of at least 10 miles (16 kilometers). **Greenstone** occurs in long, down-warped belts within the gneiss. The name reflects the presence of minerals with a green tinge, namely chlorite and epidote, formed by the metamorphic processes. In places, blobs of greenstone look remarkably similar to pillow lavas that form during

formed during the Archean Eon, a span of time from 3.8 to 2.5 billion years ago that covers about a quarter of Earth's history. The Superior Province is renowned for its mineral deposits, including iron, copper, nickel, and gold.

The igneous and metamorphic rocks of the shield must have formed at high pressure and temperature conditions deep within the crust. After hundreds of millions of years of little tectonic activity, the region has eroded to a landscape of low relief. As erosion removed the weight of the overlying material, the crust rebounded upward, exposing the igneous and metamorphic rocks from great depth (Fig. 6.5).

Ancient Geologic History

Igneous and metamorphic rocks in Voyageurs National Park in northern Minnesota show evidence of plate-tectonic processes that occurred long ago (Fig. 10.5). They provide clues to the igneous and metamorphic

FIGURE 10.5 Voyageurs National Park, Minnesota. The park is a vast region of ancient igneous and metamorphic rocks, covered in places by forests, glacial deposits, and a system of lakes. Like the Boundary Waters area next door, the park is mostly accessible by boat.

a

b

ANCIENT PLATE TECTONICS

A basic tenet of geology is that "the present is the key to the past," known as **uniformitarianism**. It helps us interpret rocks and other features by assuming that ancient geologic processes occurred in ways similar to how they occur today. Chemical and physical (or mechanical) principles involved in the geologic processes were the same in the past as in the present. For example, high temperatures are required to melt rocks today. Uniformitarianism indicates that high temperatures were also required to melt rocks in the past. But uniformitarianism does not imply that the extent or rates of melting are the same today as they were in the past.

Conditions were different during Earth's early history. The Earth was as hot as the sun when it initially formed, and it took a long time to cool down to near its present thermal state. High temperature in the distant past meant that plates of lithosphere were much thinner than they are today. Convection in the underlying asthenosphere would have been much more vigorous, so that plates moved faster. So although it's useful to compare ancient rocks and structures found in places such as Voyageurs National Park with those in modern tectonic settings, conditions and the rate of some processes then were likely different from today.

volcanic eruptions on the ocean floor (Fig. 5.11). The greenstones are thought to be from the crust of ancient oceans.

A period of mountain building, known as the Kenoran Orogeny, straddles the Archean/Proterozoic boundary (2.5 billion years ago), and involved the emplacement of granitic intrusive rocks (batholiths) accompanied by folding and uplift. A very high mountain range must have formed, perhaps because of plate subduction and continental collision. Low-silica (gabbro) dikes emplaced about 2.1 billion years ago constitute the youngest in-place rocks found in Voyageurs National Park.

Other sites in the upper midcontinent region, including Isle Royale National Park in Lake Superior and Keweenaw National Historical Park on Michigan's Upper Peninsula, contain younger Precambrian rocks. Those rocks reveal that the craton tried to rip apart about 1.1 billion years ago along the Keweenawan Rift (Chap. 3).

Younger Erosion and Glaciation

Erosion has been the dominant process in the Voyageurs National Park region for the past 2 billion years. Glaciers have scraped away younger rock and soil, exposing the very old, hard rocks beneath (Fig. 10.6). Only the deep roots of ancient mountains remain.

Glaciers take away by erosion, but they also give back. Lakebed sediment and **outwash gravel** were deposited in many national park areas as the latest ice sheet retreated and melted away about 12,000 years ago (Fig. 10.7). Unsorted glacial materials, known as **till**, occur as **ground moraines** that formed under ice sheets, **lateral moraines** deposited between glaciers and the adjacent wallrock, and **terminal moraines** dropped at the front of glaciers. **Glacial erratics**, consisting of pebble-to-bus-sized rock fragments that were embedded in ice and carried a few miles to hundreds of miles, are found in many parks. Some erratics eroded from the Voyageurs National Park region were transported as far away as Iowa.

Sand grains and rock fragments embedded in the bottom of a glacier can scrape against the underlying bedrock, forming long, parallel grooves. Such **glacial striations** preserved on rock surfaces in Voyageurs National Park suggest that ice moved in directions ranging from south to southwest across the northern Minnesota region. Many of the lakes in national park sites in the Midwest result from glacial processes. Large blocks of ice remained buried under glacial sediment after the ice sheets retreated. As the embedded ice blocks melted, the surface collapsed. The resulting depressions formed "kettles" that filled in with **kettle lakes**.

FIGURE 10.6 Effects of glaciation at Voyageurs National Park. The elongated islands and waterways reflect the carving by continental ice sheets that advanced over the region.

FIGURE 10.7 Maximum extent of continental glaciers over North America during Pleistocene ice ages. a) At times, the entire shield region was covered by ice. Old igneous and metamorphic rocks are exposed so extensively in the shield region because the south-moving ice sheets scraped away soil and younger rock layers. Glacial striations, erratics, moraines, and kettle lakes in Voyageurs National Park are evidence that the ice advanced across northern Minnesota. Mammoth Cave National Park in Kentucky lies beyond the extent of the ice sheets. **b)** Close-up of midwestern states showing extent of glaciation during the Illinoian advance of the Pleistocene Epoch, about 500,000 years ago. Many national park lands north of the Ohio River have moraines, erratics, and other evidence of Pleistocene Epoch glaciation.

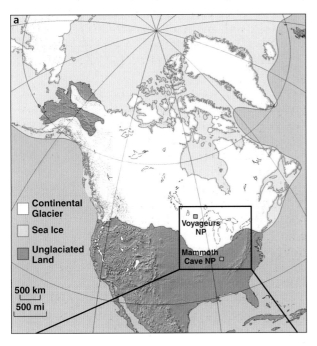

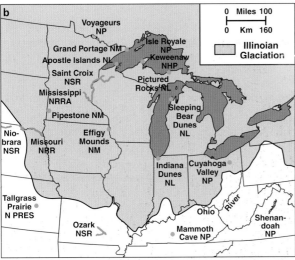

CONTINENTAL PLATFORM

The continental platform consists of a thin veneer of sedimentary layers, mostly of Paleozoic and Mesozoic age, overlying the older (Precambrian) igneous and metamorphic ("crystalline basement") rocks. Mammoth Cave, Cuyahoga Valley, and other national parks in the Midwest lie on flat sedimentary layers that were deposited as shallow seas lapped onto the low-lying craton of North America. Although the continent can bob up and down slowly over time, the main reason its low-lying interior periodically floods is that sea level rises and falls. One cause of sea level change is the accumulation or melting of polar ice caps as the temperature of Earth's atmosphere cools or warms over time (Fig. 10.8). Another cause is the changing volume of ocean basins as mid-ocean ridges swell during times of fast plate divergence and shrink when plates diverge slowly. Weathering of the platform strata and their cover of glacial, lake, and river deposits results in rich soils that make the Midwest the "breadbasket" of the United States.

The North American continent was near the equator for much of the Paleozoic and Mesozoic eras (Fig. 10.9). Shallow seas that occasionally flooded the continental interior deposited sand and mud that formed layers of sandstone and shale (Fig. 10.10). The seas contained marine organisms with shell and skeletal parts made out of calcium carbonate. In the warm seas, coral and other animals formed elaborate reefs. Some of the calcium carbonate was dissolved by seawater and then precipitated as fine lime mud, forming layers of limestone. Limestone and ancient reef material later interacted with groundwater, dissolving cavities and channels that became the passageways of Mammoth Cave National Park in Kentucky. Low, hot latitudes also produced desert conditions where giant sand dunes formed, now preserved as the spectacular sandstone layers of Zion, Arches, and other parks of the Colorado Plateau.

Precambrian basement rocks and their younger platform cover are up-warped in places, exposing some of the older sedimentary layers in broad **domes**, such as the Cincinnati Arch in Ohio and Kentucky and the Nashville Dome in Tennessee (Fig. 10.11). In other places the basement is warped downward, forming **basins**; young sedimentary layers are found in the centers of the Michigan and Illinois basins.

The landscapes of some national parks on the continental platform were formed by the interaction of the sedimentary rock layers with water flowing both above

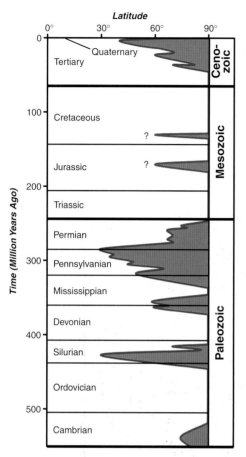

Latitude

FIGURE 10.8 Major advance and retreat of polar ice caps during the Phanerozoic Eon (past 543 million years). The graph shows that at times ice extended as far as 30° latitude. At other times, there were no polar ice caps. Times of cold temperature correspond to ice ages and sea level drop, while warm temperatures cause ice to melt and shallow seas to advance over continental platforms.

FIGURE 10.9 Position of the North American continent during the early Paleozoic (a), late Paleozoic (b), and middle Mesozoic (c) eras. Lying near the equator, the continental interior was the site of reef development and the deposition of limestone when shallow seas advanced, as well as the accumulation of desert sands when seas retreated.

(a) Late Cambrian (~510 Million Years Ago)

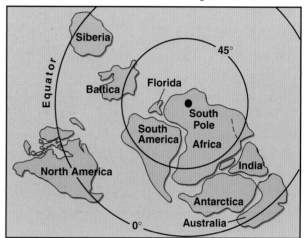

(b) Pennsylvanian (~300 Million Years Ago)

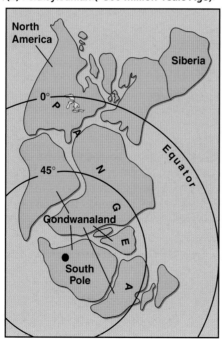

(c) Late Jurassic (~150 Million Years Ago)

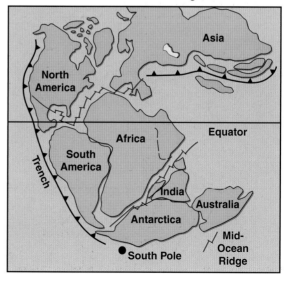

FIGURE 10.10 Shallow seas advancing over the North American Craton deposited continental platform strata. For a few hundred million years, the continent straddled the equator. Warm-water carbonate rocks, including limestone layers and coral reefs, formed where shallow seas flooded the platform region. **a)** The continent was narrower during the early part of the Paleozoic Era, and subsided along passive continental margins (Fig. 10.3). **b)** By the middle part of the Paleozoic Era, plate convergence led to the development of mountain ranges (Antler and Acadian orogens) along the former passive margins. Shallow seas continued to flood the platform region, depositing more limestone and reef material as well as sandstone and shale layers from erosion of the rising mountains (as in, for example, Catskill deltas). The limestone responsible for caves in Mammoth Cave and Wind Cave national parks was deposited during the Mississippian Period, about 350 to 325 million years ago. **c)** Intense terrane accretion and mountain building continued in the Cordilleran region of western North America during the Mesozoic Era. The Western Interior Seaway continued to deposit strata over regions that were later uplifted as the Colorado Plateau and Laramide uplifts. Regions in front of the rising mountains that received large volumes of sediment are known as foreland basins.

(a) Early Cambrian (~540 Million Years Ago)

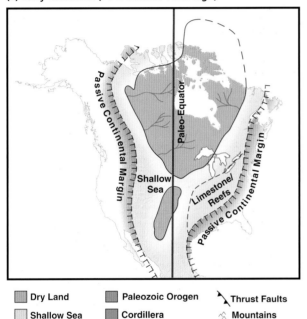

(b) Late Devonian (~360 Million Years Ago)

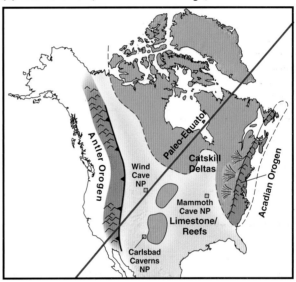

(c) Late Cretaceous (~90 Million Years Ago)

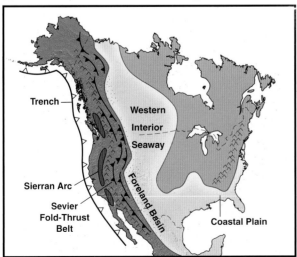

FIGURE 10.11 Broad structures on the North American continental platform. a) Domes (also called arches) form where the Precambrian crystalline basement and its cover of platform strata are warped upward like circular anticlines (Fig. 2.8a); erosion exposes older strata at the center. Basins are circular synclines (Fig. 2.8b) with younger strata near the center. **b)** Cross section along line I–I′ in Fig. 10.11a. Mammoth Cave National Park lies on gently dipping Mississippian limestone between the Nashville Dome and Illinois Basin. The vertical scale of the cross section is greatly exaggerated; the northwestward tilt (dip) of layers is only about ½° in the Mammoth Cave region.

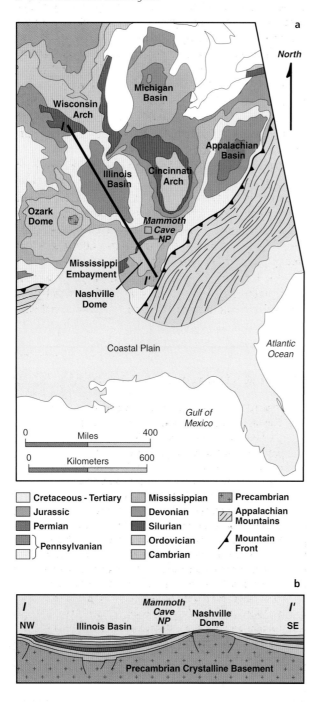

☐ Cretaceous - Tertiary	Mississippian	Precambrian
Jurassic	Devonian	Appalachian Mountains
Permian	Silurian	
Pennsylvanian	Ordovician	Mountain Front
	Cambrian	

and below the surface. Mammoth Cave National Park in Kentucky shows extraordinary features developed as water flowing through pore spaces and underground cavities and streams dissolved limestone within the platform strata. Effects of water flowing across the surface through gullies, small streams, and larger rivers are well displayed at Badlands National Park in South Dakota, Theodore Roosevelt National Park in North Dakota, and Agate Fossil Beds National Monument in Nebraska.

Cave Features

Much of the rock layering on the continental platform was deposited when North America straddled the equator (Figs. 10.9 and 10.10). Warm, shallow seas that occasionally covered the craton deposited limestone, an ideal material for the development of cave systems. Limestone is a rock formed from the hard parts of corals or shellfish, or the accumulation of fine lime mud out of water (normally seawater). Such materials are composed primarily of calcium carbonate (the mineral calcite—$CaCO_3$). The layering also includes **dolomite**, a rock similar to limestone composed of the mineral also called dolomite, a calcium-magnesium carbonate ($CaMg[CO_3]_2$). Dolomite forms where groundwater

FIGURE 10.12 Development of cave systems in platform strata. The nearly horizontal sedimentary layers develop perpendicular sets of vertical cracks (joints). Flow of surface water and groundwater tends to erode the rock along both the sedimentary layers and joints. Limestone dissolves readily, enhancing the vertical cracks and forming cave systems along the horizontal layering.

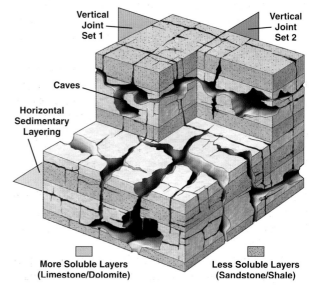

FIGURE 10.13 Mammoth Cave National Park, Kentucky. The cave features developed in limestone layers that formed as shallow seas covered the craton and deposited continental platform strata.

moving through limestone replaces some of the calcium with magnesium. Groundwater moving through the crack and pore spaces in limestone or dolomite can dissolve some of the carbonate minerals, creating narrow passageways that develop into caves (Fig. 10.12).

As water contacts air inside a cave, it releases carbon dioxide. This gas causes some of the calcium carbonate in the water to precipitate out as **travertine**, a dense form of limestone. Travertine deposits develop the spectacular formations seen in caves, known collectively as **speleothems**. Because speleothems require slow accumulations of calcium carbonate, they can reveal clues to changes in climate over the past several thousand years. Speleothems include **flowstone** (formed from sheets of water draping over cave walls), **stalactites** (drippings from ceilings), **stalagmites** (accumulations on floors), **columns** (stalactites and stalagmites merging over time), and **helictites** (small drippings forming "twisted-finger" forms from ceilings and slanted walls).

Mammoth Cave National Park is in western Kentucky, the most extensive region of karst topography in the United States. The term **karst**, meaning "stony ground," comes from a limestone region of northern Italy and Slovenia. Karst regions are characterized by features formed by the solution of limestone, including caves, sinkholes, springs, and disappearing streams. The limestone layers of Mammoth Cave National Park formed in a shallow, inland sea about 350 to 325 million years ago, during the Mississippian Period of the late Paleozoic Era (just after the late Devonian Period; see Fig. 10.10b).

Mammoth Cave National Park contains the longest cave system in the world, with 346 miles (557 kilometers) of mapped passageways, 12 miles (20 kilometers) of which are open to park visitors for general access (Fig. 10.13). The Mammoth Cave region lies between the Illinois Basin and Nashville Dome; strata have a very gentle tilt of about ½° northwestward (Fig. 10.11). The park is in the Chester Upland, composed of deep valleys dissolved from Mississippian-age limestone and dolomite, separated by ridges capped by insoluble layers of sandstone. The cave system is so extensive because capping sandstone holds the topography stable over a large area, while groundwater dissolves out passageways in the underlying limestone.

A **sinkhole** is a funnel-shaped depression in the land, formed as groundwater dissolves underlying limestone. Sinkholes in Mammoth Cave National Park are up to 1,000 feet (300 meters) across and 100 feet (30 meters) deep. Mammoth Cave is nearly horizontal, with many winding passageways dissolved out by surface water flowing down sinkholes.

The Green River flows through the heart of Mammoth Cave National Park. The level of the river controls the character of caves in the park. As the Green River Valley cuts downward, it loses some of its water to lower and lower cave systems, so that upper caves are left high and dry. Underground streams run through some of the lower-level passageways.

The flanks of some uplifted structures expose platform strata that were deposited as shallow seas lapped across the craton prior to the uplift. Two National Park

WATER FLOWING THROUGH ROCKS

Groundwater flows in different manners through two distinct zones. Cavities and pore spaces in the **phreatic zone** are completely filled with water, while they are only partially saturated in the overlying **vadose zone**. The **water table** is the surface separating the two zones. In the vadose zone, water flows downhill through the steepest path available—vertical cracks are good candidates, as are pore spaces or interfaces between tilted layers. Flow of water in the phreatic zone is determined by pressure differences between connected water bodies underground, and can be upward as well as downward. Limestone cave systems form as water dissolves calcium carbonate in both the phreatic and vadose zones. Cave deposits (speleothems) form as groundwater interacts with air in the vadose zone and precipitates some of the calcium carbonate.

FIGURE 10.14 Jewel Cave National Monument in the Black Hills region of South Dakota. The cave passageways and colorful deposits have developed in Mississippian limestone, the same age of platform strata found at Mammoth Cave National Park in Kentucky.

System sites in the Black Hills region of South Dakota showcase cave systems: Wind Cave National Park and Jewel Cave National Monument (Fig. 10.14). Both caves are carved out of Mississippian limestone, the same strata that developed Mammoth Cave in Kentucky. Wind Cave includes 108 miles (173 kilometers) of passageways, making it the fourth longest in the United States and sixth longest in the world. Caves "breathe" as barometric pressure of the surface air changes, causing wind to blow in and out of cave entrances—hence the name

"Wind Cave." Jewel Cave is about 127 miles (205 kilometers) long, ranking it second longest in the United States and third longest in the world. Some of the calcite (calcium carbonate) crystals are shaped like jewels.

Carlsbad Caverns National Park in New Mexico contains more than 40 caves. The best known is Carlsbad Cavern, at 1,028 feet (313 meters) the ninth deepest and at 19.1 miles (30.7 kilometers) the twentieth longest in the United States (Fig. 10.15). Lechuguilla Cave is the deepest in the United States (1,567 feet; 478 meters), and at 109 miles (176 kilometers) it is the fifth longest in the world. Carlsbad Cavern rocks were formed within an approximately 250-million-year-old Permian-reef system, similar to those in the shallow seas surrounding the Bahamas off Florida and the Great Barrier Reef of Australia today.

Surface Features

Many features on national park lands result from forces within the Earth deforming rock layers, heaving up mountain ranges, or spewing forth volcanic material. But some of the most beautiful and dramatic features in parks result from surface processes—the actions of wind and water—eroding flat-lying platform and foreland basin strata. Erosion can sculpt kaleidoscopic canyons and intriguing rock formations, and can sometimes expose fossil evidence of earlier forms of life on Earth.

National park lands on the prairies east of the Rocky Mountains include dramatic, colorful landscapes

FIGURE 10.15 Carlsbad Caverns National Park, New Mexico. The cave system developed in Permian-age limestone reef deposits.

FUELING CAVE FORMATION

The presence of oil deposits was important to the genesis of the Carlsbad Caverns National Park cave system. The Permian Basin was created when Gondwanaland collided with North America during the late Paleozoic Era, forming the Applachian-Ouachita-Marathon collisional mountain range (Chap. 6). Sandstone layers were deposited in the basin under fairly normal salinity. After the basin was cut off from the open ocean, a thick layer of **gypsum** formed. The gypsum layer provided insulation, so that temperatures rose to a point where oil formed from organic material in the basin sediments. Sulfuric acid from the petroleum deposits in the sandstones seeped upward and dissolved calcium carbonate within overlying limestone layers, forming the caves.

known as badlands. People visiting Badlands National Park in South Dakota or Agate Fossil Beds National Monument in Nebraska may not immediately see how those parks relate to plate tectonics, because the plate movement occurred so long ago. During the early part of the Tertiary Period, plates were converging all along the western edge of the continent, forming volcanoes and uplifting mountain ranges (Chap. 5). Sand and mud from the rising and eroding Rocky Mountains covered tropical lowlands that looked much like the savannas of eastern Africa. Volcanic ash carried eastward by the wind covered the craton region, coating lakes and stream valleys where animals grazed. Ground and surface waters altered some of the ash into clay minerals, and oxidized sand and shale layers into bright colors. Erosion of those soft materials created the unique **badlands topography** displayed in the parks, and unearthed fossil plants and animals that tell us about life that existed on those ancient savannas.

Badlands National Park lies about 50 miles (80 kilometers) east of the Black Hills of South Dakota. The oldest rocks in the park are sedimentary layers deposited about 100 million years ago during the Cretaceous Period, when shallow seas covered the western interior of the continent (Fig. 10.10c). A foreland basin adjacent to the rising mountains was filled with shallow marine layers, as well as nonmarine stream (fluvial), lake (lacustrian), and wind-blown (eolian) deposits. Most of the strata in the park are Eocene- to Oligocene-age (~37-to-23-million-year-old) silt and clay layers with some beds of volcanic ash that

FIGURE 10.17 Theodore Roosevelt National Park, North Dakota. The badlands topography developed as erosion carved up ancient river floodplain deposits.

erupted from volcanoes far to the west. The Oligocene is known as the "Golden Age of Mammals." Badlands National Park preserves excellent fossils from that epoch, including saber-toothed cats, camels, and giant pigs. Uplift in the Pliocene Epoch (~ 5 to 2 million years ago) led to ongoing erosion and development of the badlands topography (Fig. 10.16).

Theodore Roosevelt National Park was established in 1978 from part of the former president's Elkhorn Ranch in North Dakota. Teddy Roosevelt was instrumental in developing policy to manage natural resources in the public's interest; many national park lands benefited from his conservation efforts. The park lies along the Little Missouri River and contains Paleocene (~60-million-year-old) sedimentary layers that were deposited on ancient river floodplains. Marshy conditions resulted in organic-rich material that was later altered to lignite coal and petrified wood. Some volcanic ash deposits within the layering have been changed to a type of clay called bentonite. Pliocene uplift and increased erosion were followed by Pleistocene glaciation. Continued erosion of the soft strata has developed buttes, tablelands, and valleys characteristic of badlands topography (Fig. 10.17).

Agate Fossil Beds National Monument is in the valley of the Niobrara River in western Nebraska. The site was important to the development of the science of paleontology in the early 1900s. During the Miocene Epoch, about 19 million years ago, the region was a tropical lowland teeming with rhinoceros and ancient relatives of horses. When the climate dried up, animals were confined to small watering holes. Silt, sand, and

FIGURE 10.16 Badlands National Park, South Dakota. The strata are soft silt and shale layers from material eroded off the rising Rocky Mountains, with some interlayered volcanic ash layers.

FIGURE 10.18 Agate Fossil Beds National Monument, Nebraska.

volcanic ash buried animal remains under several feet (meters) of river and wind-blown sediment (Fig. 10.18).

DEFORMED CRATON

The bulk of the North American Craton has experienced very little internal deformation for the past several hundred million years, developing only broad basins and domes (Fig. 10.11). But along its edges, particularly on the west, the crust has in places been deformed and uplifted. The zones of cratonic deformation include national park lands that expose the continental platform and foreland basin layering on the Colorado Plateau, as well as deformed sedimentary layers in foreland fold-and-thrust belts and old igneous and metamorphic rocks within basement uplifts.

FIGURE 10.19 Moonrise over colorful sandstone layers in Zion National Park, Utah.

Colorado Plateau

The Colorado Plateau is a part of the North American Craton that has been broadly uplifting over the past several million years, yet internally has been only mildly deformed (Fig. 10.1). As the rocks rise to higher elevation, streams carve canyons and weather the sedimentary rock layers into colorful iron oxide minerals (Fig. 10.19).

DEPOSITIONAL SETTING. The Colorado Plateau region was part of a passive continental margin on the western edge of North America during the Paleozoic Era (Figs. 10.3 and 10.10a), and remained so until the middle part of the Jurassic Period of the Mesozoic Era. The sedimentary layering of Grand Canyon National Park, Petrified Forest National Park, and Walnut Canyon National Monument, as well as the lower part of the layering at Canyonlands National Park and Cedar Breaks National Monument, were deposited along the ancient passive margin (Chap. 4). Early- to middle-Jurassic strata of other parks—including Zion, Capitol Reef, Navajo, Arches, Grand Staircase–Escalante, and many others—were deposited farther up on the continental platform. From the late Jurassic Period onward, a convergent plate boundary was active along the western margin of the continent. Younger strata in the Colorado Plateau region were deposited in foreland basins associated with the convergent tectonic activity. The basins are filled with lake, river, and wind-blown deposits on land, as well as marine strata formed when shallow seas lapped up on the continental platform (see the Western Interior Seaway in Fig. 10.10c).

Chap. 4 shows how some parks on the Colorado Plateau, particularly Grand Canyon National Park in Arizona, represent exhumation of elements of the ancient passive continental margin (Figs. 4.17 to 4.22). From bottom to top, the rocks seen on the walls of the Grand Canyon are Precambrian igneous and metamorphic basement, younger Precambrian rift valley strata that represent the breakup of an ancient continent, and then Paleozoic passive margin strata deposited along the western edge of North America. But the Colorado Plateau is so broad that certain parks, particularly those in southern Utah and southwestern Colorado, reveal thin platform strata deposited some distance up on the craton (Fig. 10.20).

A comparison of the sequences of rock layers, known as stratigraphic columns, for some parks on the Colorado Plateau is presented in Chap. 4 (Fig. 4.24). The Grand Canyon lies on the western edge of the plateau, where the Colorado River has cut entirely through Paleozoic passive margin strata into older rocks below. The entire overlying sequence of Mesozoic strata has eroded away in the vicinity of Grand Canyon National Park. Farther up on the Colorado Plateau, erosion has not been so deep. Parks there reveal strata deposited during the Mesozoic Era, including foreland basin strata derived from rising mountain ranges.

Sedimentary layers in Zion National Park are often red because of the presence of the iron-rich mineral called hematite. The Jurassic-age Navajo Sandstone is prominent in the park (Fig. 10.21). It displays **cross bedding** developed where sand-dune layers were deposited by winds of changing direction on beaches or in deserts; cross bedding is prominent in sandstone layers in many parks on the Colorado Plateau (Fig. 10.22).

EROSIONAL FEATURES. During the Paleozoic and Mesozoic eras, shallow seas periodically transgressed coastal areas, at times extending over much of the craton. North America was at lower latitudes, where desert conditions prevailed. Strata exposed at some park lands on the Colorado Plateau include layers of sandstone deposited by wind and very shallow water (Fig. 10.23). Those sandstones are the resistant layers that result in spectacular canyons, arches, spires, and other erosional formations seen in many of the plateau's national parks and monuments (Fig. 10.24).

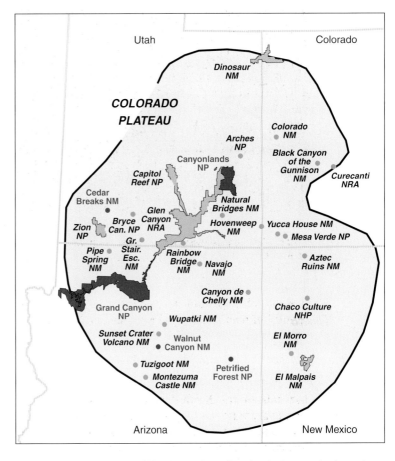

FIGURE 10.20 National park lands on the Colorado Plateau. Parks shown in bold red are included in the discussion of ancient passive continental margins in Chap. 4. Strata in the other parks relate to deposition by shallow seas up on the continental platform, or in foreland basins that received sediment from rising mountains.

FIGURE 10.21 Zion National Park, Utah. a) Checkerboard Mesa reveals frozen sand dunes of early Jurassic age. The cross-bedding characteristic of the dunes developed some distance inland from the continental edge, as sand was blown about and deposited by the wind. **b)** Mountain of the Sun, Twin Brothers, and east Temple.

a

b

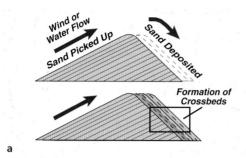

FIGURE 10.22 Development of crossbedding. a)
Wind on a beach or in a desert deposits sand layers.
A change in wind direction drapes layers with different orientation, resulting in the crossbedding. **b)**
Crossbedding in the upper De Chelly Sandstone along
White House Trail, Canyon de Chelly National
Monument, Arizona.

The colorful rock layers of Bryce Canyon National Park are mostly part of the Eocene-age Claron Formation deposited in a foreland basin. The strata are calcium-rich mudstone (shale) layers with caps of thin sandstone and conglomerate. Erosion has produced needlelike structures called **hoodoos** (Fig. 10.25). The hard capping units of sandstone and conglomerate erode very slowly. Where hard layers are breached, erosion into the softer layers underneath proceeds rapidly, while the capping units elsewhere remain high, developing the hoodoos.

Arches National Park, located in the Paradox Basin of east-central Utah, has the greatest concentration of natural rock arches in the world (Fig. 10.26). Its rock strata span the range from the Pennsylvanian portion of the late Paleozoic Era (about 300 million years ago) to near the end of the Mesozoic Era during the Cretaceous Period 85 million years ago. Arches have developed in three geologic formations, all deposited during the middle portion of the Jurassic Period: Carmel Formation about 170 million years ago, Entrada Sandstone 160 million years ago, and Curtis Formation 155 million years ago. Close inspection shows that those layers are **eolian** sandstones; they were deposited by wind and look like frozen sand dunes.

The arches developed within a clastic (sandstone and shale) sequence deposited over a thick layer of salt. During the late Pennsylvanian Period, about 300 million years ago when North America was near the equator (Fig. 10.9b), an embayment into the continent

FIGURE 10.23 Colorado National Monument, Colorado.
a) Fruita Canyon from Rim Rock Drive. The dark rocks at the bottom of the canyon are Precambrian metamorphic rocks. Red layers are Triassic-age Chinle Formation, overlain by the Wingate Sandstone and caprock of the Kayenta Formation. **b)** Balanced Rock on the rim of Fruita Canyon is made mostly of Wingate Sandstone, resting on red layers of the Chinle Formation, capped by a thin, resistant layer of the Kayenta Formation.

a

b

c   d

FIGURE 10.24 Erosional features in parks on the Colorado Plateau. a) Capitol Reef National Monument, Utah. Steep cliffs formed in resistant Wingate Sandstone layers. More gentle slopes develop on softer shales. **b)** Glen Canyon National Recreation Area, Utah. Labyrinth Canyon is a narrow slot canyon developed by erosion that occurs during flash flooding. **c)** Dinosaur National Monument, Utah/Colorado. Steep cliff face in Pennsylvanian-to-Permian-age Weber Sandstone. **d)** Wupatki National Monument, Arizona. Abundant archaeological sites are found where dwellings were constructed using steep, sandstone cliff faces that have broken along vertical sets of cracks, known as joints, that intersect at right angles.

known as the Paradox Basin began to fill with sediment. Because the embayment had restricted water flow and was in a dry climate, seawater evaporated rapidly. More seawater came in, rather than freshwater from rainfall or stream runoff, and it, too, evaporated. The process repeated, until a thick layer of salt accumulated. From the late Triassic through the late Cretaceous periods, continental platform and foreland basin strata covered the salt layer (Fig. 10.10c). Moderate deformation occurred during the early Tertiary Period, folding and faulting the strata above the salt layer. As they folded, vertical cracks (**joints**) formed in the rigid sandstone layers to accommodate stretching and squeezing. Over the past 6 million years, the Colorado River and its tributaries carved steep-sided canyons as the Colorado Plateau uplifted. Some streams cut all the way down to the salt layer. Surface runoff and groundwater dissolved some of the salt along the core regions of anticlines, causing the ground above to collapse as steep valleys. **Fins** formed where

FIGURE 10.25 Bryce Canyon National Park, Utah. The canyon is carved into soft shale layers, with thin interbeds of resistant sandstone and conglomerate. **a)** Strata on the eastern escarpment of the Paunsaugunt Plateau erode into a colorful and dramatic landscape. **b)** Hoodoos are tall spires capped by remnants of the resistant layers.

FIGURE 10.26 Arches National Park, Utah. a) The arches form through a combination of depositional, deformational, uplift, and erosional processes. Massive sandstone beds were deposited above a thick layer of salt during the Jurassic Period. Deformation caused the salt to flow into the cores of anticlines, accompanied by cracking of the massive sandstone layers along joints. During broad uplift of the Colorado Plateau, water eroded and widened the joints, forming long ridges. The ridges broke through in places, forming fins. Erosion of the core of the fins left some overlying rock suspended as arches. **b)** Delicate Arch is a remnant of sandstone perched on the northern rim of Cache Valley. The pillars and base are part of the Entrada Sandstone (about 160 million years old), while the upper part consists of sandstone layers of the 155-million-year-old Curtis Formation.

THE FLYING BUTTRESS

Some of the features in Arches National Park owe their stability to the fact that water and wind sculpted **flying buttress** designs. The flying buttress architectural form was developed because of its ability to support a large amount of weight with minimal amount of building material. This design is used in many of the churches and cathedrals of Europe.

prominent sets of vertical joints developed in the sandstones, as pressure was released when the region uplifted. Erosion along those joints produced long grooves and then deep valleys, leaving the fins behind. Further erosion formed complete holes through some of the fins, resulting in **arches**. Arches commonly occur where erosion cuts laterally through hard sandstone layers adjacent to steep canyons. **Natural bridges** have a stream or river flowing under them, while arches do not (Fig. 10.27).

a

b

FIGURE 10.27 A natural bridge is similar to an arch, but with a stream flowing underneath.
a) Rainbow Bridge National Monument, Utah. b) Natural Bridges National Monument, Utah. Owachomo
Bridge is 180 feet (55 meters) long, 106 feet (32 meters) high, and 9 feet (2.7 meters) thick.

Foreland Structures

Mount Rushmore National Memorial in the Black Hills
of South Dakota is a familiar icon to Americans (Fig.
10.28). The busts of presidents George Washington,
Thomas Jefferson, Abraham Lincoln, and Theodore
Roosevelt are carved into Precambrian granite exposed
in the center of the broad domal uplift. But why has
old crust of the North American Craton been deformed
and uplifted in such a fashion?

Other national parks, such as Rocky Mountain in
Colorado and Glacier in Montana, display breathtaking
mountain ranges formed as rocks on the edge of the
craton were compressed and shoved upward. Yet those
parks lie a considerable distance from convergent plate
boundaries that were active when the mountains
formed. So how has deformation occurred so far into
the interior of the continent?

The term **foreland** refers to the region out in front
of a developing mountainous landscape. Mountains in
the foreland that involve deformation of just the
sedimentary layering, such as those seen in Glacier
National Park, are called foreland fold-and-thrust belts.
Those that also involve the underlying, hard igneous
and metamorphic rocks are foreland basement uplifts. A
foreland basin is a region that is depressed during foreland deformation and is filled with sedimentary deposits
from the eroding mountains. Some of the younger strata
in national parks on the Colorado Plateau and east of the
frontal ranges of the Rockies are foreland basin deposits.

**FIGURE 10.28 Mount Rushmore National Memorial, South
Dakota.** The busts of the presidents are carved into the Harney
Peak Granite, part of the Precambrian crystalline basement at the
core of the Black Hills, a Laramide uplift.

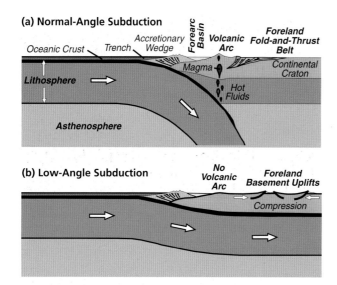

FIGURE 10.29 The subduction angle influences volcanism and the style of foreland deformation. a) Normal-angle subduction. The active continental margin morphology consists of an accretionary wedge and volcanic arc, separated by a forearc basin. Sedimentary strata are folded and thrusted toward the continental craton in a foreland fold-and-thrust belt. **b)** Low-angle subduction. The top of the plate may not extend deep enough to heat up and induce sweating, so that volcanism ceases or migrates toward the craton. The subducting plate transmits stress beneath the edge of the craton, causing hard rock of the crust to compress and break along reverse faults, forming foreland basement uplifts.

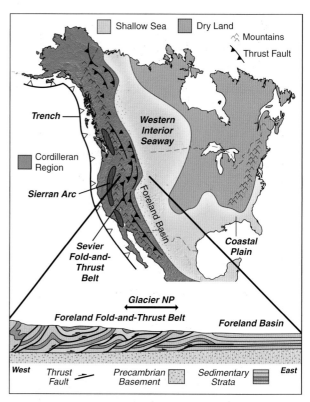

FIGURE 10.30 Geometry of foreland fold-and-thrust belt. The map shows the tectonic situation of North America during the late Cretaceous Period, approximately 90 million years ago. Sedimentary strata were folded and thrust over the underlying Precambrian basement during the Sevier Orogeny. Sediment eroded from the rising mountains was deposited in an adjacent foreland basin. Thrust faults are low-angle reverse faults. They generally detach the sedimentary cover and shove it over the basement rocks. Glacier National Park is part of a foreland fold-and-thrust belt in Montana.

During the late Mesozoic and early Cenozoic eras, western North America was a vast region of mountain building known as the **Cordillera** (Fig. 10.1). The tectonic activity was due primarily to subduction of the oceanic Farallon Plate beneath North America (Fig. 5.46a). Deformation of the edge of the craton in the foreland region was of two different styles. At times of typical, relatively steep-angle subduction, a volcanic arc developed near the West Coast (Fig. 10.29a). The hard, igneous and metamorphic basement rocks of the cratonic edge were generally not compressed and deformed, only the overlying, softer sedimentary cover. Such deformation results in the development of a foreland fold-and-thrust belt (Fig. 10.30). But at other times, the hard basement rocks were compressed and deformed, at considerable distance from where the plates were converging on the surface. Compression extended beneath the edge of the craton, where great masses of hard crust were shoved upward as foreland basement uplifts, called Laramide uplifts, in the Rocky Mountain region (Fig. 10.31).

The compression that formed the Laramide uplifts is thought to have occurred so close to the craton because the Farallon Plate was subducting at a low angle rather than more steeply (Fig. 10.29b). Such an interpretation draws from comparison with the current situation of western South America, where it is observed that the angle of subduction profoundly affects the developing landscape. Where the oceanic Nazca Plate subducts at an angle of more than 20°, active volcanoes of the Andes Mountains form. But in places where the subduction angle is less than 10°, there are no active volcanoes. Instead, large blocks of crust uplift along reverse faults, forming the Pampean Ranges of Argentina.

The landscape of western North America may have developed in a similar fashion during the late Mesozoic

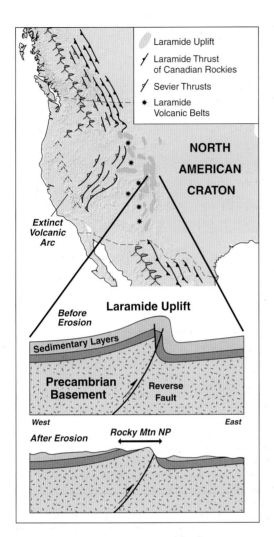

FIGURE 10.31 Foreland basement uplifts, known as Laramide uplifts, developed during compression of the edge of the North American Craton during the Laramide Orogeny 80 to 40 million years ago. They comprise the frontal ranges of the Rocky Mountains in Montana, Wyoming, South Dakota, Colorado, New Mexico, and Texas, and may have formed when a plate subducted beneath the continent at a low angle. Such foreland basement uplifts deform the hard igneous and metamorphic rocks of the Precambrian basement. The reverse faults bounding the structures are relatively steep; they differ from thrust faults, which are lower angle and generally detach the sedimentary cover and shove it over the basement rocks. Rocky Mountain National Park rests on a Laramide uplift, the Front Range in Colorado.

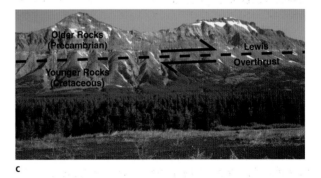

FIGURE 10.32 Glacier National Park reveals rocks uplifted and shoved eastward along low-angle thrust faults. a) The nearly horizontal layering in places involves very old rocks thrust over much younger. **b)** The Lewis Overthrust, viewed from Marias Pass. Hard, Precambrian sedimentary rocks have been thrust over softer, Cretaceous shale and sandstone. **c)** Apikuni Mountain also reveals Precambrian strata thrust over younger layers along the Lewis Overthrust.

and early Cenozoic eras. At times, the subduction angle was relatively steep, forming the ancient Cascade-Sierra volcanoes close to the coast. The foreland fold-and-thrust belt that developed is exemplified by structures found in Glacier National Park in Montana (Figs. 10.32 and 10.33). The park reveals abundant evidence of thrust

faulting. Of particular note is the Lewis Overthrust, displayed at Chief Mountain on the park's northern boundary with Waterton Lakes National Park in Canada. There, Precambrian sedimentary rocks, known as the Belt Supergroup, overlie much younger and softer Cretaceous sedimentary strata. The overlying rocks are an astonish-

a   b

FIGURE 10.33 Glacial erosion and deposition in Glacier National Park. a) U-shaped valleys reveal that glaciers were once far more extensive. **b)** The terminal moraine in the foreground was deposited in the 1800s when the Clements Glacier flowed farther down the valley.

ing 1.3 billion (1,300 million) years older than the rocks below the thrust fault, and were thrust eastward at least 50 miles (80 kilometers) over the younger rocks.

At times when the subduction angle was low, compressional forces occurred considerably closer to the craton. The contrasting Laramide uplifts form the dramatic landscapes of Rocky Mountain and other national parks along the frontal portions of the Rocky Mountains (Fig. 10.34). Most of the rocks in Rocky Mountain National Park are Precambrian metamorphic rocks. They originally were sedimentary and volcanic strata deposited between 2.0 and 1.8 billion years ago, then metamorphosed to schist and gneiss from 1.7 to 1.6 billion years

ago. Magma intruded the metamorphic rocks 1.3 billion years ago. Those older rocks were subjected to erosion and then overlain by younger sedimentary deposits starting in the early Paleozoic Era, about 500 million years ago (Fig. 10.3). Platform strata accumulated as shallow seas periodically covered the craton over the next 200 million years (Fig. 10.10a,b). Younger sedimentary layers were deposited in a foreland basin as mountains began to form in the Cordilleran region to the west (Fig. 10.30). The Precambrian basement rocks found in the park were exposed as reverse faulting formed a Laramide uplift and most of the sedimentary strata eroded (Fig. 10.31).

FIGURE 10.34 Peaks in Rocky Mountain National Park are part of a Laramide uplift. The hard igneous and metamorphic rocks were once several miles below the surface before being uplifted along reverse faults.

FURTHER READING

GENERAL

Chronic, H. 1984. *Pages of Stone, Geology of Western National Parks and Monuments: Rocky Mountains and the Western Great Plains*. Seattle: Mountaineers. 168 pp.

Chronic, H. 1988. *Pages of Stone, Geology of Western National Parks and Monuments: Grand Canyon and the Plateau Country*. Seattle: Mountaineers. 158 pp.

Howell, D. G. (editor). 1995. *Principles of Terrane Analysis: New Applications for Global Tectonics*. New York: Chapman and Hall. 245 pp.

Jennings, M. N. 1983. Karst landforms. *American Scientist*, v. 71, p. 578–586.

Klimchouk, A. B., D. C. Ford, A. N. Palmer, and W. Dreybrodt (editors). 2000. *Speleogenesis: Evolution of Karst Aquifers*. Huntsville, AL: National Speleological Society. 527 pp.

McPhee, J. 1998. Crossing the craton, in *Annals of the Former World*, Book 5. New York: Farrar, Straus and Giroux, p. 624–660.

McPhee, J. 1998. Rising from the Plains, in *Annals of the Former World*, Book 3. New York: Farrar, Straus and Giroux, p. 280–425.

Powell, J. W. 1961. *Exploration of the Colorado River and Its Canyons*. New York: Dover Publications. 397 pp.

Sprinkel, D. A., T. C. Chidsey, and P. B. Anderson (editors). 2000. *Geology of Utah's Parks and Monuments*. Salt Lake City: Utah Geological Association. 644 pp.

Wagner, S. S. 2002. *A Geology Training Manual for Grand Canyon National Park*. M.S. thesis, Oregon State University. 184 pp.

TECHNICAL

Bump, A. P. 2003. Reactivation, trishear modeling, and folded basement in Laramide uplifts: Implications for the origins of intra-continental faults. *GSA Today*, v. 13, p. 4–10.

Chumakov, N. M. 2001. Periodicity of major glaciation events and their correlation with endogenic activity of the Earth. *Doklady Earth Sciences*, v. 379, p. 507–510.

Condie, K. C. 1982. *Plate Tectonics and Crustal Evolution*, 2nd Ed. New York: Pergamon Press. 310 pp.

Crowell, J. C. 1999. *Pre-Mesozoic Ice Ages: Their Bearing on Understanding the Climate System*. Boulder, CO: Geological Society of America, Memoir 192. 106 pp.

Hamilton, W. B. 1988. Laramide crustal shortening, in *Interaction of the Rocky Mountain Foreland and the Cordilleran Thrust Belt*, edited by C. J. Schmidt and W. J. Perry, Jr. Boulder, CO: Geological Society of America, p. 27–40.

Hoffman, P. F. 1988. United Plates of America: The birth of a craton. *Annual Review of Earth and Planet Science*, v. 16, p. 543–604.

Hoffman, P. F. 1989. Precambrian geology and tectonic history of North America, in *The Geology of North America—An Overview*, vol. A, edited by A. W. Bally and A. R. Palmer. Boulder, CO: Geological Society of America, p. 447–512.

Moores, E. M., and R. J. Twiss. 1995. *Tectonics*. New York: Freeman. 415 pp.

Rodgers, J. J. W. 1994. *A History of the Earth*. Cambridge, UK: Cambridge University Press.

Strahler, A. N. 1998. *Plate Tectonics*. Cambridge, MA: Geo-Books. 554 pp.

Windley, B. F. 1995. *The Evolving Continents*, 3rd Ed. New York: Wiley. 526 pp.

WEBSITES

National Park Service: www.nps.gov

Park Geology Tour:
www2.nature.nps.gov/grd/tour/index.htm

Continental Shield:
92. Voyageurs NP:
www2.nature.nps.gov/grd/parks/voya/index.htm

Continental Platform:
93. Agate Fossil Beds NM:
www2.nature.nps.gov/grd/parks/agfo/index.htm
94. Alibates Flint Quarries NM:
www2.nature.nps.gov/grd/parks/alfl/index.htm
95. Badlands NP:
www2.nature.nps.gov/grd/parks/badl/index.htm
96. Cuyahoga Valley NP:
www2.nature.nps.gov/grd/parks/cuva/index.htm
97. Effigy Mounds NM:
www.nps.gov/efmo/index.htm
98. Indiana Dunes NL:
www2.nature.nps.gov/grd/parks/indu/index.htm
99. Lake Meredith NRA:
www.nps.gov/lamr/index.htm
100. Mammoth Cave NP:
www2.nature.nps.gov/grd/parks/maca/index.htm
101. Mississippi NR&RA:
www.nps.gov/miss/index.htm
102. Missouri NRR:
www.nps.gov/mnrr/index.htm
103. Natches Trace Parkway:
www.nps.gov/natr/index.htm
104. Niobrara NSR:
www.nps.gov/niob/index.htm
105. Ozark NSR:
www.nps.gov/ozar/index.htm

106. Pipestone NM:
www2.nature.nps.gov/grd/parks/pipe/index.htm
107. Scotts Bluff NM:
www.nps.gov/scbl/index.htm
108. Sleeping Bear Dunes NL:
www2.nature.nps.gov/grd/parks/slbe/index.htm
109. Tallgrass Prairie N Pres:
www.nps.gov/tapr/index.htm
110. Theodore Roosevelt NP:
www2.nature.nps.gov/grd/parks/thro/index.htm

Colorado Plateau:
111. Arches NP:
www2.nature.nps.gov/grd/parks/arch/index.htm
112. Black Canyon of the Gunnison NP:
www2.nature.nps.gov/grd/parks/blca/index.htm
113. Bryce Canyon NP:
www2.nature.nps.gov/grd/parks/brca/index.htm
114. Canyon de Chelly NM:
www2.nature.nps.gov/grd/parks/cach/index.htm
115. Capitol Reef NP:
www2.nature.nps.gov/grd/parks/care/index.htm
116. Colorado NM:
www2.nature.nps.gov/grd/parks/colm/index.htm
117. Curecanti NRA:
www2.nature.nps.gov/grd/parks/cure/index.htm
118. Dinosaur NM:
www2.nature.nps.gov/grd/parks/dino/index.htm
119. El Malpais NM:
www2.nature.nps.gov/grd/parks/elma/index.htm
120. El Morro NM:
www.nps.gov/elmo/index.htm
121. Glen Canyon NRA:
www2.nature.nps.gov/grd/parks/glca/index.htm
123. Mesa Verde NP:
www2.nature.nps.gov/grd/parks/meve/index.htm
124. Natural Bridges NM:
www2.nature.nps.gov/grd/parks/nabr/index.htm
125. Navajo NM:
www2.nature.nps.gov/grd/parks/nava/index.htm
126. Pipe Spring NM:
www2.nature.nps.gov/grd/parks/pisp/index.htm
127. Rainbow Bridge NM:
www.nps.gov/rabr/index.htm

128. Tuzigoot NM:
www.nps.gov/tuzi/index.htm
129. Wupatki NM:
www.nps.gov/wupa/index.htm
130. Zion NP:
www2.nature.nps.gov/grd/parks/zion/index.htm

Laramide Uplifts:
131. Bighorn Canyon NRA:
www2.nature.nps.gov/grd/parks/bica/index.htm
132. Devils Tower NM:
www2.nature.nps.gov/grd/parks/deto/index.htm
133. Florissant Fossil Beds NM:
www2.nature.nps.gov/grd/parks/flfo/index.htm
134. Great Sand Dunes NM:
www2.nature.nps.gov/grd/parks/grsa/index.htm
135. Jewel Cave NM:
www2.nature.nps.gov/grd/parks/jeca/index.htm
136. Mount Rushmore N Mem:
www2.nature.nps.gov/grd/parks/moru/index.htm
137. Rocky Mountain NP:
www2.nature.nps.gov/grd/parks/romo/index.htm
138. Wind Cave NP:
www2.nature.nps.gov/grd/parks/wica/index.htm

Foreland Fold-and-Thrust Belt:
139. Glacier NP:
www2.nature.nps.gov/grd/parks/glac/index.htm

Alaska:
140. Yukon-Charley River N Pres:
www2.nature.nps.gov/grd/parks/yuch/index.htm

U.S. Geological Survey: www.usgs.gov

Geology in the Parks:
www2.nature.nps.gov/grd/usgsnps/project/home.html
Stratigraphic Sections for Parks on Colorado Plateau:
pubs.usgs.gov/gip/geotime/section.html

Bureau of Land Management:
www.blm.gov/nhp/index.htm
Environmental Education: www.blm.gov/education
122. Grand Staircase–Escalante NM:
www.ut.blm.gov/monument

World's Longest Caves:
www.pipeline.com/~caverbob/wlong.htm

11

Accreted Terranes

▲▲▲▲ ▲▲ *"Many national parks in the United States have rocks that are out of place—manufactured elsewhere, then carried long distances before being affixed to the North American continent."*

A visit to some western parks can be viewed as a trip overseas, off of the North America we know today (Fig. 11.1). Much of the landscape in those parks formed well to the south, in some cases far out across the Pacific. A large part of western North America was added to the edge of the continent during the past 200 million years (Fig. 11.2). National park lands in some of the western states, particularly Alaska and northern Washington, contain rocks that were formed elsewhere, transported via transform or convergent plate motion, and then slammed into the edge of the continent (Table 11.1).

Terranes are created in various tectonic settings. They generally have crust that is thicker than the typical oceanic crust formed at mid-ocean ridges. Island

FIGURE 11.1 Southern Pickets in North Cascades National Park, Washington. The landscape is similar to that of Alaska. It consists of rocks that formed elsewhere, and then were deformed, metamorphosed, and accreted to North America.

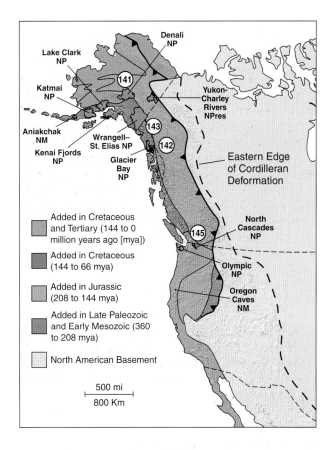

FIGURE 11.2 Westward growth of the North American continent. The Cordillera is a region of the continent that has undergone extensive deformation and terrane accretion over the past 200 million years or so. Undeformed platform and shield areas lie east of the dashed line (Chap. 10). Denali, Wrangell–St. Elias, Glacier Bay, and North Cascades national parks (numbers) are the main focus of this chapter. Aniakchak, Katmai, Lake Clark, Kenai Fjords, Olympic, and Oregon Caves are discussed with parks in active subduction zones in Chap. 5.

ALASKA	LOWER 48
141. Denali NP, AK	144. Lake Chelan NRA, WA
142. Glacier Bay NP, AK	145. North Cascades NP, WA
143. Wrangell–St. Elias NP, AK	146. Ross Lake NRA, WA

TABLE 11.1 National parks on accreted terranes.

chains formed at oceanic hotspots such as Hawai`i can smash into the edge of a continent rather than subduct. Volcanic island arcs formed at subduction zones in the Pacific Ocean move along with a plate and eventually accrete to a continent. Continental fragments sometimes break away and end up elsewhere. For example, imagine the fate of the western sliver of California and the Baja California Peninsula (Fig. 11.3). That block of continental crust is a new terrane that is made up of accretionary wedge material and volcanic arc rocks that formed during earlier plate convergence (Chap. 5). The block is being deformed as it is sheared up along the San Andreas transform fault system (Chap. 7), and is pulling apart from the rest of North America as the Gulf of California opens. At 2 inches (5 centimeters) per year, it will slide northward past the rest of North America about 2,000 miles (3,000 kilometers) in the next 60 million years. That should be just about enough for it to slam back into North America, in the region of southeastern or southern Alaska.

Many national parks in the United States have rocks that are out of place—manufactured elsewhere,

(a) Today

(b) 30 Million Years in the Future
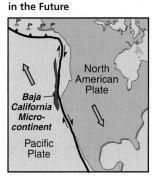

(c) 60 Million Years in the Future
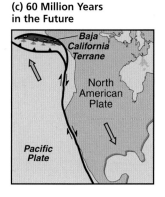

FIGURE 11.3 Fate of western and Baja California. a) The Baja California Peninsula and the western sliver of California are a block of crust (microcontinent) that is breaking away from the rest of North America. **b)** The block slides northward along the transform plate boundary between the Pacific and North American plates. **c)** The microcontinent becomes another accreted terrane as it slams into Alaska.

then carried long distances before being affixed to the North American continent. The Piedmont region of the Appalachians contains oceanic islands, continental fragments, and sedimentary, volcanic, and metamorphic rocks that were added to North America as an ancient ocean closed (Chap. 6). Parks such as Shenandoah and the Blue Ridge Parkway that border the Piedmont could be discussed here in the context of "suspect" or "accreted" terranes. Hot Springs National Park in the Ouachita Mountains of Arkansas and Gates of the Arctic in the Brooks Range of far-northern Alaska could also be considered in that vein. Other regions in the western part of the country, such as the Olympic and Klamath mountains, were also affixed to North America in fairly recent times. Olympic National Park and Oregon Caves National Monument, presented in Chap. 5, could rightfully be in this chapter. But rather than repeat material covered earlier in the book, this chapter focuses on four parks: three in Alaska, and another in Washington State. The parks are within the North American Cordillera, a broad region that has been the site of terrane accretion and other tectonic activity over the past 200 million years (Fig. 11.4).

ALASKA: A GLIMPSE OF CONTINENTAL GROWTH IN ACTION

The bulk of Alaska is material that was added to the North American continent in fairly recent geologic time. The motion between the Pacific and North American plates, coupled with the shape of the western border of the continent, explains how Alaska tends to be the collecting area for wayward terranes. Remarkably, just about all of Alaska has been assembled through terrane accretion over the past 200 million years—only a small piece of the state, on its eastern border with the Yukon Territory of Canada, is a bona fide part of the North American Craton.

Pattern of Terrane Accretion

National park lands in Alaska reveal the outward growth pattern of the North American continent (Fig. 11.5). Terranes in southern Alaska follow an arching pattern, paralleling the Gulf of Alaska (Fig. 11.6). Because terranes came crashing in from the south or southwest,

FIGURE 11.4 Western North America is a complex array of accreted terranes. Terranes discussed in the text include Chugach and Prince William (Cg), Wrangellia (W), and Yukon–Tanana (YT) in Alaska, and Northern Cascades (Ca) in Washington.

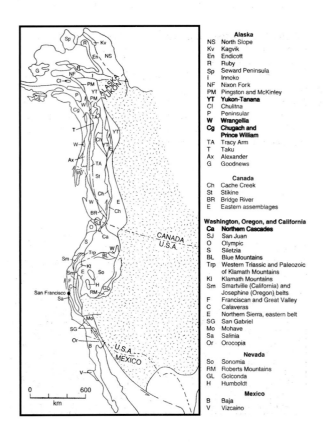

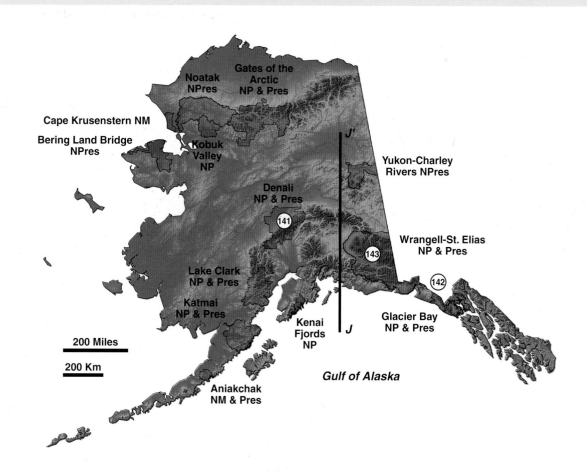

FIGURE 11.5 Shaded-relief map showing national park lands in Alaska. The rugged topography and rocks display the outward growth of the northwestern prong of North America over the past 200 million years. The mountains of Denali and Wrangell–St. Elias national parks are so high because the crust thickened as terranes came crashing in from the south. Kenai Fjords and Glacier Bay national parks lie on terranes that docked with North America more recently. Volcanoes in Aniakchak, Katmai, Lake Clark, and Wrangell–St. Elias are due to the continuing northward movement and subduction of the Pacific Plate that results in magma intruding through older terranes (Chap. 5). The park lands in the northern part of the state lie in the Brooks Range, composed of terranes added to the continent during collision with a continental fragment about 100 million years ago (Chap. 6). J–J' is the line of the simplified cross section of terrane accretion shown in Fig. 11.7.

they are generally older toward Alaska's interior and get progressively younger southward (Table 11.2).

Fig. 11.7 shows the general order of terranes from Alaska's interior southward to the Gulf of Alaska. The Yukon-Tanana Terrane began to collide with North America about 225 million years ago, and was firmly attached to the continent by 180 million years ago.

Younger rocks covering both the Yukon-Tanana and Wrangellia terranes, called the Gravina–Nutzotin assemblage, suggest that the two terranes were close by 120 million years ago. Other cover rocks suggest that the Wrangellia and Alexander terranes actually coalesced about 310 million years ago, and then combined with the Peninsular Terrane about 265 to 210 million

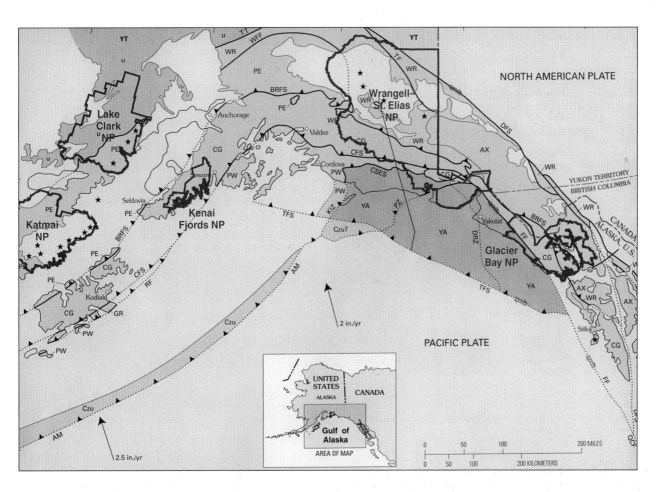

FIGURE 11.6 Accreted terranes of southern Alaska. The timing of metamorphism and deformation of rocks reveals that terranes are older toward the north, and become younger southward (Table 11.2). Stars indicate active volcanoes, which poke up through older accreted terranes.

Terranes: AX = Alexander; CG = Chugach; PE = Peninsular; PW = Prince William; WR = Wrangellia; YA = Yakutat; YT = Yukon-Tanana.

Faults: AM = Aleutian megathrust; BRFS = Border Ranges fault system; CFS = Contact fault system; CSES = Chugach–St. Elias fault system; DRZ = Dangerous River zone; DFS = Denali fault system; FF = Fairweather fault; KIZ = Kayak Island zone; PZ = Pamplona zone; QCF = Queen Charlotte fault; RF = Resurrection fault; TF = Totschunda fault; TFS = Transition fault system; TT = Talkeetna thrust; WFF = West Fork fault.

years ago, forming a crustal block sometimes referred to as the Wrangellia composite terrane. The Wrangellia composite terrane slammed into North America from 110 to 85 million years ago, followed by the Chugach Terrane about 67 million years ago, and then the Prince William Terrane by 50 million years ago. The Yakutat Terrane started to collide with North America by 25 million years ago; it is still attached to the Pacific Plate, so it continues to smash into the continent.

Denali and Wrangell–St. Elias national parks contain many of the highest mountains in North America.

The high topography of the Alaska and St. Elias ranges, which run through the two parks, is a consequence of collision of the youngest terranes from the south. The collision uplifts the region as the crust thickens and is compressed along thrust faults. On the south, Glacier Bay and Kenai Fjords national parks, as well as some of Wrangell–St. Elias National Park, lie on younger terranes that are still being deformed above the subducting oceanic crust of the Pacific Plate. The subduction leads to active volcanoes in Aniakchak National Monument and Katmai, Lake Clark, and Wrangell–

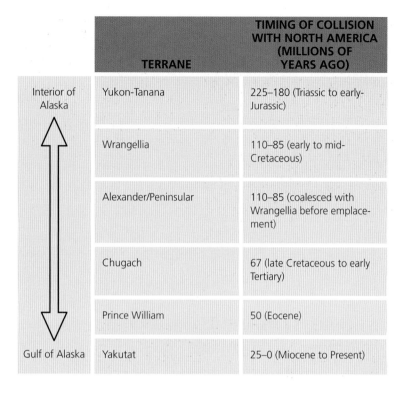

TERRANE	TIMING OF COLLISION WITH NORTH AMERICA (MILLIONS OF YEARS AGO)
Yukon-Tanana	225–180 (Triassic to early-Jurassic)
Wrangellia	110–85 (early to mid-Cretaceous)
Alexander/Peninsular	110–85 (coalesced with Wrangellia before emplacement)
Chugach	67 (late Cretaceous to early Tertiary)
Prince William	50 (Eocene)
Yakutat	25–0 (Miocene to Present)

TABLE 11.2 Sequence of terrane accretion in southern and southeastern Alaska

St. Elias national parks. The volcanoes poke up through earlier accreted terranes (Fig. 11.8). Northward subduction of the Pacific Plate also results in thrust faulting, folding, uplift, and the emplacement of igneous intrusive rock, as well as highly deformed accretionary wedge material (called **mélange**).

Accreted terranes can act as giant magnets. In the presence of Earth's magnetic field, rocks are magnetized with a certain intensity, depending on how rich they are in iron-bearing minerals. That magnetization in turn perturbs Earth's magnetic field locally. By towing an instrument, called a magnetometer, behind airplanes, cartographers have compiled an aeromagnetic map of Alaska (Fig. 11.9). A characteristic magnetic intensity pattern on the map identifies each major tectonic terrane.

Gulf of Alaska and the Denali Fault System

The Gulf of Alaska is a prominent indentation into the continent. A thick block of crust that moves northward with the Pacific Plate will eventually "dock" with North America in that region. For example, the Yakutat Terrane was more than 1,000 miles (1,600 kilometers) south of Alaska about 45 million years ago (Fig. 11.10). It moved northward and eventually crashed into the embayment as Alaska's newest accreted terrane. It continues to converge with the continent, contributing to the crustal thickening and uplift of the region.

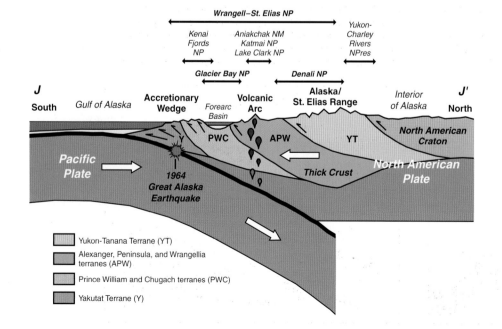

FIGURE 11.7 Simplified cross-sectional view of accreted terranes in southern Alaska. Information is projected onto line J–J' shown in Fig. 11.5. The terranes were added to the North American continent progressively from north to south (Table 11.2). Lines with double arrows show approximate extent of national park lands with respect to the various terranes. Kenai Fjords and Glacier Bay lie on relatively young terranes. Aniakchak, Katmai, Lake Clark, and Wrangell–St. Elias contain active volcanoes caused by the ongoing subduction of the Pacific Plate beneath North America (Chap. 5). The thickened crust beneath Denali and Wrangell–St. Elias contributes to the extremely high topography of the Alaska and St. Elias ranges. Yukon-Charley Rivers National Preserve lies on the small prong of the North American Craton that extends into Alaska.

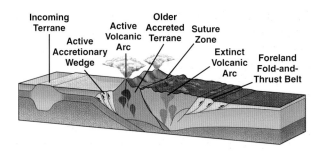

FIGURE 11.8 Three-dimensional geometry of terrane accretion. The incoming terrane approaches as part of a plate subducting beneath an active accretionary wedge. The active volcanic arc develops on crust of an older accreted terrane. Extinct volcanic arcs on still-older accreted terranes reflect the positions of earlier subduction zones. Suture zones mark the boundaries between different terranes. Foreland fold-and-thrust belts can develop some distance inland from the active volcanic arc (Chap. 10).

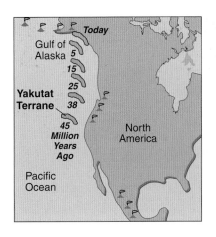

FIGURE 11.10 Emplacement of Yakutat Terrane. As the Pacific Plate moved northward, the block of crust known as the Yakutat Terrane eventually moved into the Gulf of Alaska, where it slammed into North America. The collision continues, contributing to the ongoing deformation, earthquakes, and uplift of southern Alaska (Fig. 11.7).

As terranes move up the west coast of North America, the shape of the Gulf of Alaska results in a counterclockwise rotation along the arching Denali Fault system (Fig. 11.11). The system absorbs some of the relative motion between the Pacific and North American plates. In western British Columbia and the southeastern Alaska Panhandle, plate motion is mostly transform, manifested by right-lateral motion along the Queen Charlotte and Fairweather faults. But farther north, the motion is transferred to the Denali Fault system, extending from the transform plate boundary, arching across the Yukon Territory and south-central Alaska, through Denali National Park, then westward all the way to the Bering Sea.

Geologic Features in Parks

DENALI NATIONAL PARK. At 20,320 feet (6,194 meters), Mount McKinley (Denali) towers over the landscape of Denali National Park (Fig. 11.12). *Denali* means

FIGURE 11.9 Aeromagnetic map of Alaska. Colors represent different levels of magnetic intensity, which show a strong correlation to Alaska's various accreted terranes (Figs. 11.2, 11.4, and 11.6). The trend of the Denali Fault system is also apparent (Fig. 11.11). Numbers refer to parks indicated in Table 11.1.

FIGURE 11.11 National park lands relative to the Gulf of Alaska and the Denali Fault system. Terranes moving northward with the Pacific Plate encounter the indentation in North America outlined by the Gulf of Alaska. As they crash into the continent, they rotate counterclockwise along the Denali Fault system.

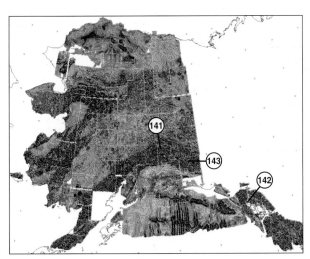

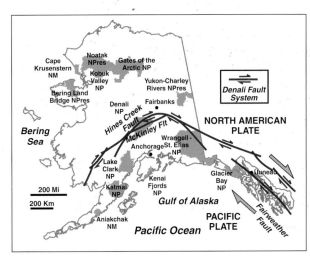

FIGURE 11.12 Denali National Park, Alaska. The highest mountain in North America, Denali (also known as Mount McKinley), consists of granite that intruded through metamorphic rocks of older accreted terranes.

TERRANES HAVE THEIR OWN COMPASSES!

One key to tracking the origin and path of accreted terranes is **paleomagnetism**. Magnetic minerals, such as magnetite, act a bit like compass needles, orienting themselves according to the magnetic field that was present at the time a rock formed. Today, the needle of a compass points toward the magnetic North Pole. The needle is also inclined upward or downward, depending on the latitude relative to the magnetic pole. Near the equator, a compass needle is horizontal. But at the magnetic North Pole, it points straight downward, and at the magnetic South Pole, it points straight upward. In between, the needle points somewhere between those extremes, depending on the magnetic latitude. (A counterbalance on the end of your compass needle is necessary to keep it horizontal.) A rock such as basalt is rich in iron-bearing minerals. When basalt forms, the magnetic inclination is frozen into its magnetic minerals, so that they act as ancient compass needles. We can determine the approximate latitude where a rock formed by examining the ancient magnetic properties frozen into the rock.

Paleomagnetic studies reveal that rocks in and near Wrangell–St. Elias National Park formed at about 25° to 30° latitude, compared to the approximate 60° where they're found today. The implication is that they were transported about 2,000 miles (3,000 kilometers) northward before being accreted to North America (Fig. 11.10).

"The Great One" in the Athabaska language. There is about 18,000 feet (5,500 meters) vertical relief between Wonder Lake, in the northern part of the park, and the top of Mount McKinley. The Wickersham Wall on the north side of the mountain has just about the same amount of relief, making it one of the most impressive vertical features on the planet. The mountain is still rising at about 0.04 inch (1 millimeter) per year.

Mount McKinley is part of the Alaska Range, which is about 600 miles (1,000 kilometers) long. The highest part of the range is the 60-mile (100-kilometer) segment that runs from southwest to northeast across Denali National Park, including mounts Dall (8,756 feet; 2,669 meters), Russell (11,670 feet; 3,557 meters); Foraker (17,400 feet; 5,303 meters), McKinley, Silverthorne (13,220 feet; 4,029 meters), Deception (11,768 feet; 3,587 meters), and Mather (12,123 feet; 3,695 meters). The mountains attain such lofty heights because of thrust faulting and crustal thickening, two consequences of compression as plates converge and terranes come crashing into the continent.

Uplift of the Alaska Range began about 60 million years ago, and the mountains are still rising. The uplift is coincident with emplacement of the Chugach, Prince William, and Yakutat terranes (Table 11.2), as well as the ongoing subduction of the Pacific Plate beneath southern Alaska (Fig. 11.7). Compression of the terranes leads to thrust faulting, which shoves blocks of crust upward (Fig. 2.7b). The crust thickens as rocks of the various terranes are shoved beneath southern Alaska; their buoyancy leads to the additional uplift necessary to achieve isostatic equilibrium (Fig. 2.6b). Add to that the fact that the Alaska Range has a core of hard granitic rocks that do not erode easily, and you end up with the highest mountains in all of North America.

Structure. The overall structure of Denali National Park relates to three features: 1) the Denali right-lateral, strike-slip fault system; 2) uplift of the Alaska Range; and 3) a system of thrust faults south of the park.

The Denali Fault system splits into two strands in the Denali National Park region: the Hines Creek Fault on the north, and the McKinley Fault on the south (Fig. 11.11). Both faults have right-lateral motion, with the north side of each fault moving eastward, relative to the south side.

The Hines Creek Fault cuts across the northern part of the park. It separates the 350-million-year-old schist of the Yukon-Tanana Terrane on the north, from the 56-million-year-old granite found on Mount McKinley. The southern block also includes coarse sandstone and shale that was deposited in the deep

FIGURE 11.13 The McKinley Fault strand of the Denali Fault system in Denali National Park. The dark rock on the left is composed of deep marine sandstone and shale layers known as flysch. On the right are light-colored layers of Devonian (350-million-year-old) limestone. The stream valley follows the McKinley Fault, which offsets those two rock masses.

FIGURE 11.14 View from Mount Healy Overlook Trail in Denali National Park. Rocks in foreground are part of the Yukon-Tanana Terrane.

ocean between converging terranes, known as **flysch**. The Hines Creek Fault crosses the Alaska Railroad near Denali National Park Headquarters, and extends westward through valleys for about 50 miles (80 kilometers). The McKinley Fault runs through the central part of the park (Fig. 11.13). The Foraker Granite, which formed about 38 million years ago, has been displaced 25 miles (40 kilometers) horizontally by the fault. Over the past 38 million years, the average slip rate on the fault has been about 0.04 inch (1 millimeter) per year.

Both strands of the Denali Fault system are north of Mount McKinley and the Alaska Range. Substantial reverse-fault motion on the McKinley Fault thrusts Mount McKinley upward and to the north. The Wickersham Wall is one effect of very recent uplift along the McKinley Fault. Another major structure, the Talkeetna Fault, lies south of the Alaska Range. Along that fault, the Talkeetna Mountains were thrust northward about 100 million years ago, during the mid-Cretaceous Period.

Accreted Terranes. Denali National Park includes three main accreted terranes: land north of the Hines Creek Fault, land between the Hines Creek and McKinley faults, and land south of the McKinley Fault. The terranes are distinguished by characteristic rock types, ages, and degrees of metamorphism.

On the north, the Yukon-Tanana Terrane consists of Paleozoic metamorphic rocks (schist and phyllite) that were placed on Alaska by the mid-Mesozoic Period (about 180 million years ago). It encompasses Mount Healy, Mount Wright, the Wyoming Hills, and

the Kantishna Hills, and includes exposures near park headquarters (Fig. 11.14). Those are some of the oldest rocks in the park, consisting of highly deformed Precambrian and Paleozoic sedimentary and volcanic rocks that have been metamorphosed to schists, quartzites, and marbles. The Toklat River flows northward across the Yukon-Tanana Terrane, skirting the east side of the Kantishna Hills (Fig. 11.15).

The region between the two strands of the Denali Fault system is a composite of the Pingston, Windy, and Dillinger terranes, and part of the McKinley Terrane. Its Paleozoic and Mesozoic sedimentary and

FIGURE 11.15 Toklat River in Denali National Park. The river is fed by glaciers coming off the Alaska Range, and flows across the Yukon-Tanana Terrane on the north side of the park.

FIGURE 11.16 Pillow basalt in Denali National Park. The circular blobs formed as fluid lava poured out on an ancient seafloor about 200 million years ago (see Fig. 5.11). They have since been accreted to North America and uplifted as part of the McKinley Terrane.

FIGURE 11.17 Glaciers in Denali National Park. The Alaska Range is an effective moisture barrier, resulting in more massive glaciers on the wet, south side of the park compared to the dryer north side. **a)** Ruth Glacier on the south side of the Alaska Range. **b)** East Fork of Toklat Glacier on the north side of the range.

a

b

igneous rocks are intensely folded and faulted. The McKinley Terrane is less metamorphosed than the others, and includes beautiful examples of pillow lavas (Fig. 11.16).

On the south side of the park, the Talkeetna Superterrane is an amalgamation that includes the Alexander, Maclaren, Peninsular, and Wrangellia terranes. Deep-sea flysch deposits in the vicinity of the McKinley Fault were highly deformed as the Talkeetna Superterrane crashed into Alaska. The smaller Talkeetna and Honolulu faults were active during the collision. South of the park, the Chugach and Prince William terranes were recently accreted, initiating uplift of the Alaska Range and Mount McKinley.

Intrusive Rocks. The granitic rocks in Denali National Park were formed during an earlier phase of subduction of the Pacific Plate beneath North America. As terranes are added to Alaska, the subduction boundary moves progressively southward. Dehydration of the oceanic crust of the subducting Pacific Plate

GLACIERS AS CLIMATE INDICATORS

It's fun to go to a national park such as Yellowstone, Yosemite, Glacier, or Crater Lake and see snow on the ground in July or August. Those parks do not have extensive glaciers now, but broad, U-shaped valleys indicate they did in the past (Figs. 9.20 and 10.33). What has changed? The climate has warmed up. In the past, snow on the ground in the summer would still be there by the next winter, and would get buried beneath more snow. Lower layers of the thickening pile would compact and turn into solid ice. On a slope, the ice would slide downhill, giving birth to glaciers. The higher elevations in Denali, Wrangell–St. Elias, Glacier Bay, and North Cascades national parks are so cold that more snow falls than melts during a year, adding new ice to glaciers.

Glaciers are sensitive indicators of climate change. They move as rivers of ice down to lower elevations, where warmer temperatures result in more melting than accumulation. More precipitation and colder temperatures tend to make glaciers advance farther downslope. The study of glaciers in national parks reveals that the zone where ice melting exceeds accumulation has been retreating uphill, indicating that the climate has warmed over the last century.

FIGURE 11.18 Wrangell–St. Elias National Park. Mount St. Elias is the second-highest peak in the United States. It consists of metamorphosed sedimentary and volcanic rocks of the Chugach Terrane.

leads to the development of volcanic arcs on top of the earlier accreted terranes (Fig. 11.8). The current arc volcanism is occurring in Aniakchak National Monument, Katmai National Park, Lake Clark National Park, and Wrangell–St. Elias National Park (Fig. 11.7). The granite found in the Alaska Range is from magma intruded into older terrane rocks, when the subduction zone was farther north. Most of the volcanic arc material has since eroded away, exposing granite on Mount McKinley and other peaks in Denali National Park. Associated lava flows are preserved in the park on the fringes of the mountains.

Glaciers. The largest glaciers in Denali National Park run down the south side of the range, including Kahiltna, Ruth, and Eldridge glaciers, each 20 to 30 miles (30 to 50 kilometers) long (Fig. 11.17). The mountains are so high that they form a weather barrier. Moist air moving northward from the Gulf of Alaska drops rain and snow as it rises. Little moisture remains for the north side of the range, which consequently has far less snow and ice available to feed glaciers. The Muldrow Glacier on the north side of Mount McKinley follows the surface trace of the McKinley Fault, below the Wickersham Wall.

WRANGELL–ST. ELIAS NATIONAL PARK. Wrangell–St. Elias National Park and Preserve is the largest unit of the National Park Service, encompassing an area of 20,707 square miles (53,608 square kilometers), making it larger than the country of Switzerland. And like Switzerland, the park has some impressively high mountains and extensive glaciers. It includes Mount St. Elias (Fig. 11.18), at 18,008 feet (5,489 meters) the second highest mountain in the United States, as well as a number of other peaks over 14,000 feet (4,300 meters). The various mountain ranges in the park trend in a northwest-southeast direction, paralleling accreted terranes and the fault zones that separate them. The park also has several active volcanoes (Chap. 5).

The Wrangell–St. Elias National Park area is important to the concept of "exotic" or "accreted" terranes. In fact, the term *Wrangellia* was a big part of the discussion about how continents grow over time. In the 1960s and 1970s, geologists recognized that rocks in the region of the park were dramatically different from nearby rocks of similar age. The idea of **tectono-stratigraphic**, or "suspect," terranes thus evolved. It suggested that large blocks of ground were made elsewhere, then transported some distance, perhaps colliding and amalgamating with other terranes during the journey. At some point, the whole assemblage "accreted" to the North American continent. One important piece of ground was known as Wrangellia, which makes up about half of Wrangell–St. Elias National Park. The Wrangellia Terrane covers much of southeastern Alaska, and fragments of the terrane are found as far south as Idaho (Fig. 11.19). It initially formed as a volcanic island arc near the equator from late Paleozoic (Pennsylvanian) to early Mesozoic (Triassic) time, and is characterized by slates and marbles, as well as gabbros and basalts that have been metamorphosed to various degrees.

Seven distinct terranes are found in Wrangell–St. Elias National Park and Preserve (Fig. 11.6). Each terrane can be recognized by its characteristic assemblage of rocks of distinct age and type. In some cases, the rocks of two or more terranes are covered by a characteristic sequence of sedimentary or volcanic layers not found elsewhere on the continent. The covering stratigraphic sequence demonstrates that those terranes coalesced elsewhere, and then were transported northward before docking with North America.

The assembly of Wrangell–St. Elias National Park terranes follows the north-to-south progression shown in Fig. 11.7: Yukon-Tanana (known locally as the Windy Terrane), Wrangellia, Alexander, Chugach, Prince William, and Yakutat. The terranes were added between the Denali Fault system on the north and the Fairweather Fault on the south (Fig. 11.11). The Fairweather Fault is a continuation of the Queen Charlotte Fault, the transform plate boundary between the Pacific and North American plates.

FIGURE 11.19 Distribution of Wrangellia Terrane. The green pattern shows that rocks of the same age, with similar structures and metamorphic facies, are found in several blocks that are exposed from Denali National Park in Alaska, through British Columbia and all the way into western Idaho.

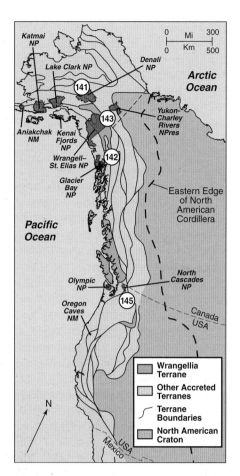

GLACIER BAY NATIONAL PARK. Glacier Bay National Park is a wonderful place to see rivers of ice flowing down the slopes of high coastal mountains and breaking off, or **calving**, into the Gulf of Alaska (Fig. 11.20). The park, on Alaska's southeast panhandle about 50 miles (80 kilometers) west of Juneau, is accessible only by plane or boat. Mount Fairweather, at 15,300 feet (4,663 meters), is the highest peak in the park. Virtually all of the park was covered by ice just 200 years ago. Warming and retreat of glaciers since then have exposed the "bedrock" geology in places. But quite a bit of ice remains in the park, including the Brandy Icefield and Glacier, the Takinsha Icefield, and the Northern Highlands.

The park's mountain ranges and narrow valleys trend southeast-northwest, parallel to the coastline. From west to east, they include the St. Elias/Fairweather, Beartrack, Excursion, Takinsha, and Chilkat ranges. Major valleys between the ranges are the Brandy Icefield and Brandy Glacier, as well as Glacier Bay. The Border Ranges

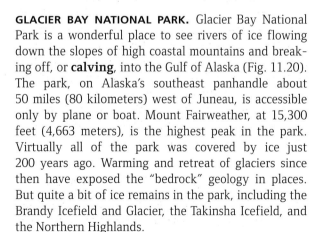

FIGURE 11.20 Glacier Bay National Park. Glaciers break up into icebergs (calve) as they enter the sea. Mountains in the background are parts of terranes that have recently been accreted to North America.

Fault runs from the St. Elias Range, on the northwest side of the park, southeastward to the Alexander Archipelago. It represents the late Mesozoic plate boundary, or **suture zone** (Fig. 11.8).

Accreted terranes in Glacier Bay National Park contain rocks of Precambrian and Paleozoic age. Some traveled great distances, perhaps from the South Pacific region. Terrane docking occurred in Mesozoic and early to mid-Tertiary time (Table 11.2). Three major terranes in the park, from east to west, include the Wrangellia, Alexander, and Chugach terranes (Figs. 11.6 and 11.7). The terranes moved northward and were placed onto North America via both strike-slip (right-lateral) and thrusting (southwest-to-northeast) motion.

The region of Glacier Bay National Park is tectonically active, as the Pacific Plate continues to slip northward and slide slightly eastward under the edge of North America. Coastal parts of the Fairweather Range are being uplifted at the rate of about ¼ inch (0.7 centimeter) per year. A magnitude 8 earthquake in 1958 triggered a large landslide that dropped material from about 3,000 feet (1,000 meters) above the steep walls of Lituya Bay. It caused a giant wave that washed across the bay at about 125 miles per hour (200 kilometers per hour) and sloshed back and forth for about 20 minutes. The gigantic wave toppled trees as high as 1,720 feet (525 meters) up the slope on the other side of the bay!

The Coast Ranges Batholith is an extensive region of igneous intrusive (plutonic) rocks extending from Washington State to Alaska (Fig. 11.21). Those rocks were formed when the entire West Coast was a subducting plate boundary. Similar to the granitic rocks found in Yosemite, Sequoia, and Kings Canyon national parks in the Sierra Nevada (Chap. 5), rocks in Wrangell–St. Elias, Glacier Bay, and North Cascades national parks display the eroded remnants of vast magma-chamber systems that formed beneath the ancient volcanoes.

ACCRETED TERRANES IN THE LOWER 48 STATES

The Cascade Mountain Range is a dramatic example of volcanic features resulting from plate convergence and subduction (Chap. 5). The Cascade Mountain region in northern California, Oregon, and southern Washington is covered by very young lava flows, cinders, pumice, mudflows, and other volcanic materials. Lassen Volcanic and Crater Lake national parks, and Mount St. Helens National Volcanic Monument, highlight composite volcanoes and the dramatic impact that recent eruptions have had on the landscape. But farther north in Washington, Mount Rainier National Park begins to reveal some of the older rocks that form the base for the volcanic material. North Cascades National Park is composed entirely of the older rocks. Together with the adjacent Ross Lake and Lake Chelan national recreation areas, the North Cascade Mountains present an opportunity to study some of the history of terrane accretion that contributed to the westward expansion of the North American continent.

Westward Building of the Pacific Northwest

Why are the North Cascade Mountains so high, and why do they contain rocks so different from those encountered elsewhere in the Cascade Mountains?

The Olympic and Cascade mountain ranges represent only the very latest of the events that have been shaping the landscape of the Pacific Northwest. In fact, 100 million years ago, the edge of the North American continent was in the far eastern portion of Washington

FIGURE 11.21 Origin of the Coast Ranges Batholith. a) One hundred million years ago, the entire West Coast was a subducting plate boundary, with a volcanic arc forming on the edge of North America (Fig. 5.46). **b)** Where the subduction ceased, the volcanoes eroded, exposing vast areas of cooled magma chambers, including the Pensinular, Sierra Nevada, and Coast Ranges batholiths. Granitic rocks in North Cascades, Glacier Bay, and Wrangell–St. Elias national parks are part of the Coast Ranges Batholith.

(a) 100 Million Years Ago

(b) Today

(a) >100 Million Years Ago

(b) 100 Million Years Ago

(c) 100–50 Million Years Ago

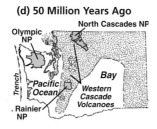

(d) 50 Million Years Ago

FIGURE 11.22 Westward growth of the North American continent in Washington State. a) The West Coast of North America was in the position of eastern Washington prior to 100 million years ago. **b)** The Kootenay Arc, a volcanic mountain chain similar to today's Cascades, developed along the continental edge. **c)** Subduction off the West Coast (Okanogan Trench) resulted in another volcanic arc and added more material to North America. **d)** After "exotic crust" of the North Cascades crashed into the continent, the trench jumped farther west. The Olympics and Cascades are the latest in a series of subduction-related mountain ranges comprising the Pacific Northwest and are represented by the landscapes of Olympic and Mount Rainier national parks (Chap. 5). North Cascades National Park lies on older, accreted-terrane rocks.

State (Fig. 11.22a). Similar to the modern situation, oceanic crust was subducting eastward, building a volcanic arc and accretionary wedge (Fig. 11.22b). Over time, fragments of thick continental crust and oceanic islands were added to the continental edge. Rocks of North Cascades National Park are not volcanic as are those of the modern Cascades; they are much older material (an accreted terrane) that crashed into North America. With addition of more and more material, the subduction boundary moved progressively westward (Fig. 11.22c). Its present position developed about 50 million years ago (Fig. 11.22d).

As subduction zones jumped westward, they formed new mountain ranges. Over time, the mountains eroded downward as the continent grew westward. Mountain ranges in western Idaho and in Oregon and Washington are progressively younger in a westward direction. Their rocks are the deep portions of the volcanic and coastal ranges that formed along the an-

FIGURE 11.23 North Cascades National Park. a) At 9,127 feet (2,782 meters), Mount Shuksan is the highest point in the park. The Alpine scenery is the result of thrust faulting as terranes came crashing in. **b)** Snow-covered Mount Baker, looking west from the park. The active volcano results from the ongoing subduction of the Juan de Fuca Plate beneath North America. It pokes up through Tertiary-age granite that intrudes metamorphic rocks of older accreted terranes. The two peaks in the middle ground are the Twin Sisters. **c)** Challenger Glacier.

a

b

c

cient subduction zones. The Coast Ranges and Cascade Mountains are the youngest set of subduction-related mountain ranges; in time, the subduction zone might again jump westward, stranding remnants of those mountains as new mountains form in the region that is currently offshore. Eventually the entire Juan de Fuca Plate will subduct, so that the San Andreas and Queen Charlotte faults will connect as a transform boundary along the entire West Coast (Fig. 11.3).

NORTH CASCADES NATIONAL PARK. In the late-Mesozoic and Cenozoic Eras, the North American Plate has moved westward. Convergence with oceanic plates led to subduction and collision with terranes of thicker crust. North Cascades National Park contains accreted terranes that were swept onto the edge of North America, as well as accretionary wedge and volcanic arc rocks associated with subduction.

The northern Cascade region looks more like the Alps than other parts of the Cascade Mountain Range (Fig. 11.23a), with elevations over 9,000 feet (2,750 meters). Such heights result from a combination of thrust faulting and crustal thickening, both consequences of compressional forces due to plate convergence. North Cascades National Park lies east of two active composite volcanoes, Mount Baker and Glacier Peak, that rest upon much older rocks (Fig. 11.23b). Massive glaciers that were far more extensive than the ones currently in the park carved the mountains of North Cascades National Park (Fig. 11.23c). While the glaciers have eroded the old igneous and metamorphic rocks in the park to sharp ridges, the nearby Cascade volcanoes are smooth cones because they continue to erupt.

Terranes in North Cascades National Park abut the older continental craton. Those old rocks, known as the Yellow Aster Complex, have been dated as forming about 1.65 billion (1,650 million) years ago (Fig. 11.24). They consist of igneous and metamorphic rocks that underwent several phases of tectonic activity, and were highly deformed along thrust and strike-slip faults. Newer terranes in the park were attached to North America from late Mesozoic to early Tertiary time.

The metamorphic facies of rocks is an important indicator of the park's geologic history. A particular facies is a collection of minerals that formed under certain temperature and pressure conditions that help determine the approximate depth to which a rock was buried. High temperatures and pressures on rocks as they were pushed down ancient subduction zones led to development of the metamorphic rocks in North Cascades National Park (Fig. 11.25). Magma that

FIGURE 11.24 Basement of the North American Craton in North Cascades National Park. The Yellow Aster Complex contains rocks as old as 1.65 billion years.

METAMORPHIC FACIES: CLUES FOR THE EARTH DETECTIVE

Certain types of minerals found in metamorphic rocks are clues to the maximum depth to which rocks were buried. A **metamorphic facies** is the suite of minerals found in a rock that develop from a specific combination of temperature and pressure. Both temperature and pressure increase with depth in the Earth. By also determining the age of the rock containing the minerals, a geologist can interpret key aspects to the tectonic history of a region. For example, sedimentary layers on top of an oceanic plate can be shoved down to considerable depth at a subduction zone. As they descend into the Earth, the rocks are metamorphosed because of the higher temperatures and pressures. Certain minerals develop as the rocks are metamorphosed, depending on how deep the rocks were pushed down. The rocks may later be scraped off the subducting plate and thrust upward and exposed at the surface as the overlying material erodes. A geologist examines the rocks while mapping an accreted terrane in a national park. By looking at the suite of minerals found within the rock, the geologist can determine the depth to which the original rock was subducted. If minerals suitable for dating analysis are present, the timing of subduction can also be determined. A geologist can interpret, for example, that the Wrangellia Terrane in Wrangell–St. Elias National Park was part of a subducting plate boundary before being accreted to North America.

formed as water rose from the subducting plate led to emplacement of the granitic rocks. As rocks were uplifted, sedimentary layers and other rock on top eroded, revealing the array of igneous and metamorphic rocks beneath.

The core area of the North Cascade Mountains is an accreted terrane containing banded gneiss and mica schist formed from the metamorphism of sandstone and shale, marble derived from limestone, and greenstones that were once lava flows beneath the ocean. Those rocks are well exposed along North Cascades Highway, which runs between the north and south units of the park. Less-metamorphosed and less-deformed Paleozoic sedimentary strata occur on the east and west flanks of the range, outside the park.

a

FIGURE 11.25 Metamorphic rocks in North Cascades National Park. a) Outcropping of Mesozoic-age Skagit Gneiss. **b)** Close-up of the same rock, which is a banded gneiss. The bands are alternating zones of dark minerals (hornblende and biotite) and lighter ones (quartz and feldspar). The rock probably started as sand and mud layers that compacted into sandstone and shale, and then were metamorphosed as the rock was subjected to high temperature and pressure. The banding is a form of foliation in a metamorphic rock that develops perpendicular to the direction of maximum compression.

b

FURTHER READING

GENERAL

Alt, D. P., and D. W. Hyndman. 1984. *Roadside Geology of Washington*. Missoula, MT: Mountain Press. 288 pp.

Alt, D., and D. W. Hyndman. 1995. *Northwest Exposures: The Geologic Story of the Northwest*. Missoula, MT: Mountain Press. 443 pp.

Chronic, H. 1986. *Pages of Stone: Geology of Western National Parks and Monuments, 2: Sierra Nevada, Cascades, and Pacific Coast*. Seattle: Mountaineers. 170 pp.

Collier, M. 1997. *The Geology of Denali National Park and Preserve*. Anchorage: Alaska Natural History Association. 48 pp.

Davidow, B. 1993 (September). The high one. *Earth*, p. 44–51.

Gilbert, W. 1979. *A Geologic Guide to Mt. McKinley National Park*. Anchorage: Alaska Natural History Association. 52 pp.

Howell, D. G. 1985. Terranes. *Scientific American*, v. 253, p. 116–125.

Howell, D. G. 1995. *Principles of Terrane Analysis: New Applications for Global Tectonics*, 2nd Ed. New York: Chapman and Hall. 245 pp.

Jones, D. L., et al. 1982. The growth of western North America. *Scientific American*, v. 247, p. 70–84.

McPhee, J. 1998. In suspect terrain, in *Annals of the Former World* 2. New York: Farrar, Straus and Giroux, p. 147–275.

West, S. 1981. Alaska: The fragmented frontier. *Science News*, v. 1, p. 10–11.

West, S. 1982. A patchwork Earth. *Science*, v. 82, p. 46–51.

Winkler, G. R., E. M. MacKevett, Jr., G. Plafker, D. H. Richter, D. Rosenkraus, and H. S. Schmall. 2000. *A Geologic Guide to Wrangell-Saint Elias National Park and Preserve,*

Alaska: A Tectonic Collage of Northbound Terranes. Washington: U.S. Geological Survey, Professional Paper 1616. 166 pp.

TECHNICAL

Brease, P., and A. Till. 1995. *The Geology and Glacial History of Denali National Park and Vicinity*. Fairbanks: Geological Society of America, Field Trip Guide Book #9, Cordilleran Section Meeting. 93 pp.

Brown, E. H. 1987. Structural geology and accretionary history of the northwest Cascades system, Washington and British Columbia. *Geological Society of America Bulletin*, v. 99, p. 201–214.

Burns, T. R. 1983. Model for the origin of the Yakutat block, an accreting terrane in the northern Gulf of Alaska. *Geology*, v. 11, p. 718–721.

Condie, K. C. 1982. *Plate Tectonics and Crustal Evolution*, 2nd Ed. New York: Pergamon Press. 310 pp.

Coney, P., D. Jones, and J. Monger. 1980. Cordilleran suspect terranes. *Nature*, v. 288, p. 329–333.

Csejety, B., Jr., et al. 1982. The Cenozoic Denali Fault system and the Cretaceous accretionary development of southern Alaska. *Journal of Geophysical Research*, v. 87, p. 3741–3754.

Hillhouse, J. W. 1987. Accretion of southern Alaska. *Tectonophysics*, v. 139, p. 107–122.

Howell, D. G. 1989. *Tectonics of Suspect Terranes*. New York: Chapman and Hall. 232 pp.

Howell, D. G., G. W. Moore, and T. J. Wiley. 1987. Tectonics and basin evolution of western North America—An overview, in *Geology and Resource Potential of the Continental Margin of Western North America and Adjacent Ocean Basins—Beaufort Sea to Baja California*, edited by D. W. Scholl, A. Grantz, and J. G. Vedder. Houston: Circum-Pacific Council for Energy and Mineral Resources, Earth Science Series, vol. 6, p. 1–15.

Hudson, T. 1983. Calc-alkaline plutonism along the Pacific rim of southern Alaska, in *Circum Pacific Plutonic Terranes*, edited by J. A. Roddick. Boulder, CO: Geological Society of America Memoir 159, p. 159–169.

Jones, D. L., N. J. Silberling, W. Gilbert, and P. Coney. 1982. Character, distribution, and the tectonic significance of accretionary terranes in the central Alaska Range. *Journal of Geophysical Research*, v. 87, p. 3709–3717.

Jones, D. L., N. J. Silberling, and J. W. Hillhouse. 1977. Wrangellia—A displaced continental block in northwestern North America. *Canadian Journal of Earth Sciences*, v. 14, p. 2565–2577.

Moores, E. M., and R. J. Twiss. 1995. *Tectonics*. New York: Freeman. 415 pp.

Plafker, G. 1965. Tectonic deformation associated with the 1964 Alaska earthquake. *Science*, v. 148, p. 1675–1687.

Plafker, G., and H. C. Berg (editors). 1994. *The Geology of Alaska*. Boulder, CO: Geological Society of America, Decade of North American Geology, vol. G-1. 1066 pp.

Reed, B. L., and M. A. Lanphere. 1974. Offset plutons and history of movement along the McKinley segment of the Denali Fault system, Alaska. *Geological Society of America Bulletin*, v. 85, p. 1883–1892.

Richards, M. A., D. L. Jones, R. A. Duncan, and D. J. DePaolo. 1991. A mantle plume initiation model for the formation of Wrangellia and other oceanic flood basalt plateaus. *Science*, v. 254, p. 263–267.

Richter, D. H., and N. A. Matson, Jr. 1971. Quaternary faulting in the eastern Alaska Range. *Geological Society of American Bulletin*, v. 82, p. 1529–1540.

Strahler, A. N. 1998. *Plate Tectonics*. Cambridge, MA: Geo-Books. 554 pp.

Stone, D. B., and W. K. Wallace. 1987. A geological framework of Alaska. *Episodes*, v. 10, p. 283–289.

WEBSITES

National Park Service: www.nps.gov

Park Geology Tour:
www2.nature.nps.gov/grd/tour/index.htm

Accreted Terranes:
Alaska:
141. Denali NP:
www2.nature.nps.gov/grd/parks/dena/index.htm
142. Glacier Bay NP:
www2.nature.nps.gov/grd/parks/glba/index.htm
143. Wrangell–St. Elias NP:
www2.nature.nps.gov/grd/parks/wrst/index.htm

Lower 48:
144. Lake Chelan NRA:
www.nps.gov/lach/index.htm
145. North Cascades NP:
www2.nature.nps.gov/grd/parks/noca/index.htm
146. Ross Lake NP: www.nps.gov/rola/index.htm

U.S. Geological Survey: www.usgs.gov

Geology in the Parks:
www2.nature.nps.gov/grd/usgsnps/project/home.html
Accreted Terranes:
walrus.wr.usgs.gov/infobank/programs/html/school/keypage/accreted_terrane.html
Origin of Accreted Terranes:
walrus.wr.usgs.gov/infobank/programs/html/school/moviepage/06.01.19.html

ANSWERS TO QUESTIONS

Part I: Earth Systems and Our National Parks

1. What role does geology play in national parks?

An area is established as a national park, monument, seashore, or other unit of the National Park Service because it displays something special about the cultural or natural history of the United States. The role of the National Park Service is to protect those features and make them accessible to the public. Geological features are an important part of this heritage, not only because they help us understand the Earth's history, but also because they are the landscapes upon which our country's cultural and natural history take place.

2. Why is tectonic activity concentrated over such narrow zones on Earth's surface?

Tectonic activity (earthquakes, volcanic eruptions, the formation of mountains) is associated with the movement of large plates of Earth's outer, rigid shell. Most of the activity is due to interactions along the boundaries of plates, or because a plate moves over a hotspot rising from Earth's deep interior.

3. Why are there plates?

The Earth is a peculiar planet. Its mantle (the zone between the thin crust and the heavy core) has hard and soft zones. Both temperature and pressure increase as you go deeper into the Earth. The crust and outermost mantle are so cold they form the solid lithosphere. But the higher temperature just below that results in a somewhat softer layer, the asthenosphere. Still deeper in the mantle, it's even hotter, but pressure is so great that the lower mantle is a hard solid. These unique properties of Earth's mantle create a situation where plates of hard lithosphere ride over the softer layer of asthenosphere.

4. Why do plates move about relative to one another?

The Earth can be viewed as a giant heat engine. It was originally molten and remains hot because radioactive elements decay to nonradioactive forms, generating heat in the process. One of the ways that the heat escapes upward is through flow of the soft asthenosphere, an extremely slow version of the convection currents in a pot of boiling water or simmering pudding. The rigid plates of lithosphere ride about on the convecting asthenosphere, ripping apart at divergent plate boundaries, crashing together at convergent plate boundaries, and sliding past one another at transform plate boundaries.

5. How fast do plates move?

Look at your fingernails and watch them grow. That will give you an idea of how fast lithospheric plates move relative to one another—about a fraction of an inch to a few inches per year! That doesn't seem like much, but over time it adds up. For example, moving at about 2 inches (5 centimeters) per year, in our lifetime the Pacific Plate moves 10 to 15 feet (3–5 meters) past the North American Plate along the San Andreas Fault, a transform plate boundary in California. As Europe and Africa move away from North and South America at about 1½ inches (4 centimeters) per year, the Atlantic Ocean has opened to a width of 4,000 miles (6,000 kilometers) in the past 150 million years!

6. What are hotspots?

Hotspots are plumes of hot material rising from deep within Earth's mantle. As a hotspot encounters the bottom of a moving plate, it melts its way through, forming a line of volcanoes, such as the Hawaiian Islands, on the plate's surface.

7. What causes volcanic activity?

Volcanic activity occurs above places in the Earth where pressure, temperature, and the presence of water conspire to melt rock. There are two main ways Earth materials melt: 1) hot mantle rising and decompressing; and 2) water flowing through hot rock. The first type of melting occurs at diverging plate boundaries (mid-ocean ridges, continental rift zones) and hotspots, where mantle beneath plates is hot but remains solid because it is under great pressure. As the hot mantle rises, a sudden drop in pressure causes melting, much as taking the lid off a pressure cooker causes superheated water to flash to steam. The second mechanism results in volcanism at convergent plate boundaries. As one plate dives (subducts) beneath the other, it heats up and releases hot fluids (similar to the way that people sweat when they get hot). The fluids (mostly water) rise, wetting hot rock in their path and causing some of it to melt.

8. What cause earthquakes?

Two things are necessary for earthquakes to occur: 1) motion within the Earth that stresses and deforms material; and 2) material that deforms by breaking rather than flowing. Moving plates of cold, brittle lithosphere produce most earthquakes, especially where they are contorted and grind against one another along their boundaries.

9. How do mountain ranges form?

Most mountain ranges are long and narrow because they form at plate boundaries or hotspots.

The largest mountain range on Earth is the system of mid-ocean ridges, a line of volcanoes formed on the seafloor as plates diverge. Plate divergence sometimes rips a continent apart, forming long mountain ranges separated by deep rift valleys. Where plates converge along an active continental margin, the one with thin oceanic crust descends (subducts) beneath the continent, forming two parallel mountain ranges. One is a coastal range consisting of material squeezed up out of the sea, the other a volcanic chain farther inland above where hot water rises from the descending plate. The highest mountains on Earth, the Himalayas, are forming where thick blocks of continental crust (India and Asia) collide as a result of plate convergence. Sheared-up mountain ridges and valleys form in a narrow zone where one plate slides past another at a transform plate boundary. A mountain chain of volcanoes forms where a plate rides over a hotspot; the volcanoes get older and lower as the plate moves them away from the hotspot.

10. Why is there so much more tectonic activity in the western United States than in the East?

Tectonic activity commonly occurs along the boundaries of large, moving plates of Earth's outer shell. The Pacific Coast of the United States coincides with plate boundaries and thus has earthquakes, volcanic eruptions, and developing mountain ranges. The Atlantic and Gulf coasts, though initially formed from plate boundary activity, are now far from any plate boundaries and accompanying tectonic events.

Part II: Divergent Plate Boundaries

1. Why is North America above sea level, whereas the surrounding ocean floors are below?

Like other continents, North America has thick crust, compared to the thin crust beneath the adjacent Atlantic and Pacific oceans. Crust is light material that "floats" on the underlying, heavier mantle. Continental crust is thick and buoyant like a beach ball; it therefore sticks up higher than thin oceanic crust.

2. If the crust in the Basin and Range Province is so thin, then why is the topography so high?

All else being equal, thin crust would result in low topography. But in active continental rift zones, such as East Africa and the Basin and Range Province, the lithospheric plate is also thin. The hot, buoyant asthenosphere rises from below and expands like a hot-air balloon, elevating the region.

3. Why are there so many types of volcanoes and volcanic products in continental rift zones?

Thick, silica-rich continental crust caps the lithospheric plate at a continental rift zone. Magma originating from the mantle must initially melt its way through that crust in order to reach the surface. The magma becomes enriched in silica because high-silica minerals tend to melt first. Some of the lava that initially pours out at continental rift zones is thick and pasty, cooling to light-colored rocks (rhyolite and dacite) and steep-sided lava domes and composite volcanoes. As rifting progresses, the crust thins, becomes depleted in silica, and has better-developed conduits, so that magma that reaches the surface is the pure, low-silica basalt that melted from the mantle. Most of the eruptions at continental rift zones produce fluid, dark-colored lava that spreads out as low-profile shield volcanoes, or erupts in fountains to form cinder cones. If rifting continues and the continent completely rips apart, only thin, basaltic oceanic crust caps the diverging plates—lava that surfaces at mid-ocean ridges is pure basalt.

4. Why are there long, north-south mountain ranges and valleys in the Basin and Range Province?

As the cold, brittle upper crust pulls apart, it breaks along normal faults. The rift valleys are the blocks of crust that drop downward along the faults, while the mountain ranges are the blocks that remain high. The "basins" (rift valleys) and "ranges" trend north-south because the direction of stretching (plate divergence) is east-west.

5. Why is Death Valley the lowest point in North America?

Death Valley is a classic example of a continental rift valley. The deepest lakes in the world form in such valleys, including Lake Baikal in Siberia (the world's deepest), lakes Tanganyika and Malawi in East Africa (the second and fourth deepest), and Lake Tahoe in the California and Nevada portion of the Basin and Range Province (the eighth deepest). Such lakes form where rift valleys drop down

too fast to be completely filled with sediment. Instead, a lot of water collects in the depression. But what if not much water is available? The lowest point on Earth, the Dead Sea, is in a rift valley in the very dry Middle East. Similarly, the Mojave Desert of Southern California is so hot and dry that water evaporates before it can accumulate in Death Valley, the lowest point in North America.

6. What types of earthquakes occur in the Basin and Range Province?

Earthquakes beneath the Basin and Range Province are shallow and relatively small because the underlying, hot asthenosphere is shallow. The lower part of the crust is so hot that it stretches in a ductile manner and does not produce earthquakes. Earthquakes occur in the relatively cold, brittle upper portion of the crust, which is only about 8 miles (12 kilometers) thick. Earthquakes are only up to magnitude 7.5 because the crust in the region is so hot.

7. Why don't large earthquakes or other tectonic events occur along margins of the Atlantic Ocean or Gulf of Mexico?

Tectonic activity commonly occurs along the boundaries of large, moving plates of Earth's outer shell. The Pacific Coast of the United States coincides with plate boundaries and is thus an active continental margin, with earthquakes, volcanoes, and developing mountain ranges. But the Atlantic and Gulf coasts are passive continental margins, far away from plate boundaries and accompanying tectonic activity.

8. How did the Atlantic Ocean and Gulf of Mexico form?

About 300 million years ago Europe, Africa, and South America collided with North America, forming the supercontinent of Pangea. As Pangea ripped apart 200 million years ago, Europe, Africa, and South America pulled away from North America, creating the Atlantic Ocean and Gulf of Mexico. The Atlantic and Gulf coasts show the classic form of a passive continental margin: a low-lying coastal plain, broad continental shelf, then a steep continental slope, gentle continental rise, and flat abyssal plain. This topography is a consequence of the transition from thick continental to thin oceanic crust. The high areas (coastal plain and shelf) are underlain by continental crust, while oceanic crust underlies low regions (continental rise and abyssal plain).

9. **What factors influence the position of a coastline?**

Coastlines are sensitive to three factors: sea level change, rate of subsidence of the continental margin, and supply of sediment. When worldwide temperatures are low, much of the water is frozen in the polar ice caps, dropping sea level. During those times, the shoreline is near the edge of the continental shelf, more than 100 miles (160 kilometers) from its current position. If Earth's atmosphere were to warm up, sea level would rise, flooding coastal areas as far as 70 miles (110 kilometers) inland. When North America ripped apart, its crust stretched and thinned, eventually subsiding below sea level as the Atlantic Ocean and Gulf of Mexico opened. Sediment eroded from North America is continuously deposited along the continental margin. The supply of sediment is enormous compared to the rate of subsidence, so over time the coastal plain and continental shelf have been building outward.

10. **Why is it thought that the Colorado Plateau region was once part of a passive continental margin?**

Layers of sandstone, shale, and limestone, similar to those found on modern passive margins, cover the area. From about 550 to 250 million years ago, the North American continent looked much different. Its western edge was roughly in the position occupied today by the Rocky Mountains. That edge was passive, like the present East Coast—slowly subsiding and being covered by sedimentary layers. Later, as the Colorado Plateau rose, rivers cut into and exposed the layers. In Grand Canyon National Park, the Colorado River has carved downward and exposed sedimentary layers and underlying rocks that comprise the ancient passive continental margin.

Part III: Convergent Plate Boundaries

1. **Why does an oceanic plate subduct beneath a continental plate?**

Earth's crust is less dense than the underlying mantle. Although both crust and mantle are solid, crust can be thought of as "floating" on the mantle. The crust of continents is thicker, and thus more buoyant, than crust of the ocean. If you envision Earth's mantle as a swimming pool, you might think of continental crust as a soccer ball, which sticks up higher out of the water than a smaller tennis ball (oceanic crust). A swimmer might easily bring the tennis ball to the bottom of the pool, but it would be more difficult to "subduct" the larger soccer ball. Similarly, a plate capped with thin oceanic crust can easily subduct, while one with thick continental crust is too buoyant. As plates converge in the Pacific Northwest, the oceanic Juan de Fuca Plate subducts beneath the continental North American Plate.

2. **Why are the Olympic and Cascade mountain ranges so different?**

Both the Olympic and Cascade mountain ranges are products of plate convergence as the Juan de Fuca Plate subducts beneath North America. But the ranges form at different parts of the subduction zone—one is volcanic, the other a product of crustal deformation. The Cascade Mountains lie above the position where the Juan de Fuca plate is about 50 miles (80 kilometers) deep. It's so hot at that depth that fluids, primarily water, begin to "sweat" from the oceanic crust and its cover of sediments. As the hot water rises, it melts rock in its path, producing magma. Some of the magma reaches the surface as lava flows and other volcanic products of Mount Baker, Mount Rainier, Mount St. Helens, and other Cascade Mountain peaks. Farther to the west, the top of the Juan de Fuca Plate is not nearly so deep. Its rocks are still pretty cold, so that it does not sweat fluids and hence produces no volcanoes. Instead, some of the hard ocean crust and sedimentary layers are scraped off the top of the plate and piled up as the Olympic Mountains.

3. **Are the Olympic Mountains getting higher or lower?**

It's thought that the Olympic Mountains are in a topographic steady state. The amount of uplift, due to layers added to the base of the mountains as the Juan de Fuca Plate subducts, is roughly balanced by erosion caused by wind, water, and ice. As more and more layers are added from below, the range continually rises. But erosion on the surface keeps the mountains at about the same maximum elevation of 1 to 1½ miles (1½ to 2½ kilometers). The overall height of the mountains is maintained as the Olympic Mountains recycle themselves. It is no coincidence that uplift and erosion are in balance. This is often the case where mountains are forming, because erosion increases as topography gets higher and higher. When mountains begin to form, the erosion rate is low, so that uplift exceeds erosion. As the mountains get really high, the rate of erosion exceeds uplift. A natural topographic height is established where rates of uplift and erosion are equal.

4. Why are subduction zone volcanoes so steep?

The thickening agent for magma is silica (the elements silicon and oxygen). As hot fluids (water and carbon dioxide) rise from the subducting plate, they melt minerals rich in silica first, because those minerals have low melting temperatures. The magma beneath mountains such as Mount Rainier and Lassen Peak is high in silica, and therefore very thick and pasty. When it erupts, the lava is so sluggish that it can't flow very far. Instead, it sticks to the sides and makes a steep composite volcano.

5. Why do subduction zone volcanoes sometimes explode?

Silica is the "thickening agent" in magma, much like flour added to gravy. Silica-rich magma is thick and pasty. Mixed gasses do not escape easily, and become trapped under high pressure as the magma rises. Subduction zone volcanoes can explode when the pressure is suddenly released. A landslide on the north side of Mount St. Helens in 1980 was equivalent to uncorking a champagne bottle and allowing pressurized gasses to escape violently.

6. Why is a devastating earthquake expected in the Pacific Northwest?

Subduction zones are where the largest earthquakes occur, such as the 1960 Peru-Chile earthquake and the 1964 earthquake in Alaska. The situation is analogous to a dresser drawer that's stuck. You can pull and pull on it and it won't budge, until finally you pull so hard that the drawer suddenly jerks open. Likewise, converging plates can lock together for decades or centuries, until accumulated stress is so great that the plates suddenly let go and snap along their boundary as a devastating earthquake. Much evidence—including sudden downdropping of coastal areas, great tsunami waves, and rapid deposition of sediment offshore—suggests that the last great earthquake in the Pacific Northwest occurred about 300 years ago. The geological evidence also suggests that equally large earthquakes occurred 1,100, 1,300, and 1,700 years ago. In fact, it appears that at least seven great earthquakes occurred in the past 3,000 years, separated by intervals of 200 to 1,000 years. Native American stories and archaeological sites also suggest that devastating earthquakes periodically hit the Pacific Northwest. The most straightforward interpretation is that the subducting Juan de Fuca Plate has been locked against the North American Plate for the past 300 years, and that a great earthquake is likely to strike the Northwest during the next one to two centuries, when the plates suddenly break free.

7. Why is the rock in parks in the Sierra Nevada Mountains so good for climbing?

The Sierra Nevada Mountains developed as a volcanic mountain chain above a subduction zone. Since the subduction stopped about 20 million years ago, the volcanoes have eroded away, exposing the rocks of the magma chambers below. That rock has high silica content because it formed at a subduction zone, and has coarse mineral crystals because it cooled very slowly within the Earth. It is a granitic intrusive rock that, unlike the weak volcanic rock of the Cascades, has strong cracks and mineral crystals that are easily grasped and are ideal for climbing.

8. How did the Appalachian Mountains form?

About 300 million years ago, an ancient ocean closed and Africa crashed into North America. Sedimentary strata and other rocks of the ocean were deformed, thrust over the edge of North America, and uplifted as the Appalachian Mountains.

9. How high were the Appalachians at the time of collision?

When continents collide, the effect is like a swimmer putting a beach ball under his or her belly—the swimmer will rise up considerably out of the water. Today the Himalayas are so high because the full thickness of the Indian subcontinent is shoving beneath Asia. In their prime, the Appalachians were probably at least as high as the Alps, and maybe even the Himalayas. Three hundred million years ago, eastern North America may have had peaks as lofty as Mount Everest or K-2, extending to nearly 30,000 feet above sea level!

10. How deep were the metamorphic rocks that are now exposed at the surface in the Appalachians?

The deeper you go into the Earth, the higher the temperature and pressure. Some of the rocks of the Appalachian Mountains were so metamorphosed that they must have been about 5 to 15 miles (10 to 25 kilometers) below the surface. The deep rocks were brought back to the surface in two ways: 1) Thrust faulting—the rocks were shoved upward along faults as the colliding continents compressed; 2) Isostatic rebound—the thick crust was weighted down by the high topography. As the mountains eroded, the thick, buoyant crust bobbed back upward, creating topography not quite as high as it was before. As much as 15 miles (25 kilometers) of rock was removed, exposing the deeply-buried rocks.

Part IV: Transform Plate Boundaries

1. **What happens to the lithosphere at transform plate boundaries?**

   Where plates slip horizontally past one another, lithosphere is neither created nor destroyed. Instead, blocks of crust are torn apart in a broad zone of shearing between the two plates.

2. **Why are such boundaries called "transform"?**

   They're called "transform" because they connect other plate boundaries in various combinations, transforming the site of plate motion. A common example is an offset connecting segments of a mid-ocean ridge. Transform movement between the plates occurs between the ridge segments, accompanied by earthquakes. Examples can be seen off the coast of the Pacific Northwest, where the Juan de Fuca Plate not only diverges from the Pacific Plate along the Juan de Fuca Ridge, but also slips laterally along transform boundaries between ridge segments.

3. **Where are prominent transform plate boundaries that disrupt continental crust?**

   Examples of transform plate boundaries that extend on land include the Anatolian Fault in Turkey, the Alpine Fault of New Zealand, the San Andreas Fault in California, and the Queen Charlotte/Fairweather Fault in western British Columbia and southeast Alaska.

4. **How destructive are earthquakes at transform plate boundaries?**

   Earthquakes at transform plate boundaries are not the largest in terms of magnitude—those occur at convergent plate boundaries. But the lithosphere can be thick and cold at transform plate boundaries, so that fair-sized earthquakes, up to about magnitude 8.5, can occur. The intensities can be devastating, because earthquakes at transform plate boundaries are always shallow.

5. **How far can blocks of crust move at transform plate boundaries?**

   The Pacific Plate slides past North America at a rate of about 2 inches (5 centimeters) per year. At that rate, a block of crust can move 300 miles (500 kilometers) in 10 million years. That's why when you go to Point Reyes National Seashore or Pinnacles National Monument, you are standing on ground that has moved 200 to 300 miles past the rest of California.

6. **Will California fall into the ocean?**

   No. But given a few million years, San Francisco and Los Angeles just might be part of a resort island off the coast of Oregon and Washington State!

7. **Is the San Andreas Fault the actual boundary between the North American and Pacific plates?**

   The San Andreas is just one of several faults that accommodate the motion between the two plates. The plate boundary is a broad zone of deformation with a width of about 60 miles (100 kilometers). Along much of the boundary, the bulk of the motion occurs along the San Andreas Fault. Point Reyes National Seashore and Golden Gate National Recreation Area are the only two park lands that are right on the San Andreas Fault. Other parks in the region—namely Channel Islands and Joshua Tree national parks, Cabrillo and Pinnacles national monuments, and Santa Monica National Recreation Area—reveal evidence of the shearing, rotation, and uplift that occurs within the broad zone of deformation between the two plates.

8. **What is meant by the "recurrence interval" of earthquakes?**

   The recurrence interval is the time between large earthquakes along a particular segment of a fault. For portions of the San Andreas Fault, the time is typically about 100 years.

9. **How much is the land surface offset during a big earthquake along the San Andreas Fault?**

   At 2 inches (5 centimeters) per year, the plates should slip 200 inches (500 centimeters), or about

17 feet (5 meters), in 100 years. A fence line now preserved at Point Reyes National Seashore is offset by about that amount, indicating the amount of movement that occurred along the San Andreas Fault during the 1906 earthquake that devastated San Francisco.

10. **Why isn't there much volcanic activity at transform plate boundaries?**

There are two circumstances that normally lead to the melting of Earth materials: 1) cold crustal material is pushed to depths where the temperature is so hot that water is driven off—the water rises and helps to melt rock in its path; and 2) hot mantle that was solid under extreme pressure deep within the Earth begins to melt as it rises and the pressure drops. In other words, there must be significant change in the vertical positions of Earth materials in order to induce melting. There can be great horizontal movement at transform plate boundaries, but things do not move up or down very much, so the conditions necessary for melting generally do not exist. There are volcanic and other igneous rocks in national park lands along the San Andreas transform plate boundary, but those rocks formed during previous tectonic episodes involving continental rifting, mid-ocean ridge development, and plate subduction.

Part V: Hotspots

1. **What causes hotspots?**

The origin of hotspots remains a puzzling question for geologists. Magmas generated at mid-ocean ridges and subduction zones result from processes in the upper mantle, within 150 miles (250 kilometers) of Earth's surface. But hotspot activity may originate from much deeper. Gasses from Hawai`ian volcanoes have relatively high concentrations of the rare elements iridium and osmium, as well as the helium-3 isotope, consistent with a deep source. It is speculated that the rising plumes may form because of bumps on the mantle-core boundary at about 1,800 miles (2,900 kilometers) depth. Another model puts the source 400 miles (700 kilometers) deep, in the region where the relatively soft asthenosphere changes to a more solid state. Under the great pressure at that depth, the mineral olivine (the major constituent of the mantle) changes its crystalline structure to spinel, giving off heat.

2. **Why does material melt at hotspots?**

Melting at hotspots is similar to the decompression melting that occurs beneath mid-ocean ridges. Deep mantle material is very hot; the only thing that keeps it solid is the enormous pressure exerted by the rock above. The pressure decreases as the hot mantle rises. Eventually, when the hot mantle reaches a shallow depth of about 30 to 60 miles (50 to 100 kilometers), the pressure is so low that the hot mantle partially melts, forming basaltic magma.

3. **Besides Hawai`i and Yellowstone, where are other hotspots in the world?**

There are several other examples of narrow plumes spewing out large volumes of magma. The best known is Iceland, where a hotspot lies directly beneath a divergent plate boundary, the Mid-Atlantic Ridge. The telltale footprint of volcanism there spreads in both eastward and westward directions from Iceland across the Atlantic Ocean floor. Another prominent hotspot extends southward across the Indian Ocean from a region of volcanism in southern India known as the Deccan Plateau. Other hotspots have formed the Galapagos Islands in the Pacific Ocean, as well as the Azores in the Atlantic.

4. **Why is Hawai`i a line of volcanic islands, rather than a long, continuous ridge of volcanic material?**

A hotspot is somewhat like the smoke rising from a chimney or the flame of a candle. With no wind, the smoke or candle flame rises vertically. But a moving plate is like the wind bending the smoke or flame. Magma rising from the Hawaiian Hotspot is deflected northwestward as it is carried by the Pacific Plate. For a while, the magma can follow the deflected route to the surface. But eventually (after a million years or so), it is easier for the magma to form a new path directly upward from the hotspot. Separate volcanic islands form, one after another, as the Pacific Plate moves over the Hawaiian Hotspot.

5. **Why do the Hawaiian Islands decrease in elevation from southeast to northwest?**

Most substances expand when heated. The increase in volume of mantle material at a hotspot causes the Pacific Ocean floor to elevate as the Pacific Plate moves over the Hawaiian Hotspot. In addition, a huge amount of volcanic material erupts onto the seafloor above the hotspot. The

Big Island of Hawai`i reaches elevations of nearly 14,000 feet (4,300 meters) above sea level. As an island moves off the hotspot, volcanic activity ceases and the region cools. The island get lower due to both thermal subsidence and erosion. In fact, the area continues to cool and subside even after the island erodes to sea level, so that a line of submerged volcanoes (the Hawaiian Ridge and Emperor Seamount Chain) extends northwestward from the Hawaiian Islands.

6. **How and why does the volcanism in Yellowstone National Park differ from that seen at parks in Hawai`i?**

Hotspot volcanism occurs where either oceanic or continental crust caps the overriding plate. The magma that originally forms as hot mantle rises and partially melts has a low-silica (basaltic) composition. The composition of lava that erupts above a hotspot depends on the materials that the original magma encounters on its way to the surface. An oceanic plate has thin crust made of basalt and gabbro (the coarse-grained equivalent of basalt). The basaltic magma that works its way to the surface through the Pacific Plate thus retains a basalt composition. The Hawaiian Islands are huge shield volcanoes made of dark, heavy lava flows. Where the rising magma must work its way through thick continental crust, it melts high-silica minerals in the process. The silica-enriched lava that erupts in the Yellowstone region is thick and pasty, resulting in explosive eruptions and light-colored, rhyolite volcanic rocks.

7. **Why have volcanoes recently erupted at Haleakalā National Park and Craters of the Moon National Monument?**

Regions directly above hotspots, such as Hawai`i Volcanoes and Yellowstone national parks, manifest vast amounts of volcanic activity. But activity can continue for some time after regions have moved over and away from hotspots—pockets of magma linger within the crust or mantle portion of the overriding plate. Haleakalā National Park and Craters of the Moon National Monument show some of the late-stage volcanic activity typical of hotspots.

8. **Why are the recent volcanic rocks of Craters of the Moon National Monument so different from those at Yellowstone?**

The Pacific Northwest of the United States reveals a region where lithosphere with thick continental crust has been moving over a hotspot. Similar to activity in continental rift regions, the Snake River/Yellowstone region has two phases of volcanic activity. Basaltic magma must initially melt its way through the thick, silica-rich crust, but later has a clearer path to the surface. Along the Snake River Plain in Idaho, high-silica (rhyolite) lavas are progressively younger from southwest to northeast, erupting as the North American plate moves southwestward over the Yellowstone Hotspot. Later eruptions are through a fractured crust that is depleted in silica, so that basalt covers the older rhyolite. The pattern of volcanism observed at the surface in recent times is light-colored, explosive rhyolite at Yellowstone National Park (directly above the hotspot), compared to darker, free-flowing basalt at Craters of the Moon National Monument (which has already passed over the hotspot).

9. **Why are there geysers, hot springs, and other geothermal features at Yellowstone?**

The Yellowstone Hotspot melts rock that rises and "ponds" as magma chambers about 5 miles (8 kilometers) beneath the surface. The heat rising from this molten material keeps the overlying rock hot. Rainwater and snowmelt percolate through the cracks and pores in the hot rock, eventually finding their way back to the surface as the geysers, hot springs, and mud pots of Yellowstone National Park.

10. **How do Yellowstone geysers work?**

The water 1 or 2 miles (1½ to 3 kilometers) below Yellowstone is under considerable pressure. The surrounding hot rocks are like a pressure cooker, causing the water temperature to rise well above its normal boiling point. But when the superheated water reaches a certain temperature, it flashes to steam, increasing to about 1,500 times its liquid volume. The water above is forced violently to the surface, through narrow constrictions and out of geysers such as Old Faithful. After the steam is released, it might take anywhere from a few minutes to hours, days, months, or even years until the temperature rises so much that another eruption occurs.

Part VI: Building the North American Continent

1. **How do continents form?**

When the Earth cooled enough for a shell of lithosphere to form, the crust that formed on top of the plates was thin like today's oceanic crust. Where plates converged, thicker crust formed at island arcs, where magma poured out on the surface as lava flows and was added to the base of the crust as igneous intrusions. When island arcs collided with other island arcs, rock and sediment were scraped off the top of subducting plates. Those blocks of coalesced material formed the nuclei of the continents. Over time, the small continents grew outward as material was added to their edges through sedimentation, subduction, and the collision of oceanic islands and other continental fragments. Continents are thus older in their interior regions, known as cratons, and get progressively younger outward.

2. **How old are portions of the North American continent?**

The nucleus of the North American Craton, known as the continental shield, has rocks as old as 3.4 billion (3,400 million) years. The shield is generally oldest near its center, and gets younger outward—a pattern consistent with continents starting as small nuclei that grow outward. The shield has blocks of igneous and metamorphic rocks comprising different provinces that were added to the continent during periods of mountain building known as orogenies. The ages of some of the igneous and metamorphic rocks in each province, formed during different orogenies, tell us the time that each piece was added to North America. The Superior Province has rocks 2.5 to 3.4 billion years old, surrounded by provinces with rocks 1.6 to 1.9 billion years old. Bordering the shield on its southern, western, and eastern sides is a thin cover of much younger sedimentary rocks comprising the continental platform. East of the shield, rocks of the Grenville Orogeny, formed about 1.1 billion years ago, poke up through the platform strata in places such as the Adirondack Mountains of upstate New York. Still farther east and south, the Appalachian-Ouachita-Marathon Orogeny emplaced rocks between 500 and 300 million years ago. West of the craton, the ages of igneous and metamorphic rocks indicate that terranes were emplaced over the past 200 million years, throughout the ongoing Cordilleran Orogeny.

3. **What do national park lands in the Midwest have to do with plate tectonics?**

The continental interior, or craton, is an amalgamation of igneous and metamorphic rocks formed by plate tectonic processes hundreds of millions to a couple of billion years ago. Rocks in Voyageurs

National Park in northern Minnesota reveal plate boundary activity that represents some of the early history of the North American continent. Some parks in the Midwest, such as Mammoth Cave in Kentucky and Cuyahoga in Ohio, contain layers that record the periodic flooding of parts of the craton by shallow seas, as well as the advance and retreat of continental ice sheets.

4. **Why do shallow seas advance and retreat over continents?**

Much of the area of continents is at low elevation, less than 600 feet (200 meters) above sea level. Adjacent continental shelves lie beneath less than 600 feet of water. Those areas are susceptible to two factors: 1) rise and fall of sea level; and 2) uplift and downwarp of the crust. Sea level rise or crustal downwarp could cause shallow seas to cover vast areas of a continent and deposit sediment. On the other hand, a drop in sea level or uplift of the crust would result in retreat of the seas and consequent erosion.

5. **Why does sea level rise and fall?**

A major factor is the overall temperature of Earth's atmosphere and oceans. At times of low temperature, a great deal of Earth's surface water accumulates in the polar ice caps, causing sea level to fall. With high temperatures, the ice caps melt and sea level rises. National parks on the North American craton are important because they help tell the story of the effects of a cooling Earth and the advance of continental ice sheets, as well as global warming and seas flooding much of the continent. Another factor that contributes to sea level rise is the rate of plate divergence. Mid-ocean ridges become broad and high when plates move apart fast, elevating the ocean floor and displacing ocean water onto coastal plains and continental interiors.

6. **What determines the extent of continental ice sheets?**

Broad continental glaciers advanced southward over North America in response to cooling climate and accumulation of ice over the North Polar region. Lobes of ice extended far south over the Midwest because the region is in the interior of the continent and therefore colder than coastal regions. The presence of soft sedimentary layers can make it easier for the ice to slip along its base, resulting in exaggerated lobes in places.

7. **What is an accreted terrane?**

   An accreted terrane is a block of crustal material, with a "basement" of characteristic igneous and metamorphic rocks, sometimes overlain by a distinctive sequence of sedimentary layers. The block traveled a significant distance on top of a moving plate and eventually slammed into the edge of a continent, where it remains today.

8. **How do we know where an accreted terrane came from?**

   There are ways to tell the approximate latitude where a block of crust formed. One way is to look in detail at the rocks. They may contain limestone layers, for example, or an assemblage of warm-water fossils suggesting that the sediments were deposited at low latitudes, near the equator. Another way is to examine the magnetic properties of certain minerals within the rock—the orientation of the magnetic field that was frozen into the minerals as the rock formed relates to latitude. Rocks formed far to the south, that are now found in parks in Alaska, must have traveled considerable distances with their respective terranes. Another hint as to the place of origin is a terrane's characteristic rocks. For example, the igneous and metamorphic ("basement") rocks and overlying sedimentary layers found in Alaskan parks and in North Cascades National Park in Washington State indicate pieces of the same continental fragment, known as the Wrangellia Terrane.

9. **Why is Mt. McKinley (Denali) the highest peak in North America?**

   Mount McKinley is 20,320 feet (6,194 meters) above sea level because it rests on a block of crust that thickened during plate convergence and accretionary tectonics. Peaks in other parks in Alaska, such as Mount St. Elias (18,008 feet; 5,489 meters) in Wrangell–St. Elias National Park and Mount Fairweather (15,300 feet; 4,663 meters) in Glacier Bay National Park are high for the same reason. In a sense, the very highest mountains on Earth, the Himalayas, are a response to the collision of an enormous accreted terrane: the entire Indian subcontinent!

10. **Why are metamorphic rocks so important in understanding the history of the development of the North American continent?**

    The types of minerals found in a metamorphic rock hint at the temperature and pressure that the rock experienced. Both temperature and pressure increase with depth within the Earth. Metamorphic rocks found in parks are "geobarometers." They tell us how deep a rock was at some time in the past—important information bearing on the tectonic history of accreted terranes that are now part of the North American continent.

GLOSSARY

`a`a A rough, blocky form of **basalt** lava surface.

absolute age dating Determination of actual number of years ago that a rock, mineral, fossil, or other material formed.

absolute plate motion The speed and direction a plate moves over the fixed, deep portions of the Earth.

abyssal plain A deep, flat portion of the ocean floor, some distance from the shallower **continental shelf** and **mid-ocean ridge** regions.

accreted terrane A block of crust with characteristic igneous and metamorphic rocks and sedimentary rock cover, that is bounded by faults and distinctly different than surrounding crustal blocks.

accretionary wedge A mountain range formed as sedimentary strata and hard crust are scraped off the top of a subducting plate.

active continental margin A transition from thick continental to thin oceanic crust that is at or near a current plate boundary, and is thus the site of tectonic activity such as earthquakes, volcanic eruptions, and mountain building.

Aleutian Islands A chain of volcanoes extending southwestward into the Pacific Ocean from the Alaska Peninsula, formed by **subduction** of the Pacific Plate beneath the North American Plate (see **island arc**).

alluvial fan Sedimentary material eroded from a steep mountain front, carried through a canyon, and deposited by flash floods in a fan-shaped pattern onto the flat land beyond the canyon mouth.

andesite A fine-grained, light-to-dark colored **igneous rock** with about 60% **silica**.

angular unconformity An **unconformity** in which the underlying strata are tilted at a different angle than the overlying strata.

anticline A **fold** in which the rocks layers are bent upward.

Appalachian Mountains An ancient highland region extending from the Atlantic provinces of Canada to Alabama, formed during ocean basin closure and continental collision during the **Paleozoic Era**.

arch 1. A broad, circular, upward **fold** in rock layering (**dome**). 2. A curved geological feature formed where a hard rock layer erodes completely through, leaving a continuous roof intact.

Archean Eon The portion of geologic time between 3,800 and 2,500 million years ago.

ash See **volcanic ash**.

asthenosphere Relatively soft portion of Earth's upper **mantle**. The rigid plates of **lithosphere** drift about over the flowing asthenosphere.

atoll A circular **coral reef** surrounding a lagoon.

badlands topography A harsh, barren landscape of hills and gullies developed as soft sedimentary layers erode.

banded gneiss   A **metamorphic rock** formed under high-temperature and high-pressure conditions, with parallel, alternating zones of dark and light-colored **minerals**.

barrier island   An elongate bar of sand and mud, cut off from the mainland, that develops parallel to the coastline.

basalt   A fine-grained, dark-colored **igneous rock** with about 50% **silica**.

basaltic andesite   A fine-grained **igneous rock** with about 55% **silica** (more silica than **basalt**, but less than **andesite**).

basin   1. A broad, circular, downward **fold** in rock layers (circular **syncline**). 2. A depression that accumulates sedimentary deposits.

Basin and Range Province   A region of long, north-south–trending mountain ranges and intervening valleys in the western United States, formed by ongoing continental rifting.

batholith   An extensive, igneous **intrusive rock** body, commonly with **granite** or **granodiorite** composition. It typically forms as a region of **magma chambers** cools.

Benioff zone   A region of earthquakes extending from near Earth's surface downward along a **subduction zone.**

blueschist   A **metamorphic rock** formed under conditions of relatively high pressure and low temperature, commonly within a **subduction zone.**

bomb   See **volcanic bomb**.

breccia   1. A **sedimentary rock** made up of coarse, angular fragments. (A similar rock with coarse, rounded fragments is a **conglomerate**.) 2. A **volcanic rock** with coarse, angular rock fragments encased in finer volcanic particles. 3. A zone of angular rock fragments resulting from grinding between fault blocks.

bridge   An **arch** (definition 2) with a stream flowing underneath.

brittle   Easily broken or cracked. Brittle solid material is vulnerable to discrete cracking and breaking (like peanut brittle or cold plastic).

Brooks Range   A highland region extending from east to west across northern Alaska, formed by ocean basin closure and continental collision during the **Mesozoic Era**.

buoyancy   The tendency of a block of material to rise upward because it is less dense than the surrounding material.

calcite   A mineral composed of the compound calcium carbonate ($CaCO_3$). The primary mineral com-

prising the sedimentary rock **limestone**, and its metamorphic equivalent, **marble**.

caldera   A relatively large, steep-sided crater, at least ½ mile (1 kilometer) in diameter, formed by the eruption and collapse of a volcano.

calve   To break a glacier into smaller pieces (**icebergs**) as the glacier flows into a bay or ocean.

Cascadia Subduction Zone   A region of parallel structural and volcanic mountain ranges in the Pacific Northwest of the United States, formed by plate convergence where the offshore Juan de Fuca Plate dives beneath the edge of the North American Plate.

Cenozoic Era   The portion of geologic time from 65 million years ago to the present. *Cenozoic* means "time of modern life."

chert   A **sedimentary rock** formed of microscopic crystals of the mineral **quartz**, commonly precipitated out of water in the deep ocean.

cinders   Pea-to-gravel-sized particles formed as molten material ejected during a volcanic eruption cools and solidifies in the air.

cinder cone   A relatively small, ≤ 1 mile (1½ kilometer) diameter volcano, formed as **pyroclastic** material (volcanic bombs, cinders, ash) blasts up into the air and rains down as solid particles that pile up into a cone with a slope of about 30°.

clastic sedimentary rock   A **rock** formed from the eroded fragments of other rocks.

cleavage   The breakage of a rock or mineral along planar faces.

coastal plain   The low-lying, flat portion of a continent adjacent to the ocean.

column   1. An upright structure bounded by the cracks (**joints**) formed as a **lava flow** cools. 2. A cave formation developed where a **stalactite** and a **stalagmite** merge.

collisional mountain range   A region of high topography formed as an ocean basin closes and thick crustal blocks crash into one another.

columnar jointing   Cracks formed due to the shrinking of a lava flow as it cools. If cooling is uniform, the cracks commonly result in vertical, six-sided (hexagonal) columns.

composite volcano   A large (up to 12 mile [20 kilometer] diameter) mountain formed during numerous eruptions of a variety of volcanic materials (lava flows, ash, pumice, cinders, mudflows) that pile up into a relatively steep volcano (also known as a *strato-volcano*).

compression   The decrease in volume of a material as it is squeezed.

compressional wave   The movement of seismic wave energy through a material, caused by alternating squeezing and stretching of the material. Also called *P-wave*.

conglomerate   A **sedimentary rock** formed from the compaction and cementation of sediment containing a lot of pebbles or cobbles. (If those particles are angular, the rock is called **breccia**.)

continental collision zone   A mountain region formed where an ocean basin closes and blocks of **continental crust** crash into one another.

continental craton   See **craton**.

continental crust   Outer layer of the Earth that is thick and high in **silica** compared to **oceanic crust**.

continental drift   The theory that continents are not stationary, but rather move about relative to one another.

continental platform   The part of the continental **craton** that is periodically invaded by shallow seas and covered by sedimentary deposits.

continental rift zone   A region of mountain ranges and long valleys formed as plate divergence pulls apart a plate capped by thick continental crust.

continental rise   Gently-sloping part of an ocean basin between the deep **abyssal plain** and the steep **continental slope**.

continental shelf   Shallow, flat region of an ocean basin immediately adjacent to a continent.

continental shield   The region of the continental **craton** where old igneous and metamorphic rocks are exposed at the surface.

continental slope   Relatively steep region of an ocean basin between shallow water of the **continental shelf** and deeper water of the **continental rise**.

convergent plate boundary   Region where two slabs of Earth's outer shell (**lithosphere**) move toward one another, destroying lithosphere. **Subduction zones** and **continental collision zones** are common manifestations of convergent plate boundaries.

coral reef   Buildup of the remains of the hard parts of coral and other marine organisms.

Cordillera   The broad, mountainous region of western North America formed by tectonic activity over the past 200 million years.

core   The central region of the Earth made mostly of iron.

craton   Relatively flat-lying region of a continent that has not experienced widespread tectonic activity for hundreds of millions of years.

crevasse   A nearly-vertical fissure on the surface of a **glacier**.

cross-bedding   A pattern of varying tilt of sedimentary layers developed as wind or water of changing direction deposits the layers. Commonly developed in sand dunes on a beach or in a desert.

crust   The outermost part of the Earth composed mostly of light silicate minerals.

curtain of fire   Sheetlike eruption of fluid lava through a **rift zone** on the flanks of a volcano.

dacite   A fine-grained **igneous rock** with about 65% **silica** (more silica than **andesite**, but less than **rhyolite**).

debris avalanche   Rock, soil, and other material that breaks away from, and flows down, the side of a volcano (see **volcanic landslide**).

debris flow   Mixture of mud, sand, rock, and water that roars down the flanks of a volcano (**lahar; mudflow**).

decompression melting   Change from solid to liquid that occurs as the pressure on hot, pressurized material drops. The change occurs as hot mantle material becomes shallow quickly.

deep-sea trench   See **trench**.

deposition   The accumulation of sedimentary material transported by water, wind, or ice.

dike   A sheetlike, igneous **intrusive rock** body formed where **magma** cuts across rock layers.

diorite   A coarse-grained, light-to-dark-colored **igneous rock** with about 60% **silica** (an intrusive equivalent of **andesite**).

disconformity   An **unconformity** where the underlying layers are parallel to the overlying layers.

divergent plate boundary   Region where two slabs of Earth's outer shell (**lithosphere**) are pulling apart from one another, creating new lithosphere. **Mid-ocean ridges** and **continental rift zones** are common manifestations of divergent plate boundaries.

dolomite   A sedimentary rock similar to **limestone** but with some of the calcium in the carbonate molecules replaced by magnesium.

dome   A broad, circular, upward **fold** in rock layers (a circular **anticline**).

ductile   Easily stretched or drawn. Ductile solid material flows (like silly putty or hot plastic).

earthquake   A sudden movement within the Earth that releases vibrations (**seismic waves**).

earthquake intensity   A measure of the actual effects due to earthquake shaking observed at a particular place on Earth's surface.

earthquake magnitude   A measure of the amount of seismic energy that an earthquake released. The scale is logarithmic, such that an increase in one number on the magnitude scale corresponds to a tenfold increase in the amplitude of seismic waves, and about a thirty-fold increase in the amount of energy released.

earthquake recurrence interval   The average time between earthquakes of a given **magnitude** along a particular **fault** or fault segment.

earthquake swarm   Many small **earthquakes** occurring in a small region over a short period of time, commonly associated with the movement of **magma** within the Earth.

eclogite   A metamorphic rock formed from a pre-existing rock (generally **basalt** or **gabbro**) that experienced high pressure, yet relatively low temperature.

ecology   The relationship between organisms and their environment.

ecosystem   A community of organisms and their physical environment, considered as a unit.

elastic limit   Level of stress where a rock will deform permanently, either by ductile flow or brittle failure. If by brittle failure, the seismic energy stored in the rock is released as an **earthquake**.

elastic rebound   The abrupt movement of rock as it reaches its **elastic limit** and releases **seismic waves** during an earthquake.

endemic   Found in one particular region of the Earth and nowhere else. The term describes a plant or animal species native to only one specific region.

eon   A major portion of geologic time, broken into smaller divisions called **eras**.

epoch   A division of a **period** of geologic time.

era   A division of an **eon** of geologic time, broken into smaller units called **periods**.

erosion   The breakdown of rock into smaller particles by the mechanical actions of water or wind, or by biological or chemical activity.

erratic   See **glacial erratic**.

escarpment   See **fault escarpment**.

extension   Forces pulling rock apart. Results in rocks breaking along **normal faults**.

extrusive rock   An **igneous rock** that solidified from magma that erupted on Earth's surface, forming fine-grained mineral crystals (**volcanic rock**).

fault   A break in the Earth along which the blocks on either side have moved parallel to the break.

fault-block mountains   Long, narrow ranges that moved upward relative to adjacent valleys along **faults**.

fault escarpment   A change in elevation representing the part of a **fault** extending above the surface.

feldspar   One of a group of magnesium/aluminum-silicate minerals that are common in rocks in Earth's crust.

fin   An eroded, vertical sliver of **sandstone**; further erosion through the fin could develop an **arch**.

flowstone   Sheetlike formation developed as water flows down a cave wall.

flysch   Coarse-grained sedimentary layers deposited in a deep-sea trough between converging crustal blocks.

fold   A structure formed as rock layers are bent.

foliation   The leaflike texture of some **metamorphic rocks**, developed as flat, platelike **minerals** orient themselves along planar surfaces.

foot wall   When a person stands across a fault plane, the block that is next to his or her feet.

forearc basin   A depression between the uplifted (**accretionary wedge**) and volcanic (**volcanic arc**) mountain ranges formed above a **subduction zone**.

foreland   The region in front of a mountain range, sometimes the site of deformation as a **foreland fold-and-thrust belt** and sedimentary deposition in a **foreland basin**.

foreland basement uplift   Deformation in front of the main region of mountain building, involving the hard igneous and metamorphic rocks as well as the overlying sedimentary layers.

foreland basin   Depression in front of a rising mountain range where sedimentary layers are deposited.

foreland fold-and-thrust belt   Deformation in front of the main region of mountain building, where sedimentary strata are bent into **folds** and transported laterally along **thrust faults**.

formation   A mass of rock with recognizable characteristics that make it distinguishable from surrounding masses of rock. Usually applied to a layer or sequence of layers of sedimentary rock.

fossil   The remnants of an ancient plant or animal. The original material might be replaced by rock, or a cast left in the rock, as the plant or animal decays.

Franciscan Group   Remnants of an ancient **accretionary wedge**, developed during the subduction of the Farallon Plate and preserved in western California.

fumarole   Hole in a volcanic region through which gasses rise.

gabbro A coarse-grained, dark-colored **igneous rock** with about 50% **silica** (an intrusive equivalent of **basalt**).

geology The study of the Earth.

geomorphology The study of landforms, such as mountains, valleys, and shorelines, and the processes that created them.

geosynclinal theory The idea that mountain building, volcanic eruptions, metamorphism, and other tectonic events involve the downwarping of large regions of Earth's crust into giant depressions, or **synclines**. It suggests that tectonic features are due mainly to vertical movements of Earth's crust. The theory contrasts with the modern **plate tectonics** which relies more on horizontal motions.

geothermal gradient The increase in temperature with depth within the Earth.

geyser A spout that periodically erupts water from the Earth.

glacial erratic A block of rock transported some distance from its original position on Earth's surface by a **glacier**.

glacial geology The study of landforms and deposits formed as a result of ice moving across Earth's surface.

glacial striation A long groove formed as sand, gravel, or larger rock imbedded in ice scrapes across the underlying bedrock.

glacier A naturally formed mass of ice that moves gradually downslope under the influence of gravity.

glaciology The study of masses of ice moving across Earth's surface.

gneiss A **metamorphic rock** formed from a pre-existing rock that underwent extreme increases in temperature and pressure. (Gneiss undergoes a greater amount of metamorphism than **schist**.)

GPS Global Positioning System. A network of satellites used to find the location of a point on Earth's surface.

graben A block of crust dropped downward along **normal faults**.

granite A coarse-grained, generally light-colored **igneous rock** with about 70% **silica**.

granodiorite A coarse-grained **igneous rock** with about 65% **silica** (less silica than **granite**, but more than **diorite**).

greenschist A low-to-medium-grade metamorphic rock containing the mineral chlorite.

greenstone A type of **metamorphic rock** rich in minerals such as chlorite and epidote, which give the rock a green tinge. Commonly, old (Precambrian) rocks formed from the metamorphism of ancient oceanic crust.

ground moraine See **moraine**.

gypsum A **mineral** formed as calcium sulfate precipitates out of evaporating water. Its chemical formula is $CaSO_4 \cdot 2H_2O$.

half-life The time it takes for half the remaining mass of a radioactive isotope to decay to another isotope form.

hanging wall When a person stands across a fault plane, the block that is next to his or her head.

hard collision An advanced state of ocean basin closure, when one continent extends a considerable distance beneath another and a high and broad mountain range forms.

helictite Fingerlike formation developed as water drips from the ceiling or slanted wall of a cave.

hoodoo A spire capped by a hard rock layer that resists erosion.

hornito See **spatter cone**.

horst A block of crust that remains high after **graben** blocks drop downward along **normal faults**.

hotspot A plume of hot material that rises from Earth's deep **mantle**. Chains of volcanoes form on the surface of a plate that rides over a hotspot.

hot spring A place where hot water escapes from the ground.

hydration melting Change from solid to liquid that occurs when a material is heated and loses water (dehydrates). The change occurs where a plate subducts and loses water as it heats up; the water rises and wets (hydrates) hot rock in its path, causing some of the rock to melt.

hydrothermal alteration A change in the mineral composition of rock induced by hot water.

iceberg A large, floating block of ice, commonly formed as a glacier or ice sheet flows into a bay or the ocean and breaks apart (**calves**).

igneous rock A **rock** that solidified from molten fluid (**magma**).

inner core The zone of dense, solid iron and nickel extending from 3,200 miles (5,100 kilometers) depth to the Earth's center at 4,000 miles (6,300 kilometers).

intensity See **earthquake intensity**.

intrusive rock An **igneous rock** that solidified from **magma** that cooled within the Earth, generally forming coarse-grained mineral crystals (**plutonic rock**).

island arc A curved chain of volcanoes formed where a plate capped by oceanic crust **subducts** beneath another plate capped by oceanic crust.

isostasy   A condition whereby a buoyant force pushing Earth materials up is balanced by an equal gravitational force (weight) pushing downward.

isostatic equilibrium   A state of **isostasy**, achieved when upward and downward forces are equal.

isostatic rebound   The upward movement of a region that occurs when erosion removes some of the weight of a mountain range or an ice sheet melts away.

isostatic uplift   Movement caused by **isostatic rebound**.

joint   A crack in a rock, with slight opening but no significant lateral movement of the blocks of rocks on either side of the crack. (A crack involving significant lateral movement is a **fault**.)

Keweenawan Rift   An ancient tear in the North American continent that occurred 1.1 billion years ago, extending from the Lake Superior region to Kansas.

lahar   1. A mass of solid material and water that moves swiftly downslope, commonly as a result of volcanic processes (**volcanic mudflow**). 2. The deposits of a volcanic mudflow.

Laramide Orogeny   A period of mountain building in the western United States from about 80 million to 40 million years ago.

Laramide uplift   A block of hard crust and overlying sedimentary strata shoved upward along a **reverse fault** during the **Laramide Orogeny**, forming one of the frontal mountain ranges of the Rocky Mountains.

lateral moraine   See **moraine**.

lava   Hot, molten rock (**magma**) that poured out on Earth's surface or beneath the sea.

lava dome   A relatively small volcano, ≤ 1 mile (1½ kilometers) in diameter, formed by the eruption of sticky, silica-rich (**rhyolite**, **dacite**) lava around a central vent.

lava flow   1. Molten material (**magma**) pouring out on Earth's surface. 2. Hard rock formed from magma that erupted from the summit of a volcano or from fissures along its sides.

lava shield   A gently sloping buildup of successive flows of fluid **lava**, not extensive enough to be considered a **shield volcano**.

lava tube   A long, commonly smooth cave formed where lava crusts over and the remaining lava flows out.

left-lateral, strike-slip fault   See **strike-slip fault**.

limestone   A **sedimentary rock** formed from the compaction and cementation of calcium carbonate (the mineral **calcite**) precipitated out of water.

lithosphere   The rigid outer shell of the Earth, composed of the outermost **mantle** and **crust**. The lithosphere is broken into plates that move over the underlying, softer **asthenosphere**.

longshore current   The flow of water parallel to a coastline.

lower mantle   The hard-solid part of the Earth between the softer **asthenosphere** and liquid **outer core**.

magma   Hot, liquid rock that may contain some gas and solid material.

magma chamber   An accumulation of molten rock below Earth's surface.

magnitude   See **earthquake magnitude**.

mantle   The portion of the Earth between the **crust** and **core**, made of **silicates** rich in iron and magnesium.

Marathon Mountains   A highland region in west Texas, formed by the same ocean basin closure and continental collision that formed the Appalachian Mountains during the Paleozoic Era.

marble   A **metamorphic rock** formed from a **limestone** layer that underwent extreme increases in temperature and pressure.

mélange   A chaotic mixture of broken and jumbled rock, commonly formed within the **accretionary wedge** at a **subduction zone**.

Mesozoic Era   The portion of geologic time between 248 and 65 million years ago. *Mesozoic* means "time of middle life."

metamorphic facies   A distinctive suite of minerals formed as a rock is subjected to a particular combination of temperature and pressure conditions.

metamorphic rock   A rock formed through the recrystallization of a preexisting rock, while the rock was still solid.

metasandstone   A **metamorphic rock** formed from a **sandstone** layer that experienced relatively small increases in temperature and pressure. (Metasandstone has a lesser amount of metamorphism than **quartzite**.)

microcontinent   A small block of **continental crust** that broke off a larger continent.

mid-ocean ridge   An undersea mountain range formed from volcanic activity where lithospheric plates diverge.

mineral   A naturally occurring, inorganic solid with specific chemical composition and crystalline structure.

moraine   Rock and other sedimentary material deposited on the sides or at the terminus of a **glacier**. A *ground moraine* forms under a glacier. A *lateral moraine* develops at the sides of a glacier, between the wall rock and the glacial ice. A *recessional moraine* forms as a glacier melts back and leaves rock, sediment, and other debris behind. A *terminal moraine* develops where the front end of a glacier melts and deposits rock, sediment, and other debris.

mountain range   A region of high topography formed by crustal deformation or volcanic processes.

mudflow   See **volcanic mudflow** or **lahar**.

mud pot   A **hot spring** with water that has incorporated a lot of earth material on its way to the surface.

mylonite   Fine-grained rock developed by the shearing of rock within a fault zone.

non-clastic sedimentary rock   A **sedimentary rock** formed through chemical precipitation of material from a solution, or through biological activity.

nonconformity   An **unconformity** where sedimentary layers overly igneous or metamorphic rocks.

normal fault   A break in the Earth along which the **hanging wall** has moved down, relative to the **foot wall**. Formed when the rocks are subjected to tension (pulled apart).

obsidian   An **igneous rock** with glassy texture, formed when **magma** cools so quickly that mineral crystals do not have enough time to develop.

oceanic crust   Outer layer of the Earth that is thin and low in **silica** compared to **continental crust**.

ophiolite   The rock sequence comprising **oceanic crust** and the uppermost **mantle**. From bottom to top, an ophiolite consists of the igneous rocks **peridotite**, **gabbro**, and **basalt**, overlain by a deep-ocean sedimentary rock layer.

orogeny   A geological event that forms mountains.

Ouachita Mountains   A highland region in western Arkansas and eastern Oklahoma, formed by the same ocean basin closure and continental collision that formed the Appalachian Mountains during the Paleozoic Era.

outer core   The zone of dense, liquid iron and nickel extending from 1,800 miles (2,900 kilometers) to 3,200 miles (5,100 kilometers) depth between the solid **lower mantle** and solid **inner core**.

outwash gravel   Coarse sedimentary material deposited by water flowing from the front of a melting glacier.

pāhoehoe   A smooth, ropy form of **basalt** lava surface.

paleomagnetism   The study of the direction and intensity of Earth's magnetic field that was frozen into rocks as they formed. Paleomagnetism can be used to date the time certain rocks formed, and in some cases determine the latitude at which they formed. (See also **remnant magnetism**.)

Paleozoic Era   The portion of geologic time between 543 and 248 million years ago. *Paleozoic* means "time of early life."

Pangea   The large continent that formed during the late **Paleozoic Era**, as an ocean closed and most of the continental crustal mass coalesced. This continent broke into the modern continents during the **Mesozoic Era** as the Atlantic Ocean opened up.

passive continental margin   A transition from thick continental to thin oceanic crust that is far from an active plate boundary, and thus lacks significant tectonic activity such as earthquakes, volcanic eruptions, and mountain building.

pencil cleavage   A property of some **slate** whereby it breaks along **cleavage** planes and original bedding planes, resulting in long pieces of rock that look like pencils.

peridotite   A coarse-grained, olive-green **igneous rock** with about 40% **silica**. (It constitutes most of Earth's **mantle**.)

period   A division of an **era** of geologic time, broken into smaller units called **epochs**.

Phanerozoic Eon   The portion of geologic time between 543 million years ago and the present. It includes the **Paleozoic**, **Mesozoic**, and **Cenozic** eras.

phreatic zone   Lower portion of a groundwater system, below the **water table**, where cracks and pore spaces in the rock are completely filled with water.

phyllite   A **metamorphic rock** formed from a **shale** layer that experienced moderate increases in temperature and pressure. (Its metamorphism is between that of that of **slate** and **schist**.)

pillow lava   A rock layer (**lava flow**) with globular structures formed when magma erupted through the seafloor or flowed into the ocean.

pit crater   A small, circular depression, less than about ½ mile (1 kilometer) in diameter, formed where ground collapses as a subsurface magma chamber partially drains.

plate tectonics   The theory that features on Earth's surface result from the horizontal movements of large plates of Earth's outer shell.

platform   See **continental platform**.

plunging fold   An **anticline** or **syncline** that is tilted so that it dips into the ground.

plutonic rock   An **igneous rock** that solidified from **magma** that cooled within the Earth (**intrusive rock**).

Precambrian   The portion of the geologic time scale before the Cambrian Period. It encompasses most of geologic history, from the formation of the Earth 4.5 billion (4,500 million) years ago to the proliferation of life 543 million years ago.

primary magma   Molten material that initially forms from melting of a solid portion of the Earth. The term commonly refers to magma of **basalt** composition that melts off of the **peridotite** of the mantle. The composition of a primary magma can be altered as it rises through the crust and incorporates other materials.

Proterozoic Eon   The portion of geologic time between 2,500 and 543 million years ago. The latest part of **Precambrian** time.

pumice   A low-density **igneous rock**, rich in **silica** and pore spaces, formed from frothy **lava**.

P-wave   See **compressional wave**.

pyroclastic   Material flung into the air during a volcanic eruption.

pyroclastic flow   Mixture of hot gas, ash, and rock fragments that moves down the slopes of a volcano during an explosive eruptions at speeds of 100 miles per hour (160 kilometers per hour) or more.

pyroclastic surge   A **pyroclastic flow** that is rich in gas. Its low density allows it to flow over hills and ridges.

quartz   A **mineral** composed of pure silicon and oxygen (chemical formula SiO_2).

quartzite   A **metamorphic rock** formed from a **sandstone** layer that underwent extreme increases in temperature and pressure. (Quartzite has a greater amount of metamorphism than **metasandstone**.)

recurrence interval   See **earthquake recurrence interval**.

recessional moraine   See **moraine**.

regression   The retreat of a sea from a region.

relative age dating   Determination of where a rock, mineral, fossil, or other material formed fits into a sequence of geological events.

relative plate motion   The speed and direction of one plate compared to another at the boundary of the two plates.

remnant magnetism   The direction and intensity of Earth's magnetic field that was frozen into rocks as they formed. (See also **paleomagnetism**.)

reverse fault   A break in the Earth along which the **hanging wall** has moved up relative to the **foot wall**. Such a fault forms when the rocks are subjected to compression (pushed together).

rhyolite   A fine-grained, generally light-colored **igneous rock** with about 70% **silica**. (It is the extrusive equivalent of **granite**.)

rift valley   A depression between mountain ranges in a **continental rift zone**.

rift valley strata   Sedimentary and volcanic layers deposited in a **rift valley**.

rift zone   Elongate region on the flanks of a volcano that experiences, or has experienced, volcanic eruptions.

right-lateral, strike-slip fault   See **strike-slip fault**.

Rio Grande Rift   An arm of the **Basin and Range Province** extending from westernmost Texas, through New Mexico and into southern Colorado.

rock   A mixture of **minerals** that are cemented together in some natural way.

rock salt   A **non-clastic sedimentary rock** formed where sodium chloride (NaCl) precipitates out of water.

sandstone   A **sedimentary rock** formed from the compaction and cementation of eroded rock particles the size of sand (commonly rich in the mineral **quartz**).

schist   A **metamorphic rock** formed from a pre-existing rock that experienced relatively large increases in temperature and pressure. (Schist represents a greater amount of metamorphism than **slate**, but less than **gneiss**.)

scoria   A porous **igneous rock**, low in **silica** and rich in iron; more dense than **pumice**. **Cinders** are a type of scoria.

seamount   An undersea mountain, formed as a volcano erodes and subsides.

sedimentary basin   A depression that fills with water and accumulates deposits of rock fragments and other sedimentary materials.

sedimentary rock   A **rock** formed from the burial and cementation of eroded rock fragments, or from material precipitated through biological or chemical activity.

sedimentary wedge   An accumulation of eroded rock fragments and other sedimentary materials over a subsiding **passive continental margin**.

seismic reflection profile   An image of Earth's subsurface obtained by sending sound waves into the ground and measuring the signals that bounce off layers and other geologic features and return to the surface.

seismic wave   Sound or other vibration that moves through the Earth or along its surface, caused by an earthquake, landslide, or other natural or artificial source.

seismology The study of **earthquakes** and the waves they generate.

serpentine A mineral formed from the alteration of the minerals olivine and pyroxene that are commonly found in the **peridotite** of Earth's mantle. It forms the metamorphic rock called serpentinite.

shale A **sedimentary rock** formed from the compaction and cementation of fine mud.

shearing The change in shape of a material as it is stressed. You can demonstrate this change by holding a deck of cards in your hands and sliding them laterally; shearing occurs where the cards slip along their faces.

shear wave The movement of seismic wave energy through a material, caused by particles of the material sliding back and forth past one another. Also called *S-wave*.

shield The region of the continental **craton** where old igneous and metamorphic rocks are exposed at the surface.

shield volcano A broad, up to 60 mile (100 kilometer) diameter mountain, with gentle slopes formed through the eruption of numerous fluid (basaltic) lava flows.

Sierra Nevada A north-south trending mountain range in eastern California and its border with Nevada. The range consists of granitic and metamorphic rocks that are the remnants of an ancient **volcanic arc**.

silica An ion consisting of the elements silicon and oxygen that combines with other elements to form most of the **minerals** in Earth's **crust** and **mantle**.

silicate A compound that contains the elements silicon and oxygen.

sill A sheetlike, igneous **intrusive rock** body formed where **magma** is injected parallel to rock layers.

siltstone A **sedimentary rock** formed from the compaction and cementation of eroded rock particles finer than sand.

sinkhole A funnel-shaped depression in the land surface formed as groundwater dissolves underlying limestone.

slate A **metamorphic rock** formed from a **shale** layer that experienced relatively small increases in temperature and pressure. (Slate has a lesser amount of metamorphism than **phyllite** or **schist**.)

slaty cleavage A property of some **slate** whereby it breaks along planes that are commonly different from the orientation of the original sedimentary bedding.

soft collision An early stage ocean basin closure, when only the edges of continents interact and low mountains of uplifted sedimentary layers form.

spatter cone A mound of hardened lava, from a few to several feet (meters) across, that forms as fluid (typically **basalt**) lava erupts in blobs that fly a short distance and accumulate around a volcanic vent. (Also known as an *hornito*, Spanish for "small oven.")

speleothem The variety of cave formations developed as calcium carbonate precipitates from a solution.

spit An elongated bar of sand and mud, connected to the mainland, that develops parallel to the coastline.

stalactite Elongated formation developed on a cave ceiling as calcium carbonate precipitates from dripping water.

stalagmite Elongated formation developed on a cave floor as calcium carbonate precipitates from dripping water.

steady state A condition reached by a system, wherein rates, dimensions, and other components of the system remain constant.

stratigraphic column A diagram that illustrates the sequence of rock layers developed in a region of the Earth.

stratigraphy A branch of **geology** that studies processes that lead to the layering of Earth materials, especially **sedimentary rocks**.

strato-volcano See **composite volcano**.

strike-slip fault A break in the Earth along which the blocks on either side of the break slid horizontally past one another. The break forms when the rocks are subjected to **shearing** stresses. At a *right-lateral strike-slip fault*, the block across the fault line appears to move to the right. At a *left-lateral strike-slip fault*, the block across the fault line appears to move to the left.

subduct To extend beneath and disappear.

subduction slab earthquake An **earthquake** occurring within the plate that extends beneath another plate at a **subduction zone**.

subduction zone A **convergent plate boundary** where one plate slides deeply beneath another.

Superior Province A region of the North American **continental shield** extending from central Canada to the northern parts of Minnesota, Wisconsin, and Michigan. The region formed during early development of the North American continent in the **Archean Eon**.

superposition   The principle that, where a sequence of sedimentary layers have not been tilted beyond vertical, the oldest layers are at the bottom and the youngest layers at the top.

surface wave   A **seismic wave** that travels along Earth's surface.

suspect terrane   See **accreted terrane**.

suture zone   A region of deformation between two **accreted terranes**, between an accreted terrane and a continent, or between two continents that have collided.

S-wave   See **shear wave**.

syncline   A **fold** in which the rock layers are bent downward.

tectonics   The study of large features on Earth's surface and the internal processes that led to their formation.

tectonic setting   The type of plate boundary or hotspot responsible for the formation of the rocks and topography of a region.

tectonostratigraphic terrane   A **terrane** characterized by a distinctive sequence of geologic layering.

tephra   A mixture of gas and rock fragments that erupts into the air from a volcano.

terminal moraine   See **moraine**.

terrane   An extensive region that is bounded by faults and has distinctive geology that is considerably different from the geology of surrounding regions.

thermal subsidence   The downward movement of a region caused by the contraction of rock as it cools.

thermophilic bacteria   Organisms capable of surviving at temperatures at or above the boiling point of water.

thrust fault   A low-angle **reverse fault**.

till   Poorly sorted and poorly stratified rock fragments and other sedimentary material carried by a glacier.

tiltmeter   A device that measures the change in the slope of Earth's surface.

topographic steady state   A condition whereby the average rate of uplift of a region is balanced by the average rate of erosion.

transform plate boundary   A region where two slabs of Earth's outer shell (**lithosphere**) slide laterally past one another.

transgression   The advance of a sea over a region.

Transverse Ranges   A mountain range in southern California that runs east-west, contrary to the more north-south orientation of most ranges along the San Andreas transform plate boundary.

travertine   A dense form of **limestone** formed where calcium carbonate precipitates out of water in caves or springs.

trench   A depression on the ocean floor formed where one lithospheric plate descends (subducts) beneath another.

triple junction   A region where three plates of Earth's outer shell meet.

tsunami   A giant sea wave (sometimes mistakenly called a "tidal wave") caused by movement of the sea floor due to an earthquake, landslide, or volcanic flow.

turbidite   A sequence of thick sandstone and thin shale layers deposited by a **turbidity flow**.

turbidity flow   A current of water saturated with sediment that rushes down a continental slope, dropping the sediment on the deep floor of the ocean.

turtleback   A convex-upward surface of a mountain flank composed of old igneous and metamorphic rocks covered in places by younger metamorphic rocks. Such features, resembling turtle shells, are common in Death Valley and other regions of the **Basin and Range Province**.

unconformity   A surface representing the disruption of the normal deposition of sedimentary strata. The cause of the disruption could be that strata were never deposited, or were deposited but later eroded.

vadose zone   Upper portion of a groundwater system, above the **water table**, where cracks and pore spaces in the rock are not completely filled with water.

vesicular basalt   **Basalt** that has visible air pockets formed where gas expanded in cooling magma.

viscosity   A measure of how well a material resists flowing.

volatile fluid   Compound such as water or carbon dioxide that evaporates easily and can exist in gaseous form at Earth's surface.

volcanic arc   The chain of volcanoes that forms on the overriding plate at a **subduction zone**.

volcanic ash   Fine-grained **pyroclastic** material, blown from a volcano and carried away and deposited by winds.

volcanic bomb   A large piece of **pyroclastic** material, commonly football- to watermelon-size, that may have developed the shape of a bomb as it traveled through the air in a liquid state.

volcanic breccia   **Volcanic rock** composed of course, angular fragments within a finer-grained mass.

volcanic cinders   See **cinders**.

volcanic landslide   See **debris avalanche**.

volcanic mudflow   1. A mass of solid material and water that moves swiftly downslope, commonly as a result of volcanic processes (see **lahar**). 2. The deposits of a volcanic mudflow.

volcanic rock   An **igneous rock** that solidified from **magma** that erupted on Earth's surface (**extrusive rock**).

volcanology   The study of molten material erupted from the Earth and the products of such eruptions.

water table   The surface separating the **vadose zone**, where cracks and pore spaces of the rock are not completely filled with water, from the underlying **phreatic zone**, where they are completely saturated.

Yellowstone Plateau   A region of high elevation centered around Yellowstone National Park.

ABOUT THE AUTHOR

 **DR. ROBERT J. LILLIE** has been a geology professor at Oregon State University since 1984. Dr. Lillie was born and raised in the Cajun country of Louisiana. He earned a B.S. degree in geology from the University of Louisiana—Lafayette, and an M.S. in geophysics from the College of Oceanography at Oregon State University. He then worked for three years in the petroleum exploration industry in the Rocky Mountains before earning a Ph.D. in geophysics from Cornell University, where he worked on deep-crustal seismic reflection profiling of continental regions. Dr. Lillie's research at Oregon State University is focused on the crustal structure and tectonic evolution of mountain ranges formed by the collision of continents, including the Himalayas in India and Pakistan, and the Carpathians in Central Europe. He is the author of *Whole Earth Geophysics,* a popular textbook used in college courses throughout the United States and other countries. Since 1994, Dr. Lillie has collaborated with the National Park Service on projects aimed at educating the public in geology. He has worked as a seasonal interpretative ranger at Crater Lake and Yellowstone national parks, and has written and illustrated geology training manuals for five parks (Crater Lake, Sunset Crater, Blue Ridge Parkway, Gulf Islands, and Olympic). He supervises graduate student theses focused on the development of geology training manuals for other parks (Grand Canyon, Hawai`i Volcanoes, Redwood, and Craters of the Moon). Dr. Lillie has presented seasonal training on geology at Crater Lake, Redwood, Olympic, Yosemite, and Mount Rainier national parks, as well as workshops on plate tectonics and the national parks at annual meetings of the National Association for Interpretation. He teaches undergraduate courses covering introductory geology, oceanography, tectonics, and geophysics, as well as graduate courses in geophysics and geological writing. *Parks and Plates* is an outgrowth of a course he teaches on geology of national parks. He has earned the credentials as a Certified Interpretative Guide and Certified Interpretative Trainer from the National Association for Interpretation. Dr. Lillie is an avid cyclist. In addition to numerous bicycle tours of the Pacific Northwest and the rest of the country, he has done six tours of Ireland; three across the Alps in Germany, Switzerland, and Austria; a trip across the Czech and Slovak republics; and one across Scandinavia above the Arctic Circle. He is also an accomplished photographer and Cajun cook.

CREDITS

Photographs:

All photographs in this text are by Robert J. Lillie except for the following:

Chapter 1

12b National Park Service; **12d** Courtesy of Stacy Wagner; **20a** National Park Service; **20b** Courtesy of Robert D. Lawrence; **20c** Courtesy of Jo Ann Callahan; **20d** Courtesy of Michael J. Appel; **20e** Courtesy of Robert D. Lawrence.

Chapter 2

2.1 John Crossley, *the American Southwest* (www.american-southwest.net); **2.5c** Courtesy of Stacy S. Wagner; **6b (right)** National Aeronautics and Space Administration; **2.12** (Granite) Courtesy of Stacy S. Wagner; **2.12** (Gabbro) U.S. Geological Survey; **2.13** (Shale) Courtesy of Julie Arrington; **2.13** (Limestone) Courtesy of Stacy S. Wagner; **2.14** (Shale) Courtesy of Julie Arrington; **2.14** (Marble) U.S. Geological Survey; **2.14** (Gneiss) National Park Service; **2.16c** National Geographic Society Physical Globe, © 1979; **2.17a** U.S. Geological Survey; **2.17b** U.S. Geological Survey; **2.19a** Stephen C. Porter, from Physical Geology, by B. J. Skinner and S. C. Porter. ©1987 John Wiley and Sons, Inc., This material is used by permission of John Wiley & Sons, Inc.; **2.19d** U.S. Geological Survey.

Chapter 3

3.6c National Park Service; **3.10a** Courtesy of Robert D. Lawrence; **3.10b** Courtesy of Robert D. Lawrence; **3.10c** Courtesy of Robert D. Lawrence; **3.10d** Courtesy of Robert D. Lawrence; **3.12c** U.S. Geological Survey; **3.15c** U.S. Geological Survey; **3.20** Courtesy of Ed Buchner; **3.25a** National Park Service; **3.25b** Randall L. Milstein.

Chapter 4

4.10a National Park Service; **4.10b** National Park Service; **4.10c** National Park Service; **4.13** National Park Service; **4.21b** National Park Service; **4.22a** Courtesy of Jo Ann Callahan; **4.22b** Courtesy of Jen Jarrell-Wetz.

Chapter 5

5.16a National Park Service; **5.16b** National Park Service; **5.17a** National Park Service; **5.17b** National Park Service; **5.18c** National Park Service; **5.19a** U.S. Geological Survey; **5.19b** U.S. Geological Survey; **5.19c** U.S. Geological Survey; **5.19d** U.S. Geological Survey; **5.19f** U.S. Geological Survey; **5.28a** National Park Service; **5.29a (top)** U.S. Geological Survey; **5.29a (bottom)** U.S. Geological Survey; **5.33** Courtesy of Naaman C. Horn; **5.34a** Courtesy of Naaman C. Horn; **5.40** Photo by Michael Greene; **5.41a** National Park Service; **5.41b** U.S. Geological Survey; **5.41c** U.S. Geological Survey; **5.41d** U.S. Geological Survey; **5.42a** U.S. Geological Survey; **5.42b** U.S. Geological Survey; **5.42c** U.S. Geological Survey; **5.42d** U.S. Geological Survey; **5.43a** U.S. Geological Survey; **5.43b** U.S. Geological Survey; **5.43c** U.S. Geological Survey; **5.43d** U.S. Geological Survey; **5.44a** U.S. Geological Survey; **5.44b** U.S. Geological Survey; **5.48** National Park Service.

Chapter 6

6.1b National Aeronautics and Space Administration; **6.7** National Park Service; **6.13b** U.S. Geological Survey; **6.14e** National Aeronautics and Space Administration; **6.15b** U.S. Geological Survey; **6.22** National Park Service.

Chapter 7

7.1a U.S. Geological Survey; **7.1b** U.S. Geological Survey; **7.9a** National Park Service; **7.11a** U.S. Geological Survey; **7.11b** U.S. Geological Survey; **7.11c** U.S. Geological Survey; **7.13a** National Park Service; **7.13b** National Park Service; **7.13c** National Park Service; **7.18a** National Park Service; **7.18b** National Park Service; **7.18c** Courtesy of Robert S. Yeats; **7.19** National Park Service; **7.20a** U.S. Geological Survey; **7.20b** Courtesy of Robert D. Lawrence; **7.22a** Photo by Mary Santelmann; **7.22b** National Park Service.

Part V opening, p. 167 Courtesy of Hawaiian Volcano Observatory, U.S. Geological Survey; **p. 172** U.S. Geological Survey.

Chapter 8

8.1 U.S. Geological Survey; **8.6a** From *Physical Geology* by Brian J. Skinner and Stephen C. Porter © 1987, John Wiley & Sons, Inc. This material is used by permission of John Wiley & Sons, Inc.; **8.6b** U.S. Geological Survey; **8.6c** U.S. Geological Survey; **8.6d** Courtesy of Chris Warner, Earth Treks Climbing Center; **8.6e** National Park Service; **8.11a** U.S. Geological Survey; **8.12** U.S. Geological Survey; **8.13** National Aeronautics and Space Administration; **8.20a** U.S. Geological Survey; **8.20b** U.S. Geological Survey; **8.21a** U.S. Geological Survey; **8.21b** Courtesy of Rebecca H. Ashton; **8.23a** U.S. Geological Survey; **8.24a** U.S. Geological Survey; **8.24b** U.S. Geological Survey; **8.24c** U.S. Geological Survey; **8.27a** U.S. Geological Survey; **8.27b** U.S. Geological Survey; **8.28d** U.S. Geological Survey; **8.33a** Courtesy of Emily Larkin; **8.33b** Courtesy of Emily Larkin; **8.33c** Courtesy of Emily Larkin.

Chapter 9

9.19a U.S. Geological Survey; **9.20b** Courtesy of Robert D. Lawrence; **9.24a** Courtesy of Robert D. Lawrence; **9.24b** Courtesy Robert D. Lawrence; **9.24d** Courtesy Robert D. Lawrence; **9.25** National Park Service.

Part VI opening, p. 209 © Danny Lehman, Corbis; **p. 210** Courtesy of Robert D. Lawrence; **p. 211** Courtesy of Robert D. Lawrence.

Chapter 10

10.5a National Park Service; **10.5b** National Park Service; **10.6** National Park Service; **10.13** National Park Service; **10.14** National Park Service; **10.15** National Park Service; **10.16** Courtesy of Robert D. Lawrence; **10.17** National Park Service; **10.18** National Park Service; **10.19** Courtesy of Jo Ann Callahan; **10.21a** U.S. Geological Survey; **10.21b** U.S. Geological Survey; **10.22b** U.S. Geological Survey; **10.23a** U.S. Geological Survey; **10.23b** U.S. Geological Survey; **10.24a** U.S. Geological Survey; **10.24b** U.S. Geological Survey; **10.24c** U.S. Geological Survey; **10.25a** Courtesy of Jo Ann Callahan; **10.25b** Courtesy of Jo Ann Callahan; **10.26a** Courtesy of Randall L. Milstein; **10.26b** Courtesy of Jo Ann Callahan; **10.27a** U.S. Geological Survey; **10.27b** U.S. Geological Survey; **10.28** Courtesy of Jo Ann Callahan; **10.32a** Courtesy of C. W. Field; **10.32b** U.S. Geological Survey; **10.32c** U.S. Geological Survey; **10.33a** Courtesy of Hillary Senden; **10.33b** U.S. Geological Survey; **10.34** Courtesy of Michael J. Appel.

Chapter 11

11.1 Courtesy of Robert D. Lawrence; **11.12** Photo by Michael Greene; **11.13** National Park Service; **11.15** Photo by Michael Greene; **11.16** National Park Service; **11.17a** National Park Service; **11.17b** National Park Service; **11.18** National Park Service; **11.20a** National Park Service; **11.20b** National Park Service; **11.20c** National Park Service; **11.23a** Courtesy of Robert D. Lawrence; **11.23b** Courtesy of Robert D. Lawrence; **11.23c** Courtesy of Robert D. Lawrence; **11.24** Courtesy of Robert D. Lawrence; **11.25a** Courtesy of Robert D. Lawrence; **11.25b** Courtesy of Robert D. Lawrence.

Line art:

All line art in this text is by Robert J. Lillie except for the following:

Chapter 1

1.6c Modified by Robert J. Lillie from Physical Geology, by B. J. Skinner and S. C. Porter. ©1987 John Wiley and Sons, Inc., New York; **1.7** Modified by Robert J. Lillie from Tanya Atwater. ©2003, University of California Regents; **1.8** From World Ocean Floor Map, by Bruce C. Heezen and Marie Tharp. ©1977, Marie Tharp.

Chapter 2

Table 2.2 Modified by Robert J. Lillie from "How old is it?" by C. Zimmer. ©2001, National Geographic, v. 200, no. 3, p. 78-101; **2.4** Modified by Robert J. Lillie from *A Geology Training Manual for Grand Canyon National Park*, by Stacy S. Wagner, M. S. thesis, Oregon State University, ©2002; **2.6** Raised relief map of United States, scale 1:6,336,000, © Omni Resources, Burlington, North Carolina, www.omnimap.com; **2.16a** National Geographic Society Physical Globe, © 1979; **2.16c** National Geographic Society Physical Globe, © 1979; **2.21** From U.S. Geological Survey, Fact Sheet - 002–97, revised June, 1998, USGS OFR 98–519, "What are Volcanic Hazards?".

Part II opening, p. 50 (left) from *Earth: Portrait of a Planet*. Used by permission of Stephen Marshak and W. W. Norton & Co. Modified by Robert J. Lillie.

Chapter 3

3.9 Modified by Robert J. Lillie from *Geology of National Parks* 5th Ed., by A. G. Harris, E. Tuttle, and S. P. Tuttle, ©1995, Kendall/Hunt Pub. Comp., Dubuque, Iowa; **3.23** National Geographic Society Physical Globe, © 1979; **3.26** Modified by Robert J. Lillie from "The geologic story of Isle Royale National Park," by N. K. Huber, ©1975, U. S. Geological Survey, Bulletin 1309; **3.27** National Geographic Society Physical Globe, © 1979; **3.28** From *Earth: Portrait of a Planet*. Used by permission of Stephen Marshak and W. W. Norton & Co. Modified by Robert J. Lillie.

Chapter 4

4.4 Shaded relief map courtesy of A. Jon Kimerling, Oregon State University; **4.5c** From *Earth: Portrait of a Planet*. Used by permission of Stephen Marshak and W. W. Norton & Co. Modified by Robert J. Lillie; **4.9a** © 1995, FOTOFLITE/PORT Publishing Company; **4.12b** From *Earth: Portrait of a Planet*. Used by permission of Stephen Marshak and W. W. Norton & Co. Modified by Robert J. Lillie; **4.14** Courtesy of Ronald Blakey, Northern Arizona University. Modified by Robert J. Lillie; **4.17** From *An Introduction to Grand Canyon Geology*, by L. G. Price, © 1999, Grand Canyon Association, Grand Canyon, Arizona. Modified by Robert J. Lillie; **4.24** From U.S. Geological Survey website (http://pubs.usgs.gov/gip/geotime/session.html). Modified by Robert J. Lillie.

Chapter 5

5.3 Bernard Garcia and Robert J. Lillie; **5.11a** From *Earth: Portrait of a Planet*. Used by permission of Stephen Marshak and W. W. Norton & Co. Modified by Robert J. Lillie; **5.22a** From T. W. Sisson, J. W. Valance, and P. T. Pringle, "Progress made in understanding Mount Rainier's Hazards," EOS, Transactions American Geophysical Union, p. 113-120, 2001; **5.28b** U.S. Geological Survey; **5.28d** U.S. Geological Survey; **5.29b** Painting by Paul Rockwood, image courtesy of Crater Lake Natural History Association; **5.46** From *Earth: Portrait of a Planet*. Used by permission of Stephen Marshak and W. W. Norton & Co. Modified by Robert J. Lillie.

Chapter 6

6.8 From *Earth: Portrait of a Planet*. Used by permission of Stephen Marshak and W. W. Norton & Co. Modified by Robert J. Lillie; **6.14c** From *Earth: Portrait of a Planet*. Used by permission of Stephen Marshak and W. W. Norton & Co. Modified by Robert J. Lillie; **6.23** Williams, J. *Crustal structure and kinematic evolution of the Brooks Range, Alaska, from gravity and isostatic considerations*, M.S. thesis, Oregon State University, 142 pp., 2000; **6.24** Williams, J. *Crustal structure and kinematic evolution of the Brooks Range, Alaska, from gravity and isostatic considerations*, M. S. thesis, Oregon State University, 142 pp., 2000.

Chapter 7

7.3 From *Earth: Portrait of a Planet*. Used by permission of Steve Marshak and W. W. Norton & Co. Modified by Robert J. Lillie; **7.5** From *Earth: Portrait of a Planet*. Used by permission of Stephen Marshak and W. W. Norton & Co. Modified by Robert J. Lillie; **7.7** U.S. Geological Survey; **7.12** From map "Geology and Active Faults in the San Francisco Bay Area," Point Reyes National Seashore Association. Modified by Robert J. Lillie; **7.16** Modified by Robert J. Lillie from San Andreas: An Animated Tectonic History of Western North America and Southern California, ©1998, Tanya Atwater and the Regents of the University of California; **7.17** From *Earth: Portrait of a Planet*. Used by permission of Steve Marshak and W. W. Norton & Co. Modified by Robert J. Lillie; **7.21** Plate boundaries from The Plates Project, University of Texas Institute for Geophysics, database. Bathymetry/topography data are from Smith, W.H.F. and Sandwell, D.T., 1997. "Global sea floor topography from satellite altimetry and ship depth soundings." Science, 277: 1956-1962.

Chapter 8

8.3 National Geographic Society Physical Globe, © 1979; **8.5** From *Earth: Portrait of a Planet*. Used by permission of Stephen Marshak and W. W. Norton & Co. Modified by Robert J. Lillie; **8.11a** Modified by Robert J. Lillie from *A Dynamic Landscape Formed by the Power of Volcanoes: Geology Training Manual for Interpretive Rangers at Hawai`i Volcanoes National Park*, by Rebbecca H. Ashton, M. S. thesis, Oregon State University, © 2003; **8.17** From *Earth: Portrait of a Planet*. Used by permission of Stephen Marshak and W. W. Norton & Co. Modified by Robert J. Lillie; **8.19** Modified by Robert J. Lillie from a drawing by Jenda Johnson, 2000; **8.26** From *Earth: Portrait of a Planet*. Used by permission of Stephen Marshak and W. W. Norton & Co. Modified by Robert J. Lillie; **8.28** Modified by Robert J. Lillie from *A Dynamic Landscape Formed by the Powers of Volcanoes: Geology Training Manual for Interpretive Rangers at Hawai`i Volcanoes National Park*, by Rebecca H. Ashton, M. S. thesis, Oregon State University, © 2003; **8.31** Hawaiian Volcano Observatory of the U. S. Geological Survey; **8.32** Hawaiian Volcano Observatory of the U. S. Geological Survey. Modified by Rebecca H. Ashton and Robert J. Lillie.

Chapter 10

10.1 From *Earth: Portrait of a Planet*. Used by permission of Stephen Marshak and W. W. Norton & Co. Modified by Robert J. Lillie; **10.4** From *Earth: Portrait of a Planet*. Used by permission of Stephen Marshak and W. W. Norton & Co. Modified by Robert J. Lillie; **10.7** From *Earth: Portrait of a Planet*. Used by permission of Stephen Marshak and W. W. Norton & Co. Modified by Robert J. Lillie; **10.9** From *Earth: Portrait of a*

INDEX

Page numbers in *italics* refer to illustrations.

Superior Province, 218–19
superposition, 25
surface waves, 45
Surprise Lake, 120, *120*
"suspect" terranes, 249
suture zones, 251
synclines, 29, *29*
   in Appalachian Mountains, 30, *139*, 140
   in Keweenaw National Historic Park, 68

Tahoe, Lake, 59
Talkeetna Fault, 247
Talkeetna Superterrane, 248
Tanganyika, Lake, 50–51, 59
Tatoosh Granodiorite, 107, *109*
Tatoosh Mountains, 107, *108*
tectonic activity, *6*
   in the Caribbean region, *134*
   at Redwood National and State parks, 105
   and volcanoes, *6*, *18*, 50, 54
   see also plate tectonics
tectonic provinces, *see* geologic provinces
tectonostratigraphic terranes, 249
Tehachapi Mountains, *158*
Tehama, Mount, *114*, 115
tephra, *40*, 41, 185–86
terminal moraines, 78, 220, *236*
terrane accretion, *see* accreted terranes
Tethys Ocean, 130
Teton Fault, 59, *59*
Teton Range, 59, *59*
Texas, 14, *77*, 79
   southwest, 145
Theodore Roosevelt National Park, 41, 210, 227, *227*
thermal subsidence, 177, *178*, 180, 181, 206
thermophilic bacteria, 201
Thermus aquaticus, 201
Thielsen, Mount, *109*, 110
third pole, 129
Three-Fingered Jack, *110*
Three Sisters, 110, *110*
thrust faults, 28, *29*, *30*, 98–99, *99*, 100, *139*, *142*, 246, 253
Tibetan Plateau, 27, 129
   isostatic uplifting of, *28*, 130, 133
till, 220
tiltmeters, *189*, 190
Timpanogos Cave National Monument, *57*
Toklat Glacier, *248*
Toklat River, 247, *247*
Tomales Bay, 157–60, *158*, *160*
topographic steady state, 102–3, *103*, 105
Trans-Alaska Pipeline, 145
transform plate boundaries, *6*, *8*, 123–25, *124*, 149–66
   California national parks and, 149–63, *155*
   continental, *13*
   definition of, 5
   earthquakes and, 42, 44
   national parks at, *155*
   oceanic, *13*

   at Pacific and North American plates, *151*, *156*, *163*
   results of, *17*
   San Andreas Fault and, 17, *17*, *22*, 42, 55, 97, 124, *151*, *153*, 154, *155*, 156–61, *158*, *159*
   sheared-up mountain ranges and, 29–30, 157–61
   strike-slip faults and, 29, 154–61, *159*
   tectonic settings and, 10–20, 97
   and U.S. Virgin Islands, *153*, 164, *164*
   volcanism and, 37
transgression, 84, *85*
Transverse Ranges, *155*, 157, *157*, 161–63, *163*
   Channel Islands and, 163
   development of, *162*
travertine, *201*, 225
trenches, 94, 97, 182, *252*
Trident Volcano, 121, *121*
tsunamis, 115–17
Tucson Mountains, 58
turbidites, 100–101, *101*, 117
turtleback, *60*
Twain, Mark, 173

unconformities, 25–26
   angular, 26, *27*, 102
   in national parks, *27*
uniformitarianism, 101, 220
Union Peak, *110*
United States:
   accreted terranes in lower 48, *13*, 211, *213*, 214, 218, *218*, *241*, 251–54
   eastern, 72–80, *73*, *144*
   eastern geologic provinces in, 135–43
   relief map of, *11*, *213*
   southwestern, *82*, 84
   tectonic settings of, 73
   western, 14, 54, *55*–67, *56*, *73*, 75, *75*, *83*, 123–24
United States Geological Survey (USGS), 188, 198
uplifts, uplifting:
   of Appalachian Mountains, *143*
   of Colorado Plateau, 19, *20*, 81–83, *83*, *213*, 216
   of crust, 27, 157
   foreland basement, 218, 233, *234*, *235*
   isostatic, *see* isostatic uplift
   of rocks, 99, *99*, *102*, 105, 129, 141, *160*, *163*

vadose zone, 225
Valley and Ridge Province, 29, 134, *135*, 136, *137*, *139*, 141
   sedimentary rocks in, 140
Valley of 10,000 Smokes, 121, *121*
valleys, 157–58
   fault zones and, 157–58
   glacial, *107*
   see also rift valleys
Van Matre, Steve, 4

vesicles, 188
vesicular basalt, 188
Virgin Islands, U.S., 150
   transform plate boundary and, *153*, 164, *164*
Virgin Islands National Park, 17, 155, *164*
volatiles, *36*, 37
volcanic arcs, *12*, *17*, *30*, 90, *90*, 94, 103, 155, 240
   active, *95*, *110*
   in Alaska, 120–23, *124*, 249
   ancient, *95*, 106
   from British Columbia to Mexico, 124–25
   in California, *156*, 160
   in Caribbean Sea region, 164
   definition of, *35*
   formation of, *95*
volcanic ash, 17, *40*, 41, 61, 121, 181
volcanic bombs, 41
volcanic breccia, 161, *161*
volcanic gasses, *40*, 41, 61–62
volcanic hazards, 40–41, *40*, 105–6, 121
   downstream effects of, *107*
volcanic islands, 136, 139, 218
   erosion of, 174, 177
   formation of, *18*, 174
   stages of development for, 178–82
   see also specific volcanic islands
volcanic landslides, *see* lahars; landslides
volcanic materials, 40–41, *40*, *41*
   in Sierra Nevada Mountains, *156*
volcanic mountain chains, 90–91, *90*
   types of, *35*
volcanic rocks, 14, 17–19, *18*, 36, *150*, 155–56, *161*
   continental rifting and, 20, *52*
volcanism, 32, 34–42, *35*, 55, 177
   age dating and, 25, 181
   in Basin and Range Province, 61, 115
   at continental rift zones, 60–67, *60*, 135
   Crater Lake caldera and, 110
   high-silica, 62, *62*, 110–11, *111*, 113, *113*, *114*, 115
   low silica, 64–65, *111*, *114*, 115
   at mid-ocean ridges, 60–61, 69, *69*
   migration of, 61
   monitoring of, 121
   in northern California, 61
   at San Francisco Volcanic Field, 61
   seafloor, 105
   in southern Oregon, 61, 110–15, *111*
   stages of, *196*
   on U.S. East Coast, *75*
   on U.S. West Coast, *73*, *75*, 105, 107, *107*, 114–18
   waning and ceasing of, 174
   of Yellowstone National Park, 198–99
   see also lava, lava flows; magma; volcanoes
volcanoes, 2, 16, *16*, 50
   active, *see* active volcanoes
   active vs. inactive, 109, *110*
   activity level of, 180, *190*
   ancient, 95, 106, 123–25, *123*